MECHANICAL DESIGN SYNTHESIS

— Creative Design and Optimization

Second Edition

MECHANICAL DESIGN SYNTHESIS

SYNTHESIS

– Creative Design and Optimization

Second Edition

RAY C. JOHNSON

Higgins Professor of Mechanical Engineering
Worcester Polytechnic Institute and
Consulting Design Engineer

ROBERT E. KRIEGER PUBLISHING COMPANY
HUNTINGTON, NEW YORK
1978

Original Edition 1971 under title
Mechanical Design Synthesis with Optimization Applications
Reprint 1978 with new material

Printed and Published by
ROBERT E. KRIEGER PUBLISHING COMPANY
645 NEW YORK AVENUE
HUNTINGTON, NEW YORK 11743

Printed in the United States of America

Library of Congress Cataloging in Publication Data

Johnson, Ray C.
    Mechanical design synthesis.

    First published in 1971 under title: Mechanical design
synthesis with optimization applications; reprinted 1978
with new material.
    Includes index.
    1. Machinery—Design.   I. Title.
[TJ230.J658 1978]   621.8'15   77-10974
ISBN 0-88275-612-5

To my wife Helen

Foreword

Engineering is a creative profession. Its members, individually and collectively, have played a dominant role in our modern society which in recent years is continuing to undergo the most rapid technological advances the historians have ever recorded. In one way or another, these technological advances can be characterized by two words, design and synthesis. Within the realm of the mechanical things of today and through projections of things of tomorrow, Professor Ray C. Johnson, as academician, industrial consultant, and designer, brings to the student engineer and the practicing professional his vast experience and leadership in optimum mechanical design through his latest publication, *Mechanical Design Synthesis*.

To those who are limited in their knowledge of the field of mechanical design and synthesis, this book, written with the delicate detail of the brilliant scholar but modified with the down-to-earth logic of the practicing engineer, introduces new concepts in fairly rapid order. Every effort has been to clarify each concept through a variety of examples and problems from the author's wide contacts with industry. To those who are already familiar with the field of mechanical design and synthesis this book successfully documents the concept of the synthesis of configuration and joins it with real-world selection of materials and dimensions.   The theory and application of mechanical elements in design are bridged through distinctive procedures which were devised by the author and field tested through the classroom and consultation work before inclusion in this book.

My colleagues and I have been privileged to see firsthand the way in which Professor Johnson has pioneered in bringing about new and realistic approaches to mechanical design.   We feel that the careful study of this book, one which is as related to the field of design as the modern jet airplane is to the field of transportation, will aid both the student and practicing engineer to work more effectively toward minimizing unneeded human energies while optimizing the output of their machines for the benefit of mankind.

Donald N. Zwiep
Professor and Head
Mechanical Engineering Department
Worcester Polytechnic Institute

Preface

There are many aspects to the total process of mechanical design which must be coped with by the successful engineer. One of the first major phases is the creative synthesis of configurations, followed by another important phase, that of selecting materials and dimensions. The basic purpose of this book is to present and to illustrate some explicit techniques which are not too well known for helping to cope with some critical problems of mechanical design synthesis in these early stages of the design process.

The techniques presented are not meant to be panaceas applicable to all problems of design, since many decisions in the total picture are noncritical in nature and the art of design must be based on ingenuity and knowledge from both practical experience and engineering science. However, the techniques presented should supplement one's background, thereby enhancing the possibility of approaching at least the optimum design for which we naturally strive. Hence, the central theme in the book is *optimization*, adhering of course to the confines of realistic limitations imposed by total system requirements as normally encountered in practice.

An introduction to the book is presented in Chapter 1, describing the relationship of mechanical design synthesis to the morphology of the total design process. The role of optimization therein is described, and some introductory examples are briefly presented for mechanical design synthesis. Also, some photographs of advanced designs are presented from various industrial fields, illustrating what either has been or will be achieved over long periods of time by the process of mechanical design in a continuing effort of optimization. Perhaps some of these examples will serve as inspiration for others to improve the products upon which they are working in the continued effort of mechanical design synthesis. In fact, it is hoped that the book will provide a challenge to the reader for developing and applying explicit techniques for use in critical problems of mechanical design synthesis in the early stages of the decision-making process.

Part I of the book considers the *synthesis of configurations* phase of the design process. In Chapter 2 we consider some aspects of creativity in design, followed by a presentation in Chapters 3 and 4 of some techniques that can aid or stimulate creative activity. In Chapter 5 the problem of selecting the optimum configuration from a list of design concepts is considered, from the viewpoint of meeting total system requirements in the best possible way.

Part II of the book considers the *selection of materials and dimensions* phase of the design process. Problems of application for the explicit techniques covered are in the category of advanced design, which is introduced in Chapter 6. The optimum shape for an element is considered in Chapter 7, whereas the more general problem of optimization for selecting materials and dimensions singular in nature and limited by both discrete value and regional constraints is considered in Chapter 8. Problems of advanced design that are more complex in nature are illustrated in Chapter 9, including some examples of application of digital computers to optimization techniques and some industrial examples with results achieved. The necessity for considering total system requirements in various modes of operation is illustrated in Chapter 9, which also integrates some of the techniques presented in earlier chapters.

Because of the necessity for limiting the size of the book, an effort was made not to duplicate material which can be found in other publications. Hence, omission of certain derivations and technical material does not belittle their value for use in the total process of mechanical design synthesis. It is assumed that the design engineer is aware of such published information, and that it will be applied where appropriate in the total process of mechanical design.

The book has been designed for use in various ways. For one, it has been written for self-study by the practicing engineer and teacher interested in design, with many illustrative examples, industrial applications, and problems presented throughout the text. Also, it can be used as a college textbook, preferably for a senior or graduate-level course in the design area. The scope is broad enough to hold interest for a two-semester course, particularly if supplemented with derivations, background, and perhaps some other topics of mechanical synthesis from other publications not duplicated herein. Also, the book can be used for a one-semester course on the senior or graduate level, giving the student a broad understanding of some problems unique to mechanical design synthesis with techniques for their solution. Hence, the scope of the book together with possibility of extension from other publications provides a wide range of applicability for the text. Regardless of the particular way in which the book is used, the reader should obtain a better insight into some of the unique problems of mechanical design synthesis and how they can be explicitly coped with in practice.

The list of acknowledgements which I should make would be unreasonably long, since there have been many people who have been instrumental in providing the necessary background for this work. However, I would like to express my appreciation to Professor Kurt Hain of Germany for the inspiration which he provided when we taught a senior design course together nearly ten years ago at Yale University. This contact was my introduction to the systematics of linkages technique which is presented as an aid to creative action in Chapter 3. Also, I appreciate the many industrial contacts which I have had in my consultation design work, such as with Messrs. John P. Kieronski, Lynwood C. Rice, and Harry B. Miller, to mention a few. Without a confrontation with actual industrial problems, I am certain that the development and application of presented techniques could not have been made.

As a final note, the emphasis today to a great extent is placed on satisfaction of total system requirements in design problems, and this is exemplified in the technical work of the book. In this same general connection, I would like to mention some people from diverse fields of activity who have been optimum in various areas of the total systems picture which I have faced, and this has been an important ingredient making possible the writing of this book. First, with respect to the industrial field I particularly appreciate the many opportunities of application for mechanical design synthesis which have been provided by Mr. John H. Osgood, director of engineering at The Langston Company. Specific industrial examples of results achieved because of these opportunities are presented in sections 9-5 and 9-6 of Chapter 9. Incidentally, he also deserves the credit for calling to my attention the quotation from Oliver Wendell Holmes presented in section 6-2 of Chapter 6, as an early writing on the philosophy of the optimization technique. Secondly, with respect to the academic field, I particularly appreciate the opportunities for teaching mechanical design synthesis at Worcester Polytechnic Institute and I extend my thanks to Professor Donald N. Zwiep, our chairman of mechanical engineering. Without this opportunity, I do not believe that it would have been possible to develop and to organize the book as presented. Thirdly, outside of the engineering fields, I appreciate greatly the inspirational example, general encouragement, and personal interest shown by Dr. Harold J. Ockenga, president of Gordon College and formerly Minister of historic Park Street Church in Boston. Fourthly, to a great extent I owe my early interest in mechanical design to my father, Mr. Olaf A. Johnson, whose excellence in the machinery design field provided a stimulant for the course which I pursued. Last, but not least, I greatly appreciate my immediate family for their part in the writing of the book. In particular, my wife Helen has shown the optimum degree of patience and understanding during the long hours of development of this work. Also, I greatly appreciate the tremendous amount of help provided by Glen who served as my draftsman on figures, by Barbara for typing the manuscript, and by Carol for secretarial assistance.

Ray C. Johnson

Holden, Massachusetts

Preface to Second Edition

The second edition of the book is primarily an expansion of the first, with new material added in Chapters 3, 4, 8 and the Appendices A through D. In the first edition some of the material in the area of creative design was presented in skeleton form, because of publication size limitations, with the omissions resulting in gaps for derivations and explanations. Also, in recent years the significance of new material not previously covered has come to light. Hence, the emphasis in writing the second edition was to make the book a more complete and up-dated volume in the area of mechanical design synthesis.

The greatest modification by far is in Chapter 3, which was expanded to include more derivations and explanations, as well as to illustrate practical application with seven more examples. The systematics of linkages covered in that Chapter is a very practical tool of creative design which has received very little attention in the American endeavors. The author first became aware of its significance in 1961 when teaching a senior design course at Yale University with Dr. K. Hain of Germany, who introduced to the class the basic technique of what he called "number synthesis" with practical creative design applications including clamping mechanisms and self-adjusting brakes. In later years, the author worked with Mr. K. Towfigh on significant advancements of the theory and application to epicyclic gear trains. Finally, the author has developed the techniques further and applied them in many industrial problems, some of which are illustrated in the text. The topic is timely and practical in the area of creative design, and it has been of great interest to my students of creative design during the past fifteen years. Hence, Chapter 3 was expanded for the second edition, and you will note that Appendices A and B have been added. These Appendices present basic catalogs of linkages for application to new problems of creative design with the systematics of linkages.

Similarly, but to a lesser extent, Chapter 4 was expanded for inclusion of further explanation and illustration pertaining to two other techniques of creative design. The implication technique from optimization goals is sometimes difficult to comprehend by the reader at first. Also, the technique of using circuit diagrams in creative design requires careful study and practice to master. Hence, in both cases, additional material was included to improve the explanation and to illustrate the basic techniques with some more simple examples.

In Chapter 8, a short section is added at the end which is philosophical in nature, but very important when developing the technique of application for solution of design optimization problems. Also, Appendix C is added on Computers in Automated Optimal Design. This is a very recent area of significance in design optimization, and the brief coverage outlined in Appendix C should provide a stimulant for further study of the subject. In connection with this addition, I wish to express my appreciation to Mr. G. E. Johnson of Vanderbilt University for his review and suggestions made pertaining to Appendix C.

Finally, in recent years the decision has been made in the United States to adopt the International System of Units. Presently, many experienced engineers in this country are most familiar with the English system of units, but a transition to at least working familiarity with SI units will be necessary in the near future. Conversion of all numerical problems of the first edition from English units to SI units was first contemplated. However, for many years to come particularly in the area of engineering calculations, the American engineer will have to be familiar with both systems. This is true, since he will have to refer to reports and use information derived in the past in English units, and he will have to work on new problems in SI units. Hence, it was decided to retain the numerical problems of the first edition in English units, but to include Appendix D on SI Units in Mechanical Design as an additional item for study. Appendix D should help the engineer to develop a working familiarity with SI units so he can solve new problems of design in that system.

The second edition of the book should be of interest to engineers in industry engaged or associated in some way with a design activity. The opportunities of application for the techniques presented are countless in problems of mechanical design synthesis. Also, the second edition of the book should be of interest to College students on the senior and graduate levels, particularly in the area of mechanical engineering design. Hence, anyone connected with engineering design may profit from the second edition, including students, professors, project engineers, design engineers, managers, etc. The subject of mechanical design synthesis is a fascinating field presently under development, and it is gratifying to learn of its expanded recognition which is continually taking place in the engineering profession.

Ray C. Johnson

Holden, Massachusetts
August 1977

Contents

8 Introduction to Optimum Design

1

Introduction

1-1   BASIC TERMINOLOGY

As an appropriate start we will consider the terminology of *mechanical design synthesis*. The word *mechanical* connotes something which is physical in nature, possessing configuration, shape, materials, and dimensions. The word *design* may be treated as either a verb or noun, similar to its synonym word *plan*. However, the word *synthesis* denotes the process of putting things together. Hence, the terminology *mechanical design synthesis* designates the process of creating and selecting configurations, shapes, materials, and dimensions for something which is physical in nature. The emphasis in the book will be on this aspect of mechanical design.

1-2   BRIEF HISTORY OF MECHANICAL DESIGN

In early days the primary emphasis was on analyzing existing mechanical designs. For example, several hundred years ago the Italian physicist and astronomer Galileo studied the behavior of bodies, such as the pendulum, falling weights, projectiles, planets, and stars. Shortly thereafter, the English mathematician and natural philosopher Sir Isaac Newton studied the law of gravity and its relation to planets, the behavior of light and its refraction through a prism, as well as many other physical phenomena of interest. Hence, initially the primary emphasis was on analyzing the behavior of existing designs.

The importance of synthesizing new mechanical designs became apparent, and greater emphasis was placed in that direction. In Figure 1.1 we see a copy of a patent on the locomotive issued in 1836 to H. R. Campbell. In Figure 1.2 we see a copy of a patent on the "road engine" issued in 1895 to G. B. Seldon. In the early 1900s it was quite common in American industry for inventions to be made by a Master Mechanic, and then a draftsman was called in to copy on paper what had been synthesized experimentally in the shop. The value of synthesizing on paper first was soon realized since it was less costly and more efficient to erase rather than to remake parts in the shop. From the preceding summary, we see that early efforts in mechanical design synthesis were primarily inventive in nature and predominantly characterized by cut-and-try approaches.

Recent trends have been to apply theory where appropriate in the process of mechanical design synthesis. However, in the total picture it is empha-

H. R. CAMPBELL.

Locomotive.

Patented Feb. 5, 1836.

Figure 1.1 Drawing from an early patent on the locomotive. *Courtesy The Management Center of Cambridge.*

2 Sheets—Sheet 1.

G. B. SELDEN.
ROAD ENGINE.

No. 549,160.   Patented Nov. 5, 1895.

Figure 1.2   Drawing from an early patent on the "road engine."   *Courtesy The Management Center of Cambridge.*

sized that all of the useful ingredients, such as various aspects of art, science, engineering, practical experience, and ingenuity, must be properly blended in the design process.   Hence, a careful coordination of efforts from sources of many talents utilizing modern facilities such as digital computers is necessary for successful mechanical design synthesis in the present day and age.

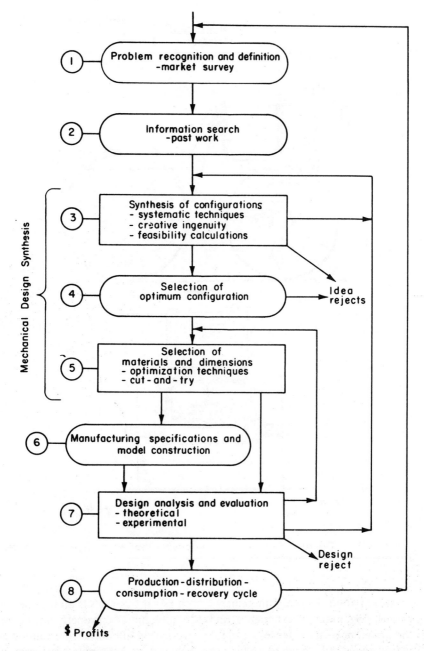

Figure 1.3   Morphology of the mechanical design process showing some typical feedback loops.

1-3   MORPHOLOGY OF THE DESIGN PROCESS

The morphology of the total process of mechanical design can be resolved into fundamental components as summarized in Figure 1.3. First, recognition and definition of the design problem leads to the establishment of the specifications which the manufactured product must satisfy. The second step is to review what has been done in the past for solving similar design problems. This information is obtained from sources such as patents, the trade literature, other publications, personal contacts, and experience.

The third step in Figure 1.3 is the synthesis of basic configurations which *might possibly* solve the problem of design. In this step of ingenuity generally there is great need for creative effort as well as for systematic techniques of synthesis. Ideas should be conceived without a consideration of final worth. Many of these will be recognized as impractical from experience and simple feasibility calculations, and they should be discarded early. In general, several ideas will remain and the most promising one must be selected in step four. The fifth step is to reach decisions on the dimensions and materials for the selected configuration. For critical problems of design it will be appropriate to apply optimization techniques in this stage. On the other hand, in many cases cut-and-try approaches in conjunction with some simple calculations of analysis will be adequate and proper from the standpoint of expediency.

Next, in step six of Figure 1.3 we are ready to complete the specifications for the making of a design prototype. Theoretical analysis and experimental evaluation in step seven are the real proofs of our design success or failure. The proof of the cook is in the pudding and not the recipe. Likewise, the real proof of the designer is in the product performance and not in the design specifications on paper. If successful in a thorough evaluation phase, the product is now ready to enter the production-distribution-consumption-recovery cycle at step eight. At this point we have reached the overall goal of our company, and profits can soon be realized.

The process of mechanical design is an iterative procedure, and in Figure 1.3 we have shown some typical feedback loops. A cardinal characteristic of a successful designer is that he never gives up. Hence, at any step in the morphology of the design process information may be obtained which makes it advisable to return to a preceding step as illustrated by a feedback loop in the figure. In fact, design success as evidenced by the production-consumption cycle often is accompanied by simultaneous activity in the design process on advanced work. Hence, repetitious application of Figure 1.3 might very well occur on a continuing basis extending over many years, resulting in tremendous advancements of product design. What can be accomplished in this long-range evolution of product design is illustrated by the present-day

Figure 1.4   The Budd Metroliner, an example of advanced design in railroad transportation (self-propelled, 2560 HP per car, over 100 mph, air-spring suspension, etc.).   *Courtesy the Budd Company.*

advancements shown in Figures 1.4 and 1.5, which should be compared with the initial concepts of the product types in Figures 1.1 and 1.2, respectively.

1-4   MEANS TO THE END AND OPTIMIZATION

Mechanical design synthesis is a decision-making process having the overall objective of successfully connecting steps 1 and 8 in the morphology of Figure 1.3. In general, there are many *possible* solutions to a problem of design, as illustrated schematically by the twelve alternatives of Figure 1.6. In fact, in step 5 alone of Figure 1.3 there are generally an infinite number of design possibilities for the selection of dimensions and materials, as will be illustrated in many of the optimum design variation studies which follow in Chapters 8 and 9. In mechanical design synthesis, the decisions made must always consider the total design process, since steps in the morphology are often dependent upon each other. Obviously, choosing the particular means in the decision-making process for reaching the desired end is not simple.

Figure 1.5   The Chevrolet Astro III experimental car, an example of advanced design in automobiles (tricycle-design configuration, rear-mounted Allison gas turbine engine, closed-circuit television rear vision, etc.).   *Courtesy The General Motors Corporation.*

With a multiplicity of possibilities, the basic objective of merely connecting steps 1 and 8 in Figure 1.3 is naturally conducive to cut-and-try techniques in the design process.   However, the stipulation that our design should be successful implies the desire for *optimization*. Hence, in our decision-making process we should strive toward the best, if we believe that significant gains can be realized in the design achieved.   In this endeavor for optimization it is assumed that the designer will adhere to all practical constraints of significance, including limit considerations on time spent and money expended.

The means to the end in the morphology of mechanical design synthesis will be made more explicit by adherence to the *central theme of optimization*. This will be illustrated many times throughout the remaining part of the book. Some of the basic characteristics of mechanical design synthesis encountered in the decision-making process will be illustrated in the introductory examples which follow in this chapter.   On the other hand, specific techniques for coping with problems of mechanical design synthesis will be covered and illustrated in the chapters which follow.

1-5   INTRODUCTORY EXAMPLE 1—IDLER GEAR MOUNTING DESIGN PROBLEM

In a machine being designed, suppose driver gear 1 must cause driven gear 3 to rotate in the same clockwise direction.   Hence, for this purpose alone it is necessary for us to introduce an idler gear 2 as shown in Figure 1.7.   At

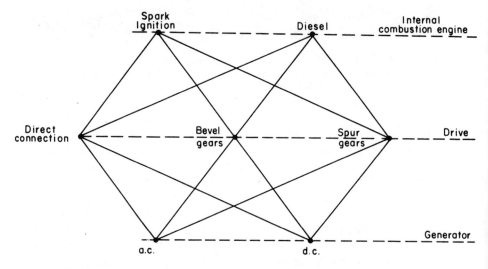

2 x 3 x 2 = 12 possible combinations of motor-generator configurations

Figure 1.6   Some strategic design alternatives for motor-generator sets.

this stage of design, power to be transmitted, $(HP)_t$, gear speeds, ω_1 and ω_3, and gear sizes, D_1 and D_3, are assumed to be specified values.   Hence, pitch line velocity V and tooth forces on the idler gear ($F_{3/2}$ and $F_{1/2}$) shown in Figure 1.7 are determined values at this stage.   The resultant of the tooth forces on the idler gear will be designated as R_{O_2}. Assume that gear pressure angle ϕ is a given value.   Also at this stage, we have some freedom in design with respect to the idler gear and its mounting.   In what follows we will make some simple variation studies which will enable us to draw some conclusions of design significance.

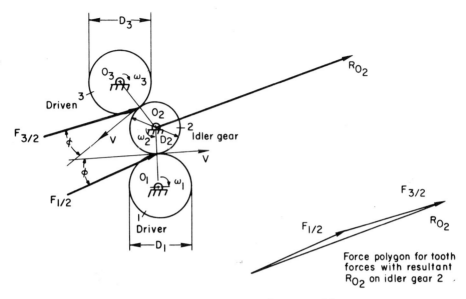

Figure 1.7   Initial design of gearset drive.

Regarding the mounting of the idler gear, two types of mountings come to mind from past experience, as shown in Figure 1.8. From the standpoints of greatest rigidity and smallest bearing loads, we could show that the straddle mounting type is the better choice. However, some overriding considerations influence our final decision. For instance, suppose that ease of removal is necessary for the idler gear because of important timing requirements in the machine. Hence, the final decision is to choose the cantilever type mounting which is better from that governing standpoint in this particular application.

We now pause to consider whether or not there are some attributes that are important objectives of optimization in the design of the idler gear and its mounting. First, from the standpoint of long-range service, we would like to maximize the life of the ball bearings in the mounting. Secondly, we realize that rigidity in mounting is highly desired for the idler gear from the standpoint of obtaining good contact for the meshing teeth.

Bearing life in hours is inversely proportional to both shaft speed ω and bearing load R_b to the fourth power, as given in proportionality equation 1.1.

$$\text{Life} \sim 1/(\omega R_b^4) \qquad (1.1)$$

Hence, for maximization of bearing life, within the limits of practicality, our mounting design should be such as to minimize both shaft speed and bearing load. Since pitch line velocity V is given, for minimization of shaft speed

Figure 1.8   Two types of mountings for idler gear, showing resultant tooth load, R_{O2}, and bearing reactions.

ω_2 in Figure 1.7 we draw the design conclusion that idler gear size D_2 should be as large as practical.   With respect to bearing load, from application of mechanics to Figure 1.8(b) we derive equation 1.2 for the critical bearing reaction force, R_L.

$$R_L = R_{O_2}\left[\frac{1}{1-(b/L)}\right] \qquad (1.2)$$

In this equation, geometric parameters L and b are shown in Figure 1.8(b).
   Thus, for minimization of bearing load R_L, the decisions of design signific-ance are as follows, the goals of which are directly drawn from an inspection of equation 1.2.   First, overhang dimension b should be as small as feasible. Next, shaft length dimension L should be as large as practical.   Finally, we

conclude that resultant tooth load R_{O_2} should be as small as possible.   What can we do to achieve this?

If we refer to the relative orientation of gears typically shown in Figures 1.7 and 1.9, we realize that an infinite number of possibilities exist and the extremes are shown in Figure 1.9(a) and (b).   For given tooth loads $F_{1/2}$ and $F_{3/2}$, the superimposed force polygons for the presented gear orientation

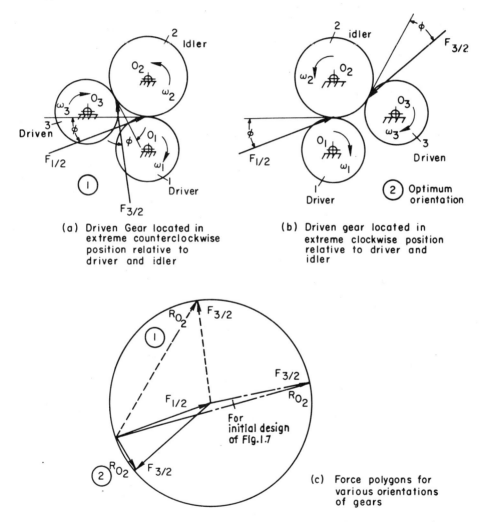

(a) Driven Gear located in
      extreme counterclockwise
      position relative to
      driver and idler

(b) Driven gear located in
      extreme clockwise position
      relative to driver and
      idler

(c) Force polygons for
      various orientations
      of gears

Figure 1.9   Extreme relative positions of gears showing optimum orientation for minimization of resultant force R_{O2}. (a) Driven gear located in extreme counterclockwise position relative to driver and idler. (b) Driven gear located in extreme clockwise position relative to driver and idler. (c) Force polygons for various orientations of gears (initial design of Figure 1.7).

possibilities are shown in Figure 1.9(c). Obviously, the optimum orientation is for the extreme position 2 of Figure 1.9(b), which minimizes magnitude of resultant force R_{O_2}.

Let us now make some decisions of design with respect to maximization of rigidity, which is essentially minimization of deflection Δ in Figure 1.8(b). From strength of materials we derive equation 1.3, which expresses deflection Δ in terms of the design parameters.

$$\Delta = \frac{R_{O_2} L b^2}{3EI} \qquad (1.3)$$

In this equation, all terms have been defined except modulus of elasticity, E, of the steel and rectangular moment of inertia, I, of the shaft cross section.

For minimization of Δ, from equation 1.3 we draw the following conclusions of design significance: First, resultant tooth load R_{O_2} should be as small as possible, and this we would achieve by the optimum orientation of gears shown in Figure 1.9(b). Also, from equation 1.3 we see that both dimensions L and b of Figure 1.8(b) should be as small as practical and that moment of inertia I should be large, which means that shaft diameter d should be as large as practical.

We will now summarize the overall decisions or goals of design reached with respect to the problem of idler-gear mounting. These are some directions in which to head from the standpoint of mechanical design synthesis.

1. Choose cantilever type mounting of Figure 1.8(b).
2. Select idler gear size D_2 of Figure 1.7 as large as reasonable.
3. Have relative orientation of gears in the extreme position shown in Figure 1.9(b).
4. Choose mounting shaft diameter d as large as practical.
5. Design for small overhang dimension b of Figure 1.8(b).
6. Regarding shaft length L of Figure 1.8(b), the variation studies made indicate that a compromise design decision much be reached. From the standpoint of bearing life, we would like to have L large. On the other hand, from the standpoint of mounting rigidity we would like to have L small. Hence, typically, a logical compromise in choice of L would be to have it as small as a 30,000-hour B-10 bearing life would allow.

The preceding decisions of design, as concluded from an overall consideration of the presented variation studies of optimization, are listed in proper chronological order for application in the design process. For greatest expediency, a graphical approach would indicate the practical extent to which we could achieve the preceding goals of design. Some simple analysis-type calculations would be necessary in the making of such a design layout drawing, together with the exercising of good engineering judgement.

In this introductory example, we have concentrated our efforts on some problems of mechanical design synthesis, particularly with respect to steps 3, 4, and 5 of the morphology in Figure 1.3.   Some points of general importance have been illustrated by the example in these areas of the design process.   Finally, it should be emphasized that a thorough analysis is recommended for the design achieved with respect to all conceivable phenomena of significance.   Such evaluation work would take place in steps 6 and 7 of the morphology in Figure 1.3.

1-6   INTRODUCTORY EXAMPLE 2—RACING CAR DESIGN PROBLEM

In the design of a high-speed racing car there are many challenging problems of optimum design.   For instance, we would like to design the fuselage shape for minimization of wind resistance and drag force.   This might very well be an interesting application for the calculus of variations.   Also, there are many other places in the design of a racing car where optimization techniques could be profitably applied.   However, space in the book would not allow us to consider the many aspects of interest.   Instead, for this introductory example we will merely consider a phase in the mechanical design synthesis, e.g., maximizing starting acceleration a_s of Figure 1.10.   This is an important goal to strive for, since we wish to get our vehicle off to a fast start ahead of our competitors in a race.

A free-body diagram of the basic configuration, having been derived from steps 3 and 4 in the morphology of Figure 1.3, is shown in Figure 1.10.   For now, at least, we will assume that both road and tire conditions are excellent and that there will be no appreciable slippage in starting.   From the application of basic dynamics we derive equation 1.4 for starting acceleration a_s. In this equation, T_s is the starting torque of the drive motor, r_s is the step-down starting gear ratio between the drive motor shaft and the wheel axle, D is the wheel diameter, W_M is the motor weight, W_B is the remaining body weight including the racing driver, and g is the gravitational acceleration constant.

$$F = (W_M + W_B)a_s/g$$
$$F(D/2) = T_s r_s$$
$$F = 2T_s r_s/D = (W_M + W_B)a_s/g$$
$$a_s = \frac{2T_s r_s g}{D(W_M + W_B)} \tag{1.4}$$

Assume that we have some freedom in design with respect to selection or design of the drive engine.   From information available, we estimate that starting torque T_s will be approximately proportional to motor weight W_M,

Figure 1.10   Free-body diagram of racer at start.

as expressed by equation 1.5.   Hence, we recognize that T_s and W_M are not independent terms in equation 1.4.

$$T_s \approx k \ W_M \tag{1.5}$$

In equation 1.5, the value for the constant of proportionality k could be estimated if desired, but this is not necessary for the mechanical design synthesis which is to follow.

By combining equations 1.4 and 1.5 we obtain equation 1.6 for starting acceleration a_s, where now the terms on the right side are independent of each other.

$$a_s \approx \frac{2kr_s g}{D\left[1 + \left(\dfrac{W_B}{W_M}\right)\right]} \tag{1.6}$$

From this developed primary design equation 1.6 we directly draw the following conclusions of design significance, for maximization of starting acceleration a_s.

1. If the choice exists, select or design the motor for maximum starting torque size.   The desirability for this decision is clear since we wish to maximize W_M as seen from equation (1.6), which means maximization of starting torque T_s (as seen from equation (1.5)).   The extent to which we could go in this direction would be governed by available space and economic limitations, for an acceptable motor size.

2. Design the transmission so starting gear ratio r_s would be as large as practical.

3. Have drive wheel diameter D as small as practical, referring to Figure 1.10.   The extent to which we could go in this direction might be governed by such considerations as required road clearance, bearing life, or perhaps other conditions of operation which have not been considered in this introductory example.

4. It would be desirable to have body design such that weight W_B would be of low value.

As in the previous example, some graphical work would be expedient for arriving at the final racing car design.   In this respect, you might work closely with your layout designer to incorporate the preceding goals of optimum design as well as practical.   The extent to which these design goals could be realized would be determined by factors in the total picture, such as considerations of space available, practical proportions, the exercising of engineering judgement, the making of some simple analysis calculations, and modes of operation other than the starting condition.   Eventually we would arrive at the design specifications, which would complete the mechanical design synthesis in step 5 of the Figure 1.3 morphology.

As for the previous example, it would be strongly recommended in steps 6 and 7 of Figure 1.3 to analyze thoroughly the design achieved for all conceivable phenomena of significance. As a specific example, we certainly would want to analyze for the validity of our assumption that no slippage would occur between tire and road on starting. If not true, a feedback loop in Figure 1.3 should be followed to include slippage effect in our variation study for optimum design.

1-7 AN INDUSTRIAL EXAMPLE OF MECHANICAL DESIGN

The morphology of mechanical design from Figure 1.3 is illustrated in Figures 1.11 through 1.14, for the design of a detachable tensile socket used in field operations such as in dredging rigs with steel wire rope. The problem of quick assembly and disassembly in field operations was recognized for steel attachments in such cases. A search of existing designs did not reveal any which satisfied all of the operational requirements. The creative phase for a new product resulted in selection of a wedge-type design shown in Figure 1.11, from the patent which was finally issued on the device. Optimization techniques were applied in eight places as an important part of selecting the final dimensions and materials. The photograph of an early model is shown in Figure 1.12, which was subsequently improved through feedback loops (see Figure 1.3); the final design is shown in Figure 1.13. Part of the experimental evaluation of the model is shown in Figure 1.14, where the socket design is shown to have successfully withstood the breaking force of the steel wire rope. Successful entry to the production-consumption cycle required extension of the prototype design to a wide range of standard sizes, which was accomplished using a digital computer.

1-8 PHILOSOPHY OF SUCCESS OR OPTIMIZATION OF ACHIEVEMENT

Success is a general goal for optimization which nearly everyone strives for in life regardless of the activity in which he is engaged. The achievement reached in any given period of time is a function of many factors, some of which are under our personal control. Hence, in the design synthesis of the path which one is to follow in life, a worthwhile objective for optimization is maximization of achievement. Let us take a look at this problem in very broad mathematical terms drawing some general philosophical conclusions. Application of this philosophy in projects of mechanical design synthesis is a very important part in the achievement of maximum success.

In general we can graphically describe the achievement reached in a given period of time as shown in Figure 1.15. The objective for optimization is

Nov. 14, 1967   R. C. JOHNSON   3,351,986

WIRE ROPE WEDGE CLAMP

Filed June 8, 1966   2 Sheets–Sheet 2

INVENTOR
RAY C. JOHNSON
By Donald G. Dalton
Attorney

Figure 1.11   Detachable tensile socket patent drawing. *Courtesy United States Steel Corporation.*

Figure 1.12   An early result from mechanical design synthesis of a detach-able tensile socket.   *Courtesy United States Steel Corporation.*

Figure 1.13   Final design for the detachable tensile socket.   *Courtesy United States Steel Corporation.*

Figure 1.14   Conclusion of a successful test in the evaluation of a detach-able tensile socket design. *Courtesy United States Steel Corporation.*

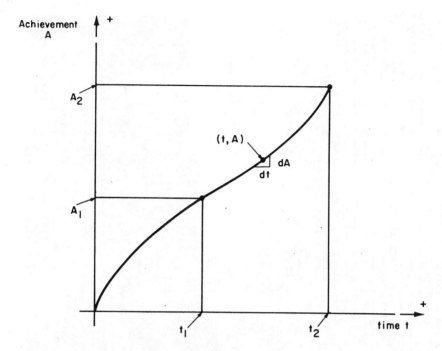

Figure 1.15   Variation of a typical function of achievement with time.

maximization of A_2. Hence, referring to the figure we describe generally the achievement function mathematically as follows:

$$A_2 = A_1 + \int_{t_1}^{t_2} \left(\frac{dA}{dt}\right) dt \tag{1.7}$$

We will assume that initial achievement A_1 at starting time t_1 as well as completion time t_2 are requirement specifications over which we have no control for the particular problem at hand. Hence, for maximization of achievement reached (A_2), from equation (1.7) we see that the basic goal is to maximize (dA/dt) at any time t within the range $t_1 \leqslant t \leqslant t_2$. To draw some philosophical conclusions of significance, we must next resolve the achievement function A into some meaningful components.

Achievement A is a composite function f of several time-dependent functions, which we will generally express as follows:

$$A = f(F_{na}, F_{ef}, F_{tr}, F_{en}, F_{at}) \tag{1.8}$$

Functions F in equation (1.8) are with respect to native ability, effort expended, training and experience, environment, and attitude, respectively, as designated by appropriate subscripts. We make no attempt to describe further the explicit character of these F functions, but we do know that they are functions of time as well as other factors.

Applying the theory of differentiation of composite functions to equation (1.8) we obtain equation (1.9) for (dA/dt), which we wish to maximize at any time t in the interval $t_1 \leqslant t \leqslant t_2$.

$$\left(\frac{dA}{dt}\right) = \left(\frac{\partial A}{\partial F_{na}}\right)\left(\frac{dF_{na}}{dt}\right) + \left(\frac{\partial A}{\partial F_{ef}}\right)\left(\frac{dF_{ef}}{dt}\right) + \left(\frac{\partial A}{\partial F_{tr}}\right)\left(\frac{dF_{tr}}{dt}\right)$$

$$+ \left(\frac{\partial A}{\partial F_{en}}\right)\left(\frac{dF_{en}}{dt}\right) + \left(\frac{\partial A}{\partial F_{at}}\right)\left(\frac{dF_{at}}{dt}\right) \quad (1.9)$$

In the right side of equation (1.9) there are two classes of terms involved. First, the $(\partial A/\partial F)$ terms are primarily problem oriented ones which are generally influenced by factors beyond your control as an individual. Hence, in this respect achievement is influenced in varying amounts by native ability, by effort expended, by training and experience, by environment, and by mental attitudes, the degrees of which depend on the particular project situation at hand. Thus, we recognize that there are many factors beyond our control as individuals which will influence the achievement reached. Secondly, the (dF/dt) terms are primarily ones which to a great extent are under your personal control as an individual. Hence, for maximization of (dA/dt), from equation (1.9) we draw the general conclusion that the (dF/dt) terms should be maximized. This, of course, assumes that you are in a healthy problem situation where the $(\partial A/\partial F)$ terms are all positive. An example of a negative term would be in what we might call an unhealthy problem situation where achievement is decreased with increased effort, because of problem conditions over which you have no control.

Since we do not know the explicit character of the F functions involved, we cannot draw explicit conclusions of optimum design for maximization of the (dF/dt) terms. Nevertheless, we have resolved the problem of maximizing achievement into some basic components to which we can associate some important philosophical goals, briefly summarized as follows, referring to the (dF/dt) terms in equation (1.9) over which we have some personal control.

1. *Native Ability.* For the most part this is an inborn factor over which we have little control. However, we can maximize the term to some extent by continually striving for good mental and physical health as related to physical exercise, proper rest, and goals of high moral conduct, all of which can affect the degree of utilization which is possible for our native abilities.

2. *Effort.* The importance of continually doing our best should always
   be kept in mind, striving for maximization of ambition and personal
   drive within proper limits.   There is no room in success for sloth and
   procrastination.
3. *Training or Experience.*   For maximization of this term we should continually be aware of the importance of increasing our knowledge and
   experience.   This can be accomplished through concentrated efforts of
   both formal and informal education, as well as on-the-job training and
   learning.
4. *Environment.*   For maximization of this term, one should be continually
   aware of his surroundings as related to the possibilities of success.   The
   development of good personal relationships as well as a consideration
   of physical surroundings and equipment followed by necessary modifications will help to maximize your achievement.
5. *Attitude.*   For maximization of this term, one should continually strive
   for a positive healthy outlook which is free of degenerating negativism.
   Be a realistic optimist, and restrict your criticism to the constructive
   type and only on matters which are significant.   The importance of
   proper mental attitude in creative work is well known, and this term is
   equally important in the more general problem of maximizing success
   in almost any activity of life.

In brief summary and conclusion, we note that achievement is a complicated function of many factors.   Many of these are beyond our control, and
we must accept the fact that they will limit the achievement possibilities in a
given problem situation.   On the other hand, many of the factors are to an
appreciable extent within our control as an individual.   Hence, we should
be aware of these factors and continually strive for the development of
characteristics which we know will maximize our achievements and success.
The basic philosophy now discussed is often overlooked by the engineer, but
nevertheless it is an extremely important part in the activity of mechanical
design synthesis.

1-9   ADVANCED DESIGN AND MECHANICAL DESIGN SYNTHESIS

In Figures 1.4 and 1.5 some industrial examples of advanced designs were
presented, illustrating the long-range evolution of product designs as related
to repetitious applications of essentially the Figure 1.3 morphology on a
continuing basis.   Some other examples of advanced designs and concepts
are presented in Figures 1.16 through 1.20 for the automotive, aircraft, and
space fields.   They also illustrate very well what will be or what has been
achieved by various industries in the evolutionary process of mechanical

design synthesis for certain product types.   Each of the examples illustrates mechanical design synthesis based on satisfaction of particular functional requirements in specific problems of design.

Successful solution to a problem of design requires a great deal of creative effort in the *synthesis of configurations* as well as much engineering effort, both practical and theoretical in nature, for the *selection of materials and dimensions*.   These two basic aspects of mechanical design synthesis will be discussed in Parts I and II, respectively, which follow in the book.   Some explicit techniques for coping with challenging problems in these areas of design will be presented and illustrated.

Figure 1.16   The Chrysler Concept 70X has many features of advanced design in the automotive field, such as the parallelogram mounting of doors, which is ideal for tight parking situations.   *Courtesy The Chrysler Corporation.*

Figure 1.17   The U.S. Supersonic Transport has many advanced design features for supersonic flight (300 passengers, 1800 mph cruising speed at 12-mile altitude, 298-foot length, 143-foot wing span, 67,000 pounds of jet engine thrust, etc.).   *Courtesy The Boeing Company.*

Figure 1.18   The S-3A carrier-based antisubmarine aircraft has many features of advanced design, resulting in a speed and range which is twice that of the earlier S-2 design.   Its basic airframe is of minimum size and weight, and its payload of sensors and avionics gives detection capabilities equal to or better than those of the land-based P-3C.   *Courtesy The Lockheed Aircraft Corporation.*

Figure 1.19   The DC-10 Jetliner has many advanced design features including the latest turbofan engines, a powerful high-lift system, and high-speed airfoil design (345 passengers, 180-foot length, nearly 20-foot fuselage diameter, 161-foot wing span, 49,000 pounds of engine thrust at take-off, etc.). *Courtesy McDonnell Douglas Corporation.*

Figure 1.20   Concept of an advanced spacecraft design.   The twelve-man space station with three decks would be designed for eventual orbit around a planet such as Mars.   *Courtesy National Aeronautics and Space Administration.*

PART I

SYNTHESIS OF CONFIGURATIONS IN CREATIVE DESIGN

2

Introduction to Creative Design

2-1 PROCESS OF CREATIVE ACTIVITY

Man will always encounter problems which demand originality in action for successful solution. As an individual he has various degrees of native ability, knowledge, and experience as basic tools from which to draw. The precipitation of creative activity is necessary to bring this background to bear on the problem at hand for obtainment of an original and successful solution. Obviously we wish to maximize the degree of creative accomplishment which we achieve for a particular problem.

Throughout a lifetime, an individual may either maximize or minimize his creative accomplishment. He may allow himself to be influenced by forces of positivism and freedom with responsibility, leading to the path of success in creative accomplishment. On the other hand, he may be overcome by the forces of negativism and subjection, leading to the path of defeat in creative accomplishment. Extreme possibilities for the path which we as individuals may take are graphically depicted in Figure 2.1.

Let us more specifically consider some of the factors which affect our degree of success with respect to creative accomplishment. The basic

Figure 2.1   Typical process of creative activity for an individual of given native ability, depicting some basic factors of influence.

philosophy applies to problems of life in general.   However, specific techniques to be presented in later chapters pertain to the solution of problems in the creative stages of mechanical design synthesis.

2-2   TRAITS FOR CREATIVE ACTIVITY

The precipitation of creative activity is influenced by many personal traits which one can develop or control at least to some extent.   These factors have been studied and published in many places, such as in references 2-1 through 2-6 of the bibliography.   Also, some of these traits are mentioned in section 1-8 of Chapter 1 for more general application than the achievement of creative activity.

The path of success for creative accomplishment is most likely followed if one properly develops some significant personal traits in his work activity. An optimistic outlook, an open-minded approach, self-confidence, constructive discontent, patience, ambitious drive, and perseverance and diligence in carrying through one's convictions are some personal attributes which are helpful for creative action.   These personal traits of positivism applied in an environment of freedom with an exercising of individual responsibility are characteristic of creative success.   Development of traits such as these are forces of influence which will tend to maximize your degree of creative accomplishment, as depicted in Figure 2.1.

On the other hand, the path of defeat for creative accomplishment is most likely followed if one develops personal traits of negative character.   A pessimistic outlook, bigotry, skepticism, disbelief, uncertainty, fear of criticism or failure, and slothfulness are some examples of personal characteristics which generally do not enhance creative activity.   They are forces of negativism and subjection which tend to depress the achievement of creative accomplishment, as depicted in Figure 2.1.   Unfortunately, if one submits to such prevailing influences from many sources of negativism, creative activity will be quenched more and more the older one becomes at a stage in life when he has the greatest wealth of experience and knowledge from which to draw.   Hence, the natural tendency is to drop off in creative activity as time progresses, because of such degenerating influences.

From personal experience, it is the belief of the author that creative activity can increase with age by a concentrated effort to overcome the forces of negativism with an emphasis on those of positivism.   Of course, this requires that one be removed as much as possible from subjection, and that he be given a great deal of freedom, which must be taken with a high degree of personal responsibility.   Fortunately, for the most part, such an environment has been available to the author in much of his work.

An encyclopedic check of history reveals that many highly creative people of the past had characteristics of positivism and they were not stymied in their work by extreme forces of negativism to which they were subjected. Two early examples are Galileo and Sir Isaac Newton as described in reference 2-7. Another good example of positivism prevailing over the forces of negativism is described in reference 2-8, as related to the creative work of Alexander Graham Bell. Also, many more contemporary examples could be cited which would attest to the validity of this philosophy. Hence, we conclude that creative accomplishment is enhanced by the development of traits of positivism in an environment of freedom with responsibility. It is depressed by succumbing to the influence of negativism and subjection.

2-3   STIMULATING CREATIVE ACTIVITY

The development of proper personal traits is not the only consideration which you should make for maximizing the possibility of creative accomplishment. There are specific techniques which we can apply that will enhance the precipitation of creative activity. Some of the stimulants or "catalysts" for creative activity are well known, such as the brainstorming technique in group effort. Others are not so well known, some of which will be presented briefly in Chapters 3 and 4. Let us first describe the general purpose and function for an aid to creative effort.

When presented with a problem requiring a creative solution, many people find it difficult to get started in the conception of ideas. We cannot always depend on the "flash-in-the-night" solution as a stroke of the genius. To move away from the initially inactive dead-center position of the otherwise dullard, mental activity of a creative nature must be stimulated and a tool for accomplishing this must be introduced. Such an aid for creative effort is a necessary link for most of us in order to generate many ideas for the solution of a problem. Even the genius can supplement his list of conceived ideas by the application of a stimulant technique for increased creative activity.

The personal trait of ambitious drive must be exercised for the *initiation* of creative activity. A concentrated effort must be made to come up with at least a poor solution to the problem. This may be enhanced by following the logical building-block approach described in Chapter 4 as an aid for creative activity. After conception of the initial solution, another stimulant technique may be introduced for the precipitation of other ideas. This basic function of the stimulant technique in creative design is depicted graphically in Figure 2.2. Oftentimes perseverance in the repetitious application of this plan can help you to generate new ideas which otherwise would not be conceived. Let us next briefly summarize some of the stimulant techniques which the author has found to be helpful in the early stages of mechanical design synthesis.

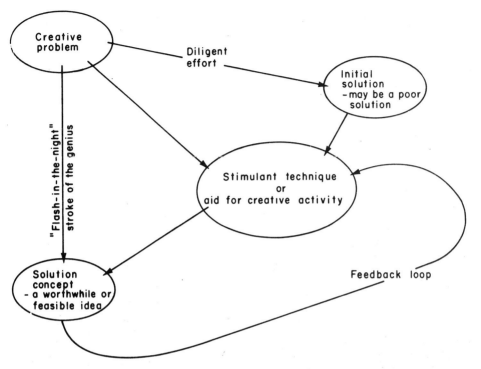

Figure 2.2   Function of a stimulant technique in creative design effort.

2-4   SUMMARY OF SOME STIMULANT TECHNIQUES

In the next two chapters we will briefly describe and illustrate some stimulant techniques for enhancing creative activity in the early stages of mechanical design synthesis.   In Chapter 3 the systematics of linkages will be described and illustrated as an aid for the creative design of mechanical devices involving linkages.   In Chapter 4 some other stimulant techniques will be described and illustrated for the creative design of mechanical devices in particular. These are the logical building-block approach, synthesis by implication from optimum design, and the use of circuit diagrams as an aid for creative activity. The technique of brainstorming will not be described since it is quite well known and it is presented in many other references.   The interested reader undoubtedly will want to refer to other sources for some other stimulant techniques which can help in the precipitation of ideas according to the Figure 2.2 plan.   A very recent publication is reference 2-10, which describes the techniques of self-interrogation, role playing, functional visualization, idea diagrams, morphological idea matrixes, and synectics, as well as brainstorming.

Finally, it is suggested that the reader be on the lookout for other stimulant techniques which according to the plan of Figure 2.2 might serve as an aid for precipitating creative activity. The author believes that there are many such possibilities which can be discovered if a conscientious effort is made in that direction.

BIBLIOGRAPHY

2-1   A. D. Moore, *Invention, Discovery, and Creativity*, Doubleday & Company, Inc., Garden City, New York, 1969, 178 pp.

2-2   J. R. M. Alger and C. V. Hays, *Creative Synthesis In Design*, Prentice-Hall, Inc., Englewood Cliffs, N.J., 1964, 92 pp.

2-3   D. H. Edel, Jr., *Introduction To Creative Design*, Prentice-Hall, Inc., Englewood Cliffs, N.J., 1967, 237 pp.

2-4   H. R. Buhl, *Creative Engineering Design*, The Iowa State University Press, Ames, Iowa, 1960, 195 pp.

2-5   E. K. Von Fange, *Professional Creativity*, Prentice-Hall, Inc., Englewood Cliffs, N.J., 1959, 260 pp.

2-6   L. Harrisberger, *Engineersmanship—a Philosophy of Design*, Brooks/Cole Publishing Company, Belmont, Calif., 1966, 149 pp.

2-7   *The World Book Encyclopedia*, Field Enterprises Educational Corporation, Chicago, Ill., 1966 ed., Vols. 8 and 14.

2-8   B. A. Weisberger, "The Age of Steel and Steam," Vol. 7, in *The Life History of the United States*, Time Incorporated, New York, 1964, page 42.

2-9   R. C. Johnson, "Optimum Design By Synthesis," presented at the 1970 Design Engineering Conference on May 14, 1970 in Chicago, Ill., sponsored by A.S.M.E., Paper No. 70–DE–8.

2-10 P. H. Hill, *The Science of Engineering Design*, Holt, Rinehart and Winston, Inc., New York, 1970, Chap. 2.

2-11 R. Barron, *Creativity and Professional Freedom*, D. Van Nostrand Co., Inc., Princeton, N.J., 1968.

2-12 G. L. Glegg, *The Design of Design*, Cambridge University Press, London, 1969, 93 pp.

2-13 W. J. Gordon, *Synectics: The Development of Creative Capacity*, Harper & Row Publishers, New York, 1961, 180 pp.

2-14 R. C. Johnson, "Design Synthesis – Aids to Creative Thinking", *Machine Design*, Penton Publishing Co., Cleveland, Nov. 15, 1973, pp. 158-163.

2-15 R. May, *The Courage To Create*, Norton Co., New York, 1975, 143 pp.

2-16 J. D. Roslansky, *Creativity*, North Holland Publishing Co., Amsterdam, 1970, 90 pp.

2-17 C. W. Taylor and F. Barron, *Scientific Creativity*, John Wiley & Sons, Inc., New York, 1963, 419 pp.

3

Systematics of
Linkages

3-1 INTRODUCTION

The systematics of linkages is a relatively unknown technique which can serve as an aid for creative activity in execution of the general plan presented in Figure 2.2. The technique has many practical applications in the creative synthesis of various types of mechanical devices, such as for mechanisms, structures, and internal-force-exerting devices. For supplementary study the interested reader should particularly consult Chapter 3 of reference 3-1, as well as references 3-2 through 3-5 and 3-10. Also, in the present book some other examples employing this technique are presented in sections 4-2 and 9-6 of Chapters 4 and 9, respectively.

Some well-known mechanical devices are sketched in Figure 3.1 together with what we will call their *associated linkage* for each case. In the figure we actually depict steps of analysis, since we use the existing device to sketch the associated linkage. On the other hand, for the work of creative synthesis the procedure will be in reverse since the associated linkage will be the basic tool used for the precipitation of new configurations. Hence, it will be helpful for us to have on hand a catalog of such basic linkages from which we can select associated linkages for creative synthesis use.

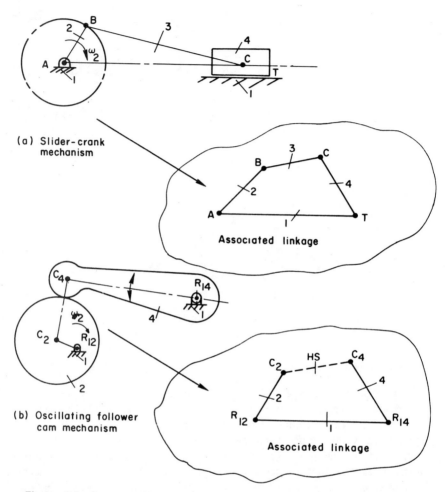

(a) Slider-crank
    mechanism

Associated linkage

(b) Oscillating follower
    cam mechanism

Associated linkage

Figure 3.1   Some well-known mechanical devices and their associated linkages.

Some basic notation used in later work is brought out in Figure 3.1. First, imaginary links in a mechanical device will be depicted by dashed lines in the associated linkage. Hence, the imaginary binary link joining the centers of curvature of surfaces in higher-pair sliding contact are shown as dashed links labeled HS in the associated linkages of Figures 3.1(b) and (c). Incidentally, in what follows we will depict equivalent higher-pair sliding in an associated linkage even if a roller follower is actually used in the mechanism, since the

R = rotation point of gears

B = base point of gears, centers of curvature of involute teeth in contact

Center of curvature of gear tooth in contact

Equivalent quadric crank mechanism shown in phantom

(c) Meshing gearset

(d) Simple structure with external load

Associated linkage

Figure 3.1 (*continued*)

only effect of the roller is friction and wear reduction. Also, the imaginary links introduced to represent the effects of forces *N* and *P* are shown with dashed lines in the associated linkage of Figure 3.1(e). Finally, if clarity is improved, lower-pair sliding joints in a mechanical device will be designated by *R* or *T* in the all pin-jointed associated linkage. *R* will designate an actual pin-jointed or hinged connection between links allowing relative rotation,

(e) Nut cracker
an internal-force
exerting device

Associated linkage

Figure 3.1 (*continued*)

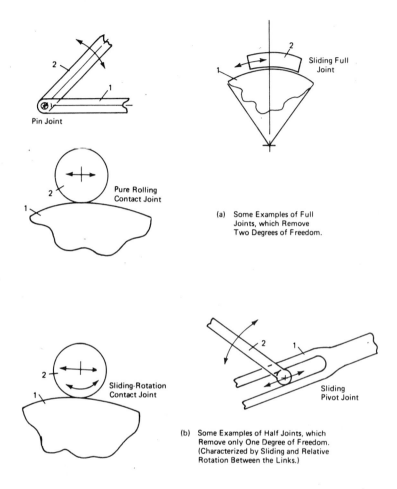

Pin Joint

Pure Rolling Contact Joint

Sliding Full Joint

(a) Some Examples of Full Joints, which Remove Two Degrees of Freedom.

Sliding-Rotation Contact Joint

Sliding Pivot Joint

(b) Some Examples of Half Joints, which Remove only One Degree of Freedom. (Characterized by Sliding and Relative Rotation Between the Links.)

Figure 3.2   Examples of Full and Half Joints for Connecting Two Links.

whereas T will designate that in the device a lower-pair sliding joint actually will be used allowing relative motion of translation between the connected links. The R terminology is employed in Figure 3.1(c), and for a meshing pair of gears we will always have a four-sided closed loop of $RR3B$ sequence with the gears rotating relative to a common frame link. Incidentally, the B's designate the base points or centers of curvature of the involute gear teeth in contact. The T terminology is illustrated in Figure 3.1(a) for the slider crank mechanism, designating in the associated linkage that the device actually has a lower-pair sliding joint allowing relative translational motion between links 1 and 4 instead of a pin-jointed connection.

Grübler's equation for plane linkages having one link fixed is as follows:

$$F = 3(L - 1) - 2J \tag{3.1}$$

In equation (3.1), F is the number of degrees of freedom for the device, L is the number of links including the frame, and J is the number of full joints. A full joint we will define as one which removes two degrees of freedom from the system, such as a hinged pin-joint, a lower-pair sliding translational joint, or higher-pair rolling contact without relative sliding. On the other hand, higher-pair sliding connections between links could be considered as half-joints since they remove only one degree of freedom from the system. Some examples of full and half joints are illustrated in Figure 3.2 for connecting links 1 and 2.

Incidentally, we define a *joint* as a connection between two links. It is also known as a *kinematic pair*, having two *elements*, one from each link. In application of Grübler's equation (3.1), be sure to count only the joints actually in use for connecting links. Hence, any inactive joints not affecting constraint of the linkage should be ignored and should not be counted in the J number.

Regarding F in Grübler's equation (3.1), the *number of degrees of freedom* which a mechanical system has is defined as the number of *independent* co-ordinates required to define uniquely the position of the links, assuming that the dimensions of the links themselves have been established. Thus, for the simple devices of Figure 3.1 the values of F are obvious from a visual inspection, and they are confirmed by the application of equation (3.1) to the linkages shown. Incidentally, in going from the Nut Cracker device to the Associated Linkage in Figure 3.1(e), the internal forces P and N have been replaced by imaginary binary links, which are two-force members of equivalent effect. This technique in reverse will be applied in the synthesis of some internal force exerting devices to follow.

It will be expedient in synthesis work to use full joints in associated linkages depicted as hinged pin-joints, even though the actual device may have other types of connections, and this is illustrated in Figures 3.1(a), (b), and (c). Incidentally, some examples of higher-pair sliding joints are illustrated in Figures 3.1(b) and (c). However, for such cases a hypothetically equivalent mechanism is obtained by introducing an imaginary binary link, HS, between the centers of curvature for the contacting surfaces, resulting in only full joints for the associated linkage as shown in the figures.

Grübler's equation (3.1) is derived quite simply as follows. Any link constrained only to planar motion has three degrees of freedom. However, fixing one of the links removes three degrees of freedom from a system of links, and each full joint removes two degrees of freedom from the system. Thus, we have

$$F = 3L - 3 - 2J = 3(L - 1) - 2J$$

Applying Grübler's equation (3.1) to the associated linkages of the mechanisms shown in Figures 3.1(a), (b), and (c) we obtain

$$F = 3(L - 1) - 2J = 3(4 - 1) - 2(4) = + 1$$

For the associated linkage of the structure in Figure 3.1(d) we obtain

$$F = 3(L - 1) - 2J = 3(3 - 1) - 2(3) = 0$$

For the associated linkage of the internal force exerting device of Figure 3.1(e) we obtain

$$F = 3(L - 1) - 2J = 3(4 - 1) - 2(5) = - 1$$

The negative value of F merely indicates that the linkage has an additional constraint beyond what is necessary to establish uniquely the position of the assemblage. Hence, with the technique previously described of replacing internal forces of a device with binary links in the Associated Linkage, $F = -1$ is characteristic of both preloaded structures and internal force exerting devices. In general, we would find that the value of F is characteristic of the function which the device performs. This fact is summarized more specifically in Table 3-1.

TABLE 3-1 CHARACTERISTIC FUNCTIONS OF F FOR MECHANICAL DEVICES

F	Function of Device
+1	Single degree-of-freedom mechanism, for converting a given input motion to a desired output motion.
0	Structure for supporting or resisting external loads.
−1	Preloaded structure with internal forces built in, *or* an internal-force-exerting device.
+2	Differential mechanism, such as for integrating two input motions to give a single output motion.

3-2 SOME BASIC LINKAGES FOR SYNTHESIS USE

An appropriately selected associated linkage often may be used as an aid for creative activity in synthesis of the basic configuration for a new device. Generally, simplicity of construction is a desired feature in realistic design situations. Hence, we would like to have available a catalog of the simplest possible basic linkages of various F values from which we may select associated linkages for synthesis use. Such linkages have been derived by combining basic types of links in different possible ways. The simplest types of individual links for

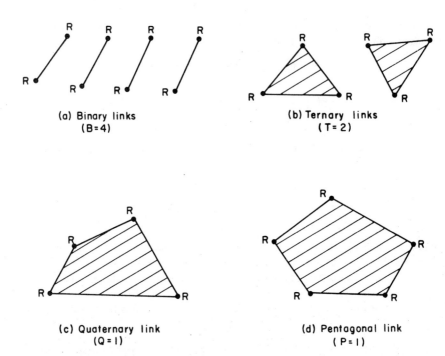

(a) Binary links
(B=4)

(b) Ternary links
(T= 2)

(c) Quaternary link
(Q = 1)

(d) Pentagonal link
(P= 1)

Figure 3.3 Some basic types of links showing elements of rotational pairs.

this purpose are shown in Figure 3.3, with the letter R designating an element of an unconnected pair. The number of such elements determine what type of link it is, designated as a binary, ternary, quaternary, or pentagonal link as shown in Figure 3.3.

The first part of our catalog will summarize the *possible link combinations* for single pin-jointed plane linkages, for the various F values of interest. The *numbers* of the various types of links in a linkage are designated by B, T, Q, and P for the binary, ternary, quaternary, and pentagonal links, respectively. Derivation of the possible link combinations follows, and the results are summarized in Table 3-2 as the first part of our catalog for synthesis use. The simplest possible link combinations are presented in the table, to the degree of complexity of eight links for $L = 8$ and considering the use of only binary, ternary, quaternary, and pentagonal links.

Derivation of Possible Link Combinations

Based on the definitions now made for the various terms, the values for L

and J must be positive whole number, *assuming planar linkages with full joints* as desired in our basic catalog. Rearrangement of equation (3.1) gives

$$J = \frac{3}{2} L - \frac{(F+3)}{2} \qquad (3.2)$$

Since the value of F must be a whole number, positive or negative, the following important conclusions are readily drawn from an examination of equation (3.2).

1. If F is an odd number $(-1, +1, +3, \text{etc.})$, L must be an even number.
2. If F is an even number $(0, +2, +4, \text{etc.})$, L must be an odd number.

The above conclusions must be true, or in light of the ŝtated restrictions on J, L and F it would be impossible to satisfy equation (3.2).

Next, based on the definitions now made for the terms, for any linkage we will have

$$L = B + T + Q + P + \cdots \qquad (3.3)$$

Also, since each binary link contributes two elements of two joints, and each ternary link contributes three elements of three joints, etc., with two elements required for each joint we obtain the following equation for J.

$$J = (2B + 3T + 4Q + 5P + \cdots)/2$$

Substituting this and equation (3.3) in equation (3.1), we obtain the following modified version of Grübler's equation for planar linkages with one link fixed.

$$F = 3(B + T + Q + P + \cdots - 1)$$
$$- 2(2B + 3T + 4Q + 5P + \cdots)/2$$
$$\therefore F = B - Q - 2P - 3 \qquad (3.4)$$

From the derivation, we realize that equation (3.4) would contain additional terms if links of greater complexity than the pentagonal type were included, but this would be getting into a range of impracticality away from the desire of simplicity in design. Also, it is interesting to note from equation (3.4) that the value of F does not depend on the number of ternary links in the linkage.

Next, if we subtract equation (3.4) from (3.3) we obtain

$$T + 2Q + 3P = L - (F + 3) \qquad (3.5)$$

The simultaneous progressive solution of this equation and equation (3.3) will explicitly determine the possible link combinations for planar linkages with full joints, in the manner illustrated as follows.

Suppose we wish to determine the possible link combinations for $F = +1$.

Then, equations (3.5) and (3.3) become as follows.

$$T + 2Q + 3P = L - 4 \tag{3.5a}$$

$$B + T + Q + P = L \tag{3.3}$$

From the definitions of the terms, we know that B, T, Q and P must be positive whole numbers, and since F is odd we know that L must be a positive even number. Hence, from equation (3.5a) the smallest number of links possible is for $L = 4$, requiring then that $T = Q = P = 0$. Thus, from equation (3.3) we conclude that $B = 4$ is the simplest possible link combination for $F = +1$, which would be for the well-known four bar linkage. The next even number to consider is for $L = 6$, and then equations (3.5a) and (3.3) become as follows.

$$T + 2Q + 3P = 6 - 4 = 2$$

$$B + T + Q + P = 6$$

The progressive simultaneous solution of these equations is facilitated using a decision tree as shown in Figure 3.4a. From the first of these equations we

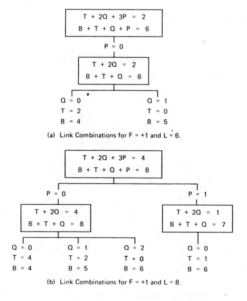

(a) Link Combinations for F = +1 and L = 6.

(b) Link Combinations for F = +1 and L = 8.

Figure 3.4   Decision Trees for Progressive Simultaneous Solutions of Equations (3.5a) and (3.3).

see that P must be 0, since a value of 1, 2, 3, etc would require a negative value of T or Q to satisfy the equation. Thus, with $P = 0$, the first equation reveals that the only possibilities for Q are $Q = 0$ or $Q = 1$, since 2, 3, etc. would require a negative value for T. Thus, in Figure 3.4(a) with $P = Q = 0$, from the first equation we conclude that $T = 2$ and from the second that $B = 4$ is then necessary. Hence, the possible link combination determined is for $L = 6$, $P = Q = 0$, $T = 2$, and $B = 4$. This is summarized as link combination II in Table 3-2. Similarly, in Figure 3.4(a) with $P = 0$, and $Q = 1$, from the first equation we conclude that $T = 0$ and from the second that $B = 5$ is then necessary. Hence, the possible link combination determined is for $L = 6$, $P = 0$, $Q = 1$, $T = 0$, and $B = 5$. This is summarized as link combination III in Table 3-2. Thus, for $F = +1$ and $L = 6$, the two possible link combinations have been determined explicitly. Finally, for $F = +1$ and $L = 8$, the next even number, there are four possible link combinations summarized as IV, V, VI and VII in Table 3-2. These are determined explicitly in a similar way, using the decision tree of Figure 3.4(b). The first two equations of Figure 3.4(b) are equations (3.5a) and (3.3) using $L = 8$, with the progressive simultaneous solution technique as illustrated for $L = 6$.

For the case of $F = 0$, equations (3.5) and (3.3) become as follows.

$$T + 2Q + 3P = L - 3 \tag{3.5b}$$

$$B + T + Q + P = L \tag{3.3}$$

As before, we realize that B, T, Q and P must be positive whole numbers, and since F is even we know that L must be a positive odd number. Thus, in a similar manner to $F = +1$, we derive the seven possible link combinations VIII through XIV summarized in Table 3-2 for $F = 0$.

For the case of $F = -1$, equations (3.5) and (3.3) become as follows.

$$T + 2Q + 3P = L - 2 \tag{3.5c}$$

$$B + T + Q + P = L \tag{3.3}$$

As before, we realize that B, T, Q and P must be positive whole numbers, and since F is odd for this case we know that L must be a positive even number. Thus, in a similar manner as previously illustrated, we derive the fourteen possible link combinations XV through XXVIII summarized in Table 3-2 for the $F = -1$ category.

Finally, for the case of $F = +2$, equations (3.5) and (3.3) become as follows.

$$T + 2Q + 3P = L - 5 \tag{3.5d}$$

$$B + T + Q + P = L \tag{3.3}$$

TABLE 3-2   POSSIBLE LINK COMBINATIONS FOR SINGLE PIN-JOINTED PLANE LINKAGES FOR $L \leqslant 8$

F	L	B	T	Q	P	Designation
+1	4	4	0	0	0	I
	6	4	2	0	0	II
	6	5	0	1	0	III
	8	4	4	0	0	IV
	8	5	2	1	0	V
	8	6	0	2	0	VI
	8	6	1	0	1	VII
0	3	3	0	0	0	VIII
	5	3	2	0	0	IX
	5	4	0	1	0	X
	7	5	1	0	1	XI
	7	3	4	0	0	XII
	7	4	2	1	0	XIII
	7	5	0	2	0	XIV
−1	2	2	0	0	0	XV
	4	2	2	0	0	XVI
	4	3	0	1	0	XVII
	6	4	1	0	1	XVIII
	6	4	0	2	0	XIX
	6	3	2	1	0	XX
	6	2	4	0	0	XXI
	8	6	0	0	2	XXII
	8	5	1	1	1	XXIII
	8	4	3	0	1	XXIV
	8	5	0	3	0	XXV
	8	4	2	2	0	XXVI
	8	3	4	1	0	XXVII
	8	2	6	0	0	XXVIII
2	5	5	0	0	0	XXIX
	7	6	0	1	0	XXX
	7	5	2	0	0	XXXI

As before, we realize that B, T, Q and P must be positive whole numbers, and since F is even for this case we know that L must be a positive odd number. Thus, in a similar manner as previously illustrated, we derive the three possible link combinations XXIX through XXXI summarized in Table 3-2 for the $F = +2$ category.

As an exercise in the basic technique illustrated specifically for $F = +1$, the interested reader should derive the possible link combinations to $L \leq 8$, for $F = 0$, $F = -1$ and $F = +2$. The answers can be checked in Table 3-2.

Derivation of Acceptable Isomers for the Link Combinations

Basic linkages for synthesis use have been determined from the thirty-one possible link combinations of Table 3-2, considering for each the possible ways in which the links may be connected relative to each other, disregarding variations obtained by dimensional changes and satisfying throughout the distribution of the characteristic function of the F value. Thus, we would consider the two linkages of Figure 3.5(a) to be of the same type, differing only in specific dimensions but having the same relative connection between the basic links of the link combination $B = 4$, $T = 2$. Likewise, for $F = +1$ synthesis use we would discard the isomer shown in Figure 3.5(b) because it contains a structural component of $F = 0$ characteristic. In fact, this particular linkage is equivalent for $F = +1$ synthesis use to the four-bar linkage which is derived in a simpler category. Thus it has no value of uniqueness for $F = +1$ synthesis use, and for that reason it is disregarded for our catalog.

The problem of deriving basic linkages for synthesis use will now be considered more specifically. To start, we define the *isomers* of a given link combination as the collection of linkages obtained having *different* relative connections between the types of links involved. Hence, in Figure 3.5 there are two isomers shown for the link combination $B = 4$ and $T = 2$. Either one, but not both, of those shown in Figure 3.5(a) would be considered as an isomer of the link combination. The other one shown would be the type of relative connection of links depicted in Figure 3.5(b).

Next, we define the *acceptable isomers* of a link combination as the isomers having the desired degree-of-freedom function distributed throughout. Thus, in Figure 3.5 only one acceptable isomer is shown for $F = +1$, which would be either one but not both of the linkages shown in Figure 3.5(a). The linkage shown in Figure 3.5(b) would not be an acceptable isomer of the link combination, since the $F = +1$ function is not distributed throughout the linkage. Only acceptable isomers will be useful for the synthesis of new devices in creative design. This requirement will be referred to as the *degree-of-freedom distribution principle*.

For synthesis use in creative design, we are interested in determining the acceptable isomers up to a preselected degree of complexity with respect to link combinations. Thus, for $L = 4$ in the $F = +1$ category of Table 3-2, we readily determine by graphical inspection that there is only one acceptable

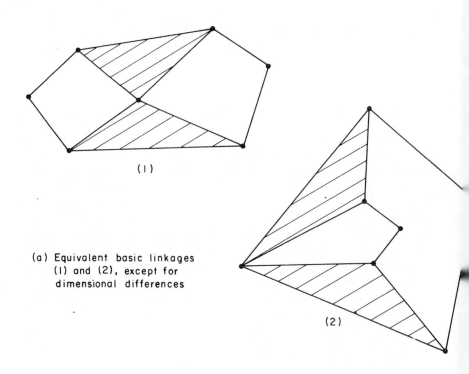

(1)

(a) Equivalent basic linkages
(1) and (2), except for
dimensional differences

(2)

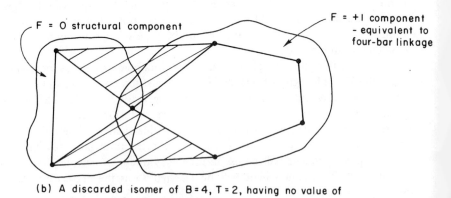

F = 0 structural component

F = +1 component
- equivalent to
four-bar linkage

(b) A discarded isomer of B = 4, T = 2, having no value of
uniqueness for F = +1 synthesis use

Figure 3.5 Some isomers of link combination II from Table 3-2, where
$L = 6$, $B = 4$, and $T = 2$.

Figure 3.6   Catalog of simplest acceptable isomers for synthesis use (see Table 3-2 for link combinations as designated above). (See Appendix A for Extended Catalog of Acceptable Isomers.)

isomer of the link combination $B = 4$, which is the four bar linkage shown at the top of Figure 3.6. On the other hand, there are two acceptable isomers for $L = 6$ in the $F = +1$ category of Table 3-2. These are shown as Watt's linkage and Stephenson's linkage at the top of Figure 3.6, both being derived from the link combination II of Table 3-2 having $B = 4$ and $T = 2$. They are readily determined by graphical manipulation of sketches using the links shown in (a) and (b) of Figure 3.3. In the same way, we can readily determine that the degree-of-freedom distribution principle cannot be satisfied for $F = +1$ with link combination III of Table 3-2, which is for $B = 5$ and $Q = 1$. To illustrate, the isomer of Figure 3.7 can readily be sketched using five binary links and one quaternary link, but since it contains the structural component it is not an acceptable isomer for $F = +1$ synthesis use. Its value for $F = +1$ use merely reverts to the four bar linkage already derived in a lower level of complexity. Hence, for $F = +1$ synthesis use, nothing new is contributed to our catalog of basic linkages by the isomer of Figure 3.7 and it is discarded from further consideration.

Acceptable Isomers from Number Sets

A number set approach can sometimes be helpful in determining the acceptable isomers of a link combination. To illustrate with a specific example, let us consider link combination II of Table 3-2 having $B = 4$ and $T = 2$ with the $F = +1$ function desired. Consider first the different ways of relative connection between the links involved, using the links of Figure 3.8 (a). Hence, let 0,1,2,3 and 4 designate the number of binary links in series emanating at an element of a ternary link for a linkage being derived. Since each ternary link has three elements of three different joints, a set of three numbers will be associated with any ternary link. Realizing that there are only four binary links available to work with in the link combination, and that we are interested in finding only the different *relative* connections possible between the links involved, we will have only the following four number sets possible for any ternary link in this example.

$$0\ 0\ 4$$

$$0\ 1\ 3$$

$$0\ 2\ 2$$

$$1\ 1\ 2$$

In the preceding number sets we note that the order in which the numbers appear makes no difference, since only the relative connections of links are sought. Hence, such numbers as 0 4 0, 4 0 0, 1 0 3, 3 1 0, etc. are not included, since the digit combination disregarding order is all we are interested in. Also,

(a)   Basic links to work with.

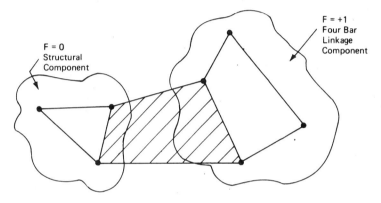

(b)   An isomer of the link combination contributing nothing new for F = +1 synthesis use.

Figure 3.7   A Link Combination $B = 5$ **and** $Q = 1$ **not Having Any Acceptable Isomers for** $F = + 1$ **Synthesis Use.**

note that for any of the four number sets the sum of digits equals four. This is true for the example since for either of the two ternary links all four binary links in some manner must emanate from the three elements of that ternary link in its connection to the other ternary link.

Incidentally, in any of the four possible number sets listed, a 0 designates that the element of the ternary link is directly connected (no binary link involved) to an element of the other ternary link in forming the joint. The digit 1 designates that the element of the ternary link is separated from an element of the other ternary link by one binary link. Similarly, the digits 2, 3 and 4 designate the number of binary links of separation between connected elements of the ternary links.

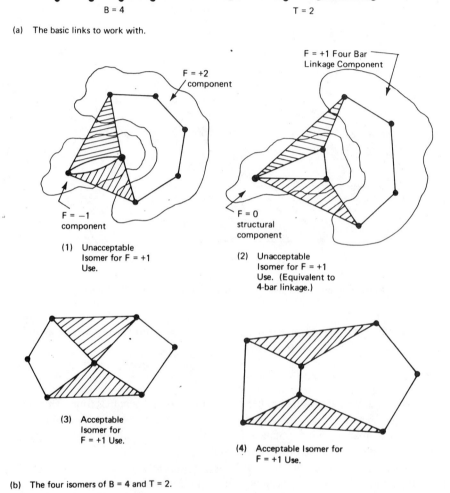

(a) The basic links to work with.

(b) The four isomers of B = 4 and T = 2.

Figure 3.8   The Isomers of Link Combination II from Table 3-2.

Continuing with the explanation of the number set approach for derivation of the isomers in the example, we realize that all elements of both ternary links must be used in their connection by binary links in various ways. Hence, since there are only two ternary links involved for the example, we require matching pairs of identical digits between two identical number sets for the derivation of an isomer. Thus, using the possible number sets previously summarized, the following four groups of matching digits in number sets are readily seen to be the only possibilities for isomers of the example.

Sketches of the four isomers associated with the four groups of matching number sets are given in Figure 3.8(b). The derivation is explicit for any group from the table of matching number sets, with a one-to-one correspondence between matched numbers and connections of elements in the isomer of the figure.

From a visual inspection of the isomers in Figure 3.8(b), it is obvious that the first two do not satisfy the degree-of-freedom distribution principle, whereas the last two are acceptable isomers for $F = +1$ synthesis use.

Similar number set approaches can be developed for the derivation of the isomers for various link combinations. The approach is particularly useful in conjunction with graphical manipulation by sketching in certain more complicated cases, such as for link combination IV in the $F = +1$ category of

Table 3-2, having $B = 4$ and $T = 4$. As a challenging exercise, the interested reader should derive the acceptable isomers for this link combination, with Appendix A used as a check at the end. If the matching number set approach is developed for aiding in the solution of this problem, there are seven groups of number sets to consider as follows.

Ternary Link				Acceptable Isomer Number
T_1	T_2	T_3	T_4	
0 0 0	0 1 1	0 1 2	0 1 2	(1)
0 0 1	0 0 1	0 1 2	0 1 2	(2)
0 0 1	0 1 1	0 1 1	1 1 1	(3)
0 0 1	0 0 2	0 1 2	0 1 1	(4)
0 1 1	0 1 1	0 1 1	0 1 1	(5), (6)
0 0 2	0 0 2	0 1 1	0 1 1	(7)
0 0 2	0 0 2	0 0 2	0 0 2	(8), (9)

The number sets summarized in the preceding table are not easily determined, and they have already been screened for derivation of only acceptable isomers of $B = 4$ and $T = 4$ for $F = +1$ use. Hence, a brief explanation should be given. The sum of number set digits in any row of the table must total eight, since each of the binary links is counted twice in the matching of identical digits. Also, such number sets as

T_1	T_2	T_3	T_4
0 0 0	0 0 0	0 0 4	0 0 4

have been eliminated at the start, since it is impossible to avoid $F = 0$ and $F = -1$ components in the matching of identical digits. Thus, for $F = +1$

use in this example, the degree-of-freedom distribution principle could not be satisfied for such groups of number sets. Finally, in the preceding table note that two acceptable isomers can be derived from each of two groups of number sets, since it is possible to match numbers two ways in each group and yet satisfy the degree-of-freedom distribution principle for $F = +1$ use.

For the suggested exercise, once you have matched the identical digits in the table, the nine acceptable isomers (1) through (9) can be sketched directly based on the one-to-one correspondence between the matched digits in the number set group and the linkage connections for ternary links. As a hint for the matching of identical digits in the table of seven number set groups, avoid the following matching conditions in particular. The matching of pairs of 0's and pairs of 1's should be avoided if the digits both occur in the same two ternary links, since such a condition will result in a structural component of $F = 0$ characteristic in the isomer. Also, the matching of more than one pair of 0's if they occur in the same two ternary links should be avoided, since such a condition will result in a preloaded structural component of $F = -1$ characteristic. Any of these conditions will result in an isomer which does not satisfy the degree-of-freedom distribution principle for $F = +1$ use, and hence it would not be an acceptable isomer and it would be discarded from further consideration.

Incidentally, another example of a Number Set Table is given in Appendix A, as an aid for finding the acceptable isomers of the $F = 0$ link combination $B = 3$ and $T = 4$.

Acceptable Isomers by Sketching Alone

Graphical manipulation of links in sketching is often most expedient and the simplest approach for the derivation of acceptable isomers. This would be the case for all of the acceptable isomers presented in Figure 3.6, which is the catalog of simplest ones for synthesis use. However, even in some of the more complicated link combinations of Table 3-2, free-hand sketching alone would be the most practical way to determine the acceptable isomers. The technique is best illustrated by example. Hence, examples are shown in Figures 3.9, 3.10 and 3.11 for link combinations arbitrarily chosen from Table 3-2 for the $F = +1$, $F = 0$ and $F = -1$ functions respectively. In each of these figures, all of the acceptable isomers are presented under (a), whereas one example of an isomer which does not satisfy the degree-of-freedom distribution principle for the particular F function desired is presented in (b).

The simplest acceptable isomers are of greatest value for synthesis use in practical design situations. Seventeen of these are summarized in Figure 3.6, which becomes the second part of our catalog. These simple basic linkages

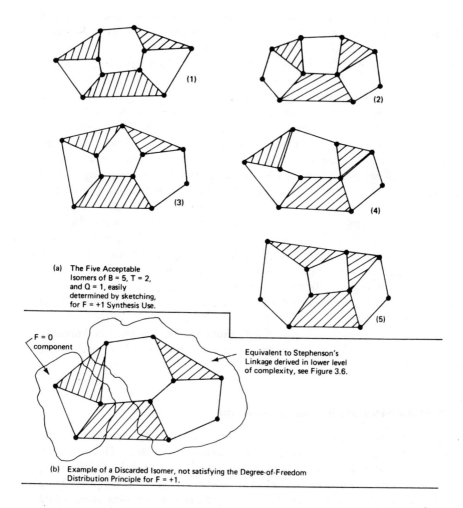

(1)

(2)

(3)

(4)

(a) The Five Acceptable
     Isomers of B = 5, T = 2,
     and Q = 1, easily
     determined by sketching,
     for F = +1 Synthesis Use.

(5)

F = 0
component

Equivalent to Stephenson's
Linkage derived in lower level
of complexity, see Figure 3.6.

(b) Example of a Discarded Isomer, not satisfying the Degree-of-Freedom
     Distribution Principle for F = +1.

**Figure 3.9   Isomers of Link Combination V in the $F = +1$ Category of
Table 3-2.**

may also be used with the logical building-block approach described in Chapter
4, as illustrated in Example 4-2. On the other hand, there are applications
where more complicated basic linkages are of direct value, such as in the synthesis of the yoke riveters in Example 3-10. In such situations the engineer
could sketch a feasible associated linkage directly from an appropriate link

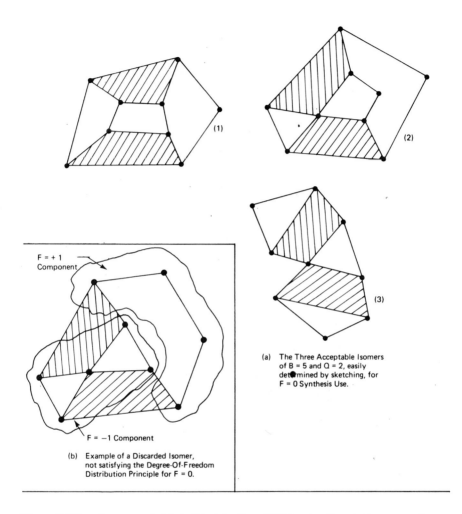

(1)

(2)

(3)

F = +1
Component

F = −1 Component

(a) The Three Acceptable Isomers
of B = 5 and Q = 2, easily
determined by sketching, for
F = 0 Synthesis Use.

(b) Example of a Discarded Isomer,
not satisfying the Degree-Of-Freedom
Distribution Principle for F = 0.

Figure 3.10   Isomers of Link Combination XIV in the $F = 0$ Category of Table 3-2.

combination taken from Table 3-2, using the techniques now described and illustrated for derivation of acceptable isomers. The interested reader should obtain some practice in the sketching of acceptable isomers from the link combinations of Table 3-2 as a good exercise. Also, an extended catalog of acceptable isomers for $F = +1$, $F = 0$, $F = -1$ and $F = +2$ is presented in Appendix A of the book.

(a) The Two Acceptable Isomers
of B = 5, T = 1, Q = 1 and
P = 1, determined by sketching,
for F = −1 Synthesis Use.

(b) Example of a Discarded Isomer, not satisfying the Degree-of-Freedom
Distribution Principle for F = −1.

Figure 3.11   Isomers of Link Combination XXIII in the $F = -1$ **Category**
of Table 3-2.

Basic linkages with higher joints may be derived by *partial shrinkage* of a
higher link in a single pin-jointed linkage. Incidentally, a *higher joint* is one
which connects more than two links, and a *higher link* is one having more
than two joint elements. Thus, if in Watt's linkage of Figure 3.12(a) we make
points F and G coincident, we derive the basic linkage having one higher joint,

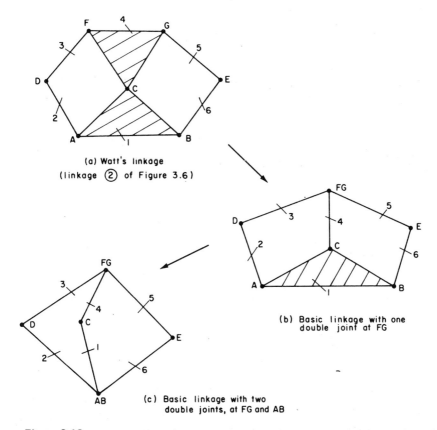

(a) Watt's linkage
(linkage ② of Figure 3.6)

(b) Basic linkage with one
    double joint at FG

(c) Basic linkage with two
    double joints, at FG and AB

Figure 3.12 Basic linkages for $F = +1$ with higher joints, as derived from Watt's linkage by partial shrinkage of links.

which is double joint FG shown in Figure 3-12(b). If in that linkage we make points A and B coincident, we derive the basic linkage having two double joints, FG and AB, as shown in Figure 3.12(c). This partial shrinkage technique for a link can be used to derive the possible basic linkages having higher joints from the single pin-jointed acceptable isomers of the catalog, and actually what is derived is merely a special dimension situation of coincident points. Also, note from Grübler's equation (3.1) that the value of F is not changed by execution of this technique, since both L and J retain their original values in the equation. In the partial shrinkage technique for derivation of acceptable isomers with higher joints, we would discard any linkage which violates the degree-of-freedom distribution principle.   Such unacceptable isomers with higher joints may be obtained from acceptable isomers of single joints, depending upon the particular way in which a link is shrunk. As an example, if link 4 in Figure 3.12(a) were partially shrunk by making points F and C coincident, a structural component of $F = 0$ would result whereas $F = +1$ is the characteristic desired.

We may partially shrink a higher link in many different ways. Specifically, the number of higher joints N_{hj} which are possible may be calculated from equation (3.6), giving the number of combinations of S_j single joints taken H_j at a time disregarding the order. Hence, for the formation of double joints use $H_j = 2$, for triple joints $H_j = 3$, etc.

$$N_{hj} = \frac{S_j!}{H_j!\,(S_j - H_j)!} \tag{3.6}$$

Thus, if a higher link such as the ternary link in Figure 3.12(b) has three single joints to start, and we are interested in the number of double joints possible by the partial shrinkage technique, from equation (3.6) we calculate $N_{hj} = 3!/[2!\,(3 - 2)!] = 3$. More specifically, the three double joints possible are obtained by making AB, BC and CA coincident points in Figure 3.12(b). Incidentally, the double joint formation AB shown in Figure 3.12(c) is the only one of the three possibilities which satisfies the degree-of-freedom distribution principle for $F = +1$. For the case of BC or CA we would form structural components of $F = 0$ characteristic, and such higher joint linkages would be discarded from further consideration.

To illustrate the partial shrinkage technique in a more complicated example, consider the acceptable isomer of $B = 5$, $T = 2$ and $Q = 1$ for $F = +1$ synthesis use as shown in Figure 3.13(a). Double joints may be formed by the partial shrinkage of ternary links 1 and/or 2 by forming coincident points AB and/or DE respectively, as shown in Figure 3.13(b). The possibilities of forming coincident points AC, BC, DF, and EF for double joints would form structural components, and these would be discarded since the degree-of-freedom distribution principle would not be satisfied for $F = +1$. Next, consider the possible ways in which we may partially shrink the quaternary link 3 of Figure 3.13(a) to form higher joints. Thus, for formation of double joints we use $H_j = 2$ and $S_j = 4$ in equation (3.6), giving $N_{hj} = 4!/[2!\,(4 - 2)!] = 6$. The six possible coincident points for formation of double joints with the quaternary link in Figure 3.13(a) are GC, CF, FH, HG, GF, and CH. To facilitate the determination of these possible double joints, we could use the circle diagram shown in Figure 3.13(c) with the single joints G, C, F and H of the link shown as points on the circle diagram. The double joint possibilities would be given by the terminal points of the six possible lines shown on this circle diagram.

In Figure 3.13(a) we readily note that formation of double joints GC, CF

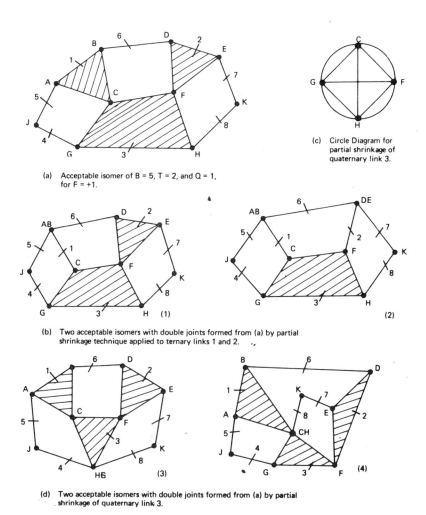

(c)   Circle Diagram for partial shrinkage of quaternary link 3.

(a)   Acceptable isomer of B = 5, T = 2, and Q = 1, for F = +1.

(b)   Two acceptable isomers with double joints formed from (a) by partial shrinkage technique applied to ternary links 1 and 2.

(d)   Two acceptable isomers with double joints formed from (a) by partial shrinkage of quaternary link 3.

Figure 3.13   Example of Some Acceptable Isomers with Higher Joints for $F = +1$ as Derived by Partial Shrinkage Technique.

and *FH* would give structural components of $F = 0$ characteristic, and these possibilities are discarded for $F = +1$ use. Formation of double joint *HG* would result in an acceptable isomer of $F = +1$. Also, in Figure 3.13(a) we note that formation of double joint *CH* would result in an acceptable isomer for $F = +1$, and by symmetry the formation of double joint *GF* would give the same type of linkage. Thus, only two acceptable isomers with one double

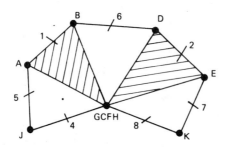

Figure 3.14   Linkage with One Triple Joint (GCFH) Formed by Total Shrinkage of Quaternary Link 3 in Figure 3.13(a). $(F = 0)$

joint are formed from partial shrinkage of the quaternary link 3 in Figure 3.13(a), and they are both shown in Figure 3.13(d).

Regarding partial shrinkage of the quaternary link 3 in Figure 3.13(a), we could form triple joints by making any three of the four joint elements of the link coincident. Hence, using $S_j = 4$ and $H_j = 3$ for formation of triple joints with the link, equation (3.6) gives $N_{hj} = 4!/[3! \, (4 - 3)!] = 4$. The four possible triple joints would be formed in Figure 3.13(a) by making GCF, GFH, CGH and CFH coincident, and these we recognize as the vertex points of the four possible triangles seen in the circle diagram of Figure 3.13(c), which could have been used to facilitate the determination. However, structural components of $F = 0$ characteristic are formed in each case, and all isomers with triple joints are discarded since they do not satisfy the degree-of-freedom distribution principle for $F = + 1$.

In conclusion for the partial shrinkage technique, the acceptable isomers having higher joints derived from Figure 3.13(a) are the four linkages shown in Figure 3.13(b) and (d) plus five more. Three of the additional ones are derived from Figure 3.13(b) (1) by making GH, HC and FG coincident, whereas the remaining two are derived from Figure 3.13(b) (2) by making GH and either CH or GF the coincident points. The interested reader should sketch these five additional acceptable isomers with higher joints as derived by the partial shrinkage technique to Figure 3.13(b).

Another technique for the derivation of linkages with higher joints would be by the *total shrinkage* of one or more higher links. Thus, as a specific example, if in Figure 3.13(a) we would make points G, C, F and H coincident the quaternary link 3 would vanish and a triple joint $GCFH$ would be formed. This is shown in Figure 3.14, with $GCFH$ joining four links and this is known as a triple joint. However, applying Grübler's equation (3.1) to the linkage of Figure 3.14, we obtain $F = 3(L - 1) - 2J = 3(7 - 1) - 2(9) = 0$. In this calculation, note that the triple joint was counted as 3 in the J value. Also, note that F has decreased from $F = +1$ in Figure 3.13(a) to $F = 0$ in Figure 3.14, and incidentally we see that the $F = 0$ characteristic is distributed throughout the linkage in this case.

In general, the total shrinkage of a link reduces the F value of the linkage by one. Referring to equation (3.1) we see that this is true, since the total shrinkage of a link reduces L by one and also in general it reduces the number of joints by one as illustrated in the preceding example. Thus, if F, L and J are the values of the terms for the original linkage, total shrinkage of n_{ts} higher links would result in $L_{ts} = L - n_{ts}$ and $J_{ts} = J - n_{ts}$ for the number of links and joints respectively of the new linkage. Applying Grübler's equation (3.1) we obtain equation (3.7) for the number of degrees-of-freedom of the new linkage after total shrinkage of n_{ts} higher links, and n_{ts} higher joints would be formed in the process.

$$F_{ts} = 3(L_{ts} - 1) - 2J_{ts} = 3(L - n_{ts} - 1) - 2(J - n_{ts})$$

$$= 3(L - 1) - 3n_{ts} - 2J + 2n_{ts} = F - n_{ts}$$

$$\therefore F_{ts} = F - n_{ts} \qquad (3.7)$$

Thus, the number of degrees-of-freedom F which a linkage originally has is reduced by the number of higher links n_{ts} which are totally shrunk to form higher joints in a new linkage. Hence, if we were interested in deriving a catalog of acceptable isomers with higher joints for $F = +1$ synthesis use by the total shrinkage technique, we would have to start with a catalog of proper F value taking equation (3.7) into account.

Incidentally, we would find that the same acceptable isomers are derived for any case by either the partial shrinkage or the total shrinkage techniques now explained. The total shrinkage technique is sometimes easier to apply, since it is not necessary to consider the many different ways in which we may partially shrink the higher links. However, it does have the disadvantage of requiring catalogs of acceptable isomers of different F values to start in order to account for the F value change by equation (3.7) in the total shrinkage procedure. From this standpoint, the partial shrinkage technique is simpler. As a suggested exercise, the interested reader should sketch acceptable isomers of various F values with higher joints from Appendix A, using both the partial shrinkage and total shrinkage techniques. A catalog of some acceptable isomers for $F = +1$ with one higher joint is given in Appendix B.

Addition and Removal of Binary Links

In many cases to be illustrated later, a mechanical device must function with different values of F during its total operation. Hence, a structure may require $F = 0$ for supporting external loads during its normal operation, whereas an adjustment feature of $F = +1$ characteristic may be desired in set-up for bringing the load surface into proper position. Similarly, an internal force exerting device such as a self-adjusting brake to be illustrated later may have three internal forces requiring $F = -1$ during the braking operation, but to accommodate shoe lining wear we might require $F = +2$ during the engagement period. We would find that release of the three forces would cause the binary links for those to vanish in the associated linkage, and F would automatically change from -1 to $+2$ in the process. In general, the value of F can readily be changed in a device by the addition or removal of binary links connected between higher links, and this could be the basis for the self-adjustment feature mentioned previously for the structure and illustrated in a later example. The effect of adding or removing a binary link is specifically illustrated in Figure 3.15, with respect to change in the F value.

The addition of a binary link between any two links of a linkage adds one link and two joints to the system. Hence, if n_{ba} designates the number of binary links added to a linkage, then the number of links would become $L_{ba} = L + n_{ba}$ and the number of joints would become $J_{ba} = J + 2n_{ba}$. Thus, the number of degrees-of-freedom for the new linkage would become F_{ba} by application of equation (3.1) as follows.

$$F_{ba} = 3(L_{ba} - 1) - 2J_{ba} = 3(L + n_{ba} - 1) - 2(J + 2n_{ba})$$

$$= 3(L - 1) - 2J - n_{ba} = F - n_{ba}$$

$$\therefore F_{ba} = F - n_{ba} \tag{3.8}$$

Incidentally, if a binary link is removed from a given linkage, merely use a negative value for the count in n_{ba}.

Referring to Figure 3.15(a), we recognize Stephenson's linkage having $F = +1$. By adding binary link 7 we obtain Figure 3.15(b), and with $n_{ba} = +1$ we see immediately from equation (3.8) that F is reduced by one. This, of course, is confirmed by application of equation (3.1) in the figure. Similarly, by removing binary link 4 from Figure 3.15(a), use $n_{ba} = -1$ and we see immediately from equation (3.8) that F is increased by one. This is confirmed by application of equation (3.1) in Figure 3.15(c).

3-3 SYNTHESIS OF MECHANICAL DEVICES

In Figure 3.1 we illustrated the one-to-one correspondence between a *mechanical device* and its *associated linkage*. This was done in a procedure of analysis by starting with the given device and sketching the basic linkage associated with it. Hence, in Figure 3.1 we sketched the associated linkages for several mechanisms

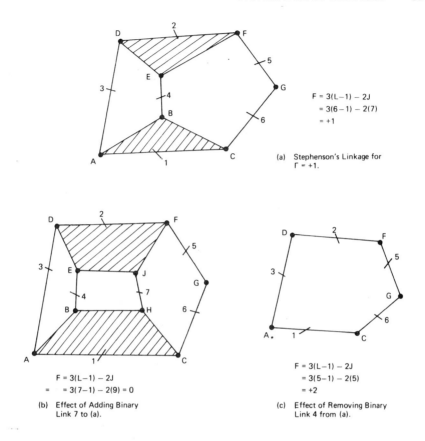

$F = 3(L-1) - 2J$
$= 3(6-1) - 2(7)$
$= +1$

(a) Stephenson's Linkage for $\Gamma = +1$.

$F = 3(L-1) - 2J$
$= \; = 3(7-1) - 2(9) = 0$

(b) Effect of Adding Binary Link 7 to (a).

$F = 3(L-1) - 2J$
$= 3(5-1) - 2(5)$
$= +2$

(c) Effect of Removing Binary Link 4 from (a).

Figure 3.15   Illustrative Example Showing Effect of Adding or Removing a Binary Link Between Higher Links.

of $F = +1$ characteristic, a structure having $F = 0$, and an internal force exerting device whose associated linkage was characterized by $F = -1$. Also, we note in Figure 3.1 that sliders, cams and gears were in the devices and that these elements had corresponding parts in the associated linkages as illustrated.

In the work of synthesis for new mechanical devices, the procedure of analysis illustrated in Figure 3.1 is merely reversed. Thus, in mechanical synthesis we start with an appropriately selected associated linkage and use it as a guide for the creative design of a new device. The procedure is based on a familiarity with the one-to-one correspondence between the mechanical device and its associated linkage, to the extent of being able to carry through the illustrated process of analysis in reverse. Hence, with much practice the clever designer will find that new mechanical devices will almost "fall together" using an appropriately selected associated linkage as a crutch for creative synthesis. The non-creative person soon becomes highly creative in the area of mechanical devices, because of this stimulant technique which will be referred to as the *systematics of linkages*. It will be illustrated many times in the remaining pages of the Chapter.

Several theorems useful for the introduction of cams, gears, sliders, and sliding pivots in the work of synthesis will now be stated.

Theorem I: In any pin jointed linkage it is permissible to replace any binary link and its two single joints with contacting surfaces having centers of curvature which do not coincide.

Theorem II: In any pin jointed linkage it is permissible to replace any single pin joint with a lower pair sliding joint or with a pure rolling contact joint, see Figure 3.2(a).

Theorem III: In any pin jointed linkage it is permissible to replace any binary link and its two single joints with a sliding pivot type of joint as illustrated in Figure 3.2(b).

To save on space, the proofs for these theorems will not be presented here although they are based on straight-forward kinematic considerations. However, the application of these theorems will be illustrated many times in the synthesis examples which follow. Incidentally, you may think of Theorem III as merely being a special case for Theorem I, involving the sliding pivot of Figure 3.2(b), which is really a special type of cam connection.

Before the presentation of some specific examples, let us make a note on the *introduction of gearsets* in the work of synthesis. The procedure is merely the reverse of what is shown in Figure 3.1(c), which should be studied carefully for the one-to-one correspondence between meshing gearset and associated linkage. In a more complicated associated linkage the meshing gearset may be only a component of the total linkage. Based on Figure 3.1(c), the characteristics of such a component may be summarized as follows:

1. The meshing gearset corresponds to a four-jointed closed loop encountering four links.

2. The four-jointed closed loop has two rotation points R and two base points B corresponding to the meshing gearset, with the sequence $RRBB$ encountered in the loop.

3. The link joining the two base points is an imaginary binary link of higher pair sliding (HS) characteristic for the gear tooth surfaces in contact whose centers of curvature are the base points.

4. The imaginary binary link HS is discarded in synthesis of the meshing gearset, by Theorem I previously stated.

5. The base points B connect *an* imaginary binary link, and hence the base points must be single joints.

6. The rotation joints R connect real links, and hence they may be either single joints or higher joints.

7. The four links encountered in the *RRBB* closed loop are the imaginary HS binary link, and three real links (the pinion, gear and center connecting frame link).

After becoming thoroughly familiar with the forementioned characteristics in conjunction with Figure 3.1(c), the reversed procedure of introducing meshing gearsets in synthesis becomes obvious. In a complicated associated linkage, be on the lookout for four-jointed closed loop components having at least one binary link between single joints. If these characteristics are met, the *RRBB* closed loop notation can be introduced and a meshing gearset corresponds to this component of the linkage if you wish. A meshing gearset can be introduced if desired at such a place in the synthesis of a mechanical device by the systematics of linkages.

Also, we should make a note regarding the general introduction of cams and followers in the synthesis of a mechanical device. By Theorem I an imaginary HS binary link will be discarded in the work of synthesis, in the reverse procedure of what is illustrated in Figure 3.1(b). Therein we see that this imaginary HS binary link connects the centers of curvature C_c and C_f of cam and follower respectively, and these points are imaginary single pin joints for the particular instant of observation of the device. Hence, in the procedure of synthesis for a new device, in any associated linkage a cam and follower can be introduced wherever a binary link connecting single joints exists. Incidentally, if desired a roller can then be introduced in the device as a final step of creative design, and this would be done merely for the reduction of friction loss.

In the introduction of general cams and followers, it is *not* necessary to have a *four*-jointed closed loop in the associated linkage. Gear teeth are special types of cams, and the four-jointed closed loop requirement was introduced since a meshing gearset requires normally a constant center distance for proper operation. Hence, for the introduction of a general cam and follower, in the associated linkage we would merely look for a binary link connecting single joints and by Theorem I this is where the cam and follower can be introduced.

Finally, we should call attention to the fact that cams, gears, sliders and sliding pivots can be introduced regardless of whether or not the device being synthesized is a mechanism. The basic requirements illustrated by the one-to-one correspondence in Figure 3.1, and now described in conjunction with Theorems I, II and III, are all that is necessary. Hence, cams, gears, sliders and sliding pivots may be introduced in structures and internal force exerting devices. This may be advantageous with respect to the introduction of adjustment features or obtainment of high mechanical advantage, and this will be illustrated in later examples.

In all applications of the systematics of linkages, it should be emphasized that the technique does not automatically by itself give you an acceptable idea for a mechanical device. It is merely a very valuable aid for guidance in the synthesis of ideas in creative design. Thus, background knowledge and good judgement must be integrated with the technique when sketching the initial configuration of the new mechanical device. Appropriate associated linkages should be selected with discretion, and all of the parts in the new device should be functional. Hence, as an example, an associated linkage which introduces non-functional structural components in an internal force exerting device generally should be abandoned at once as a bad idea from a practical standpoint. Also, in all cases, remember that simplicity of construction is a cardinal principle of good design. Hence, start by considering the simplest acceptable isomers, and increase the complexity only if the results warrant it.

In addition to keeping the forementioned practical considerations in mind when selecting an appropriate associated linkage from the catalog of acceptable isomers, in specific problems special basic requirements must also be considered. These are based on the particular problem at hand, and they become apparent after a review of the input-output requirements and perhaps after some initial effort in deriving ideas by the technique. Special basic requirements will be illustrated in many of the examples to follow.

In the remaining part of the chapter, we will present many examples illustrating creative design of various devices by the systematics of linkages. It should be emphasized that the ideas derived as sketched in the figures are generally not the final designs. Undoubtedly, we would incorporate relatively minor refinements in some cases, and the final selection of materials and dimensions might require some optimization of design. Also, in the creative design presentations which follow, for any one problem we do not claim that all of the ideas generated are good ones. Many may be bad, but hopefully some will be good. However, this is characteristic of creative design, where in the uninhibited process of creative action both good and bad ideas are generated, with the bad ones of course being discarded. The important thing illustrated in what follows is that the systematics of linkages is a practical and explicit stimulant technique which in many cases enables you to generate large numbers of ideas for the creative design of mechanical devices. Without a working knowledge of this technique, undoubtedly you would be far more stagnant in your abilities for creative action. Hence, practice should be gained by doing the many suggested exercises which follow.

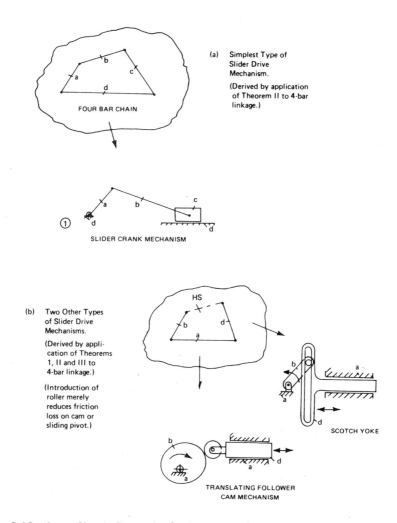

Figure 3.16   Some Simple Examples for Synthesis of Slider Drive Mechanisms.

Appropriately selected associated linkages from an available catalog of acceptable isomers are used as the aid for creative action in the synthesis of configurations for new devices. This technique of creative synthesis is best illustrated by example, and proficiency will be gained after some practice on new problems. Hence, in the remaining part of the chapter we will present several challenging examples, and suggestions will be made for worthwhile exercise by the interested reader.

3-4   SYNTHESIS OF SOME MECHANISMS OF
$F = +1$ CHARACTERISTIC

Mechanisms of $F = +1$ characteristic are probably the most commonly encountered type of application for linkages in mechanical design. Some very simple examples of synthesis for this type of problem will be presented first in Figures 3.16 through 3.18, before considering some more complicated examples for the systematics of linkages. The basic problem is the synthesis of a slider drive mechanism to start. Application of Theorems I, II and III to the four bar linkage results in the synthesis of the well-known slider drive mechanisms shown in Figure 3.16. Another introductory example is given in Figure 3.17, where Theorem II is applied to the six link level of complexity in Figure 3.6 and we drive five additional types of slider drive mechanisms. Four are from Stephenson's linkage and only one from Watt's linkage, and all five are different from the simple slider-crank mechanism of Figure 3.16. As a suggested exercise, the interested reader should apply Theorems I and III to Figure 3.17 to derive additional types of slider drive mechanisms, including ones whose motion can be cam controlled.

As a final introductory example of the synthesis technique, we could apply Theorems I and II to acceptable isomer (2) of Figure 3.9(a) to obtain the cam controlled slider drive mechanism incorporating a meshing gearset in the reverse procedure of Figure 3.1(c), and this is illustrated in Figure 3.18. The gearset in the mechanism could give special effects, such as a large amplitude of slider motion or a large mechanical advantage in the system to overcome a large slider resistance. Beyond the stage of creative design, the optimum gear ratio would be chosen to suit the particular situation at hand. As a suggested exercise, the interested reader should sketch some other slider drive mechanisms from the acceptable isomers of Figure 3.9, incorporating cams, gears, and sliding pivots if appropriate. This can be extended further if you wish by using appropriate linkages from Appendix A.

***Example* 3-1**   In the design of a new machine we wish to synthesize the simplest possible drive mechanism for oscillating two widely separated agitator shafts, as shown in Figure 3.19. The available constant-speed drive shaft and the particular space constraints are shown in the figure. Also, we note that three ground pivots are required, for the three moving links, 1, 2, and 3 relative to ground link 4. Hence, the associated linkage must contain at least one link having three or more joints, which will be the ground link in the mechanism. Thus, we refer to Figure 3.6 and select basic linkage ② as the aid for creative synthesis in Figure 3.20. As an exercise the interested reader should try his hand at the creative synthesis of some other feasible configurations from basic

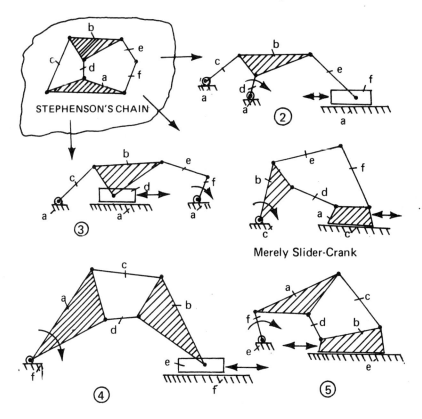

(a) Slider mechanisms derived from Stephenson's chain as the associated linkage.

Figure 3.17   Five Types of Slider Drive Mechanisms Derived for $F = +1$ from Acceptable Isomers of $L = 6$, see Figure 3.6.

linkages ② and ③ of Figure 3.6, applied as associated linkages to the problem of Figure 3.19.

***Example* 3-2** We wish to design a cam-controlled mechanism for driving a slider along fixed ways in a machine. The cam is to have constant speed and the slider is to reciprocate through a given stroke. As an exercise, sketch several mechanism configurations from associated linkages of $F = +1$ characteristic, appropriately selected from Figure 3.6 and/or Appendix A. As a more challenging example, suppose we extend the problem further by stipulating

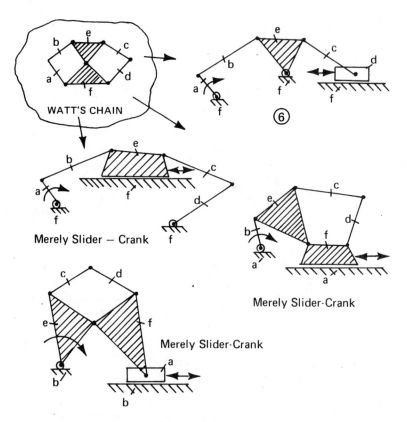

(b) Slider mechanisms derived from Watt's chain as the associated linkage.

Figure 3.17   (continued)

that the slider must be part of a rack which is to be gear driven as well as cam controlled. Recognizing that a rack is merely a gear of infinite pitch radius, we refer to the catalog of Figure 3.6 and seek a basic linkage of the $F = +1$ classification having a four-sided closed loop to be designated *RRBB* corresponding to the meshing gears, as previously discussed for Figure 3.1(c). Thus, from basic linkage ② of Figure 3.6 we derive the mechanism of Figure 3.21. Similarly, from basic linkage ③ of Figure 3.6 we derive the two mechanisms of Figure 3.22. This we can do by fixing either link 1 or 3 of the associated linkage in Figure 3.22.

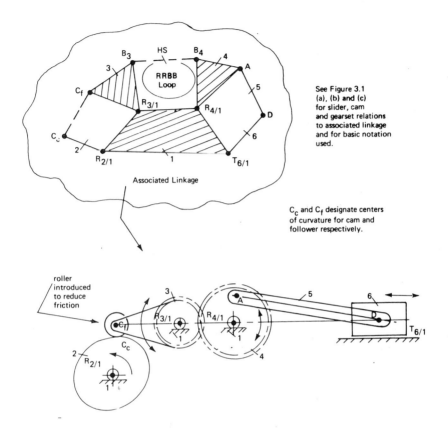

See Figure 3.1
(a), (b) and (c)
for slider, cam
and gearset relations
to associated linkage
and for basic notation
used.

C_c and C_f designate centers
of curvature for cam and
follower respectively.

Figure 3.18   Cam Controlled Slider Drive Mechanism with Meshing Gearset Derived from Figure 3.9(a) Acceptable Isomer (2) as the Associated Linkage.

***Example* 3-3**   A cam-controlled mechanism is to reciprocate a large mass *M* through given stroke *D*, and one of the initial configurations conceived is shown in Figure 3.23.   However, based on feasibility calculations, we have determined that other ideas are desired having greater mechanical advantage than can be obtained with the Figure 3.23 mechanism.   Requirements and constraints for a new mechanism are briefly summarized in the sketch of Figure 3.24, with the basic objective of synthesis being to drive connection point *P* with a cam-controlled mechanism of large mechanical advantage.   For the initial configuration of Figure 3.23, this driving component of the total mechanism would have the four-bar linkage as the associated linkage, as sketched

Figure 3.19 Input and output requirements with boundary constraints for dual-agitator synthesis (example 3-1).

in Figure 3.25. From this we see that all links of the initial configuration are functional, so for the synthesis of a new driving-mechanism component we require at least the same four types of basic links, which are: (1) frame link; (2) drive point link; (3) cam link; and (4) imaginary HS binary link.

The four-bar linkage yields no other solution besides the one shown in Figure 3.25, if the cam axis is to be fixed against translation. Hence, we proceed to the next degree of complexity in synthesis, and we can use both linkages ② and ➘ ③ of Figure 3.6 for the associated linkages of new mechanisms. Nine basic associated linkage possibilities are uncovered from Watt's linkage ② of the Figure 3.6 catalog, as determined and presented in the design tree of Figure 3.26. On the other hand, eleven basic associated linkage possibilities are uncovered from Stephenson's linkage ③ of the Figure 3.6 catalog, as determined and presented in the design tree of Figure 3.27. Review of all these possibilities revealed that four drive-mechanism configurations were feasible in synthesis to give a large mechanical advantage, and these new ideas so derived are summarized in Figures 3.28 and 3.29. The most promising one was selected as Figure 3.29(a), shown in the total mechanism setting of Figure 3.30 for large mechanical advantage or low cam force. As an exercise, the interested reader should synthesize some other mechanism configurations for the solution of the basic problem of this example.

Figure 3.20 Synthesis of a dual-agitator drive mechanism from Watt's linkage ② of the Figure 3.6 catalog.

***Example* 3-4**   Epicyclic gear trains of $F = +1$ characteristic are used in aplications where a large gear ratio is desired in a compact design taking little space. A well-known epicyclic gear train is shown in Figure 3.31, with its associated linkage sketched in a step of analysis based on the techniques now described for meshing gears. Normally, this epicyclic gear train design would

Figure 3.21 Cam-gear-slider mechanism derived from Watt's linkage ② of the catalog in Figure 3.6. (The roller follower is introduced for reduction of friction and wear.)

operate with gear 4 fixed, the input supplied to gear 1, and the output taken from arm 5. We would like to derive some other epicyclic gear train ideas using the systematics of linkages as an aid for creative design.

Figure 3.22 Cam-gear-slider mechanisms derived from Stephenson's linkage ③ of the catalog in Figure 3.6 (same notation for C_c, C_f, B_g, and B_r as in Figure 3.21).

Figure 3.23   Initial configuration conceived for cam-controlled slider drive mechanism.

Figure 3.24   Graphical summary of input and output requirements and boundary constraints.

First, for the existing epicyclic gear train *associated linkage* in Figure 3.31, we recognize some basic characteristics which will be the requirements for the selection of other acceptable isomers from our catalog. They are as follows:

1. All joints are parts of four-jointed closed loops, each of which corresponds to a meshing gearset.

2. Each four-jointed closed loop has an imaginary HS binary link connected to single joints, which are the base points of the meshing gears.

3. The *RRBB* sequence is encountered in each loop, and the rotation points may or may not be higher joints. The link joining the *R* and *B* in any loop corresponds to a gear, and of course there are two of these for each loop.

Figure 3.25   Driving-mechanism component of the initial configuration from Figure 3.23 for connecting the input to output in Figure 3.24.

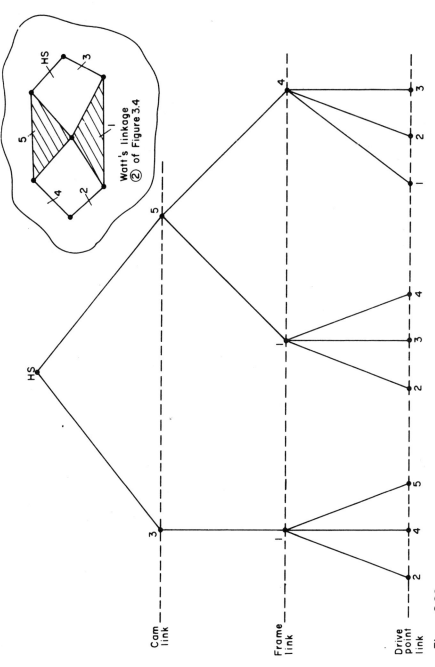

Figure 3.26 Possible combinations of functional links in Watt's linkage as the associated linkage for synthesis of drive-mechanism components in example 3-3.

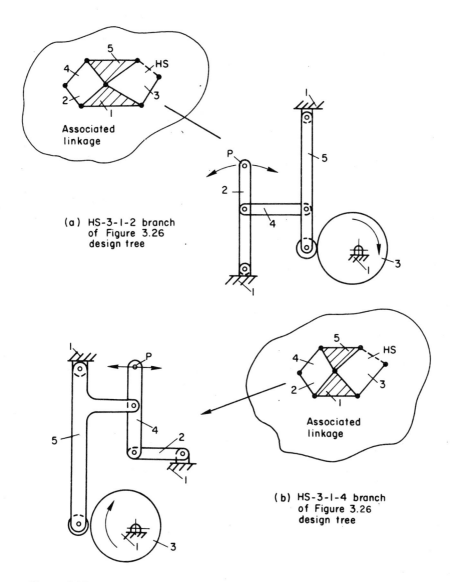

(a) HS-3-1-2 branch
of Figure 3.26
design tree

(b) HS-3-1-4 branch
of Figure 3.26
design tree

Figure 3.28   Synthesis of feasible drive-mechanism components from the design tree of Figure 3.26, for application to the setting in Figure 3.24. (Optimum dimensions are not shown in sketches.)

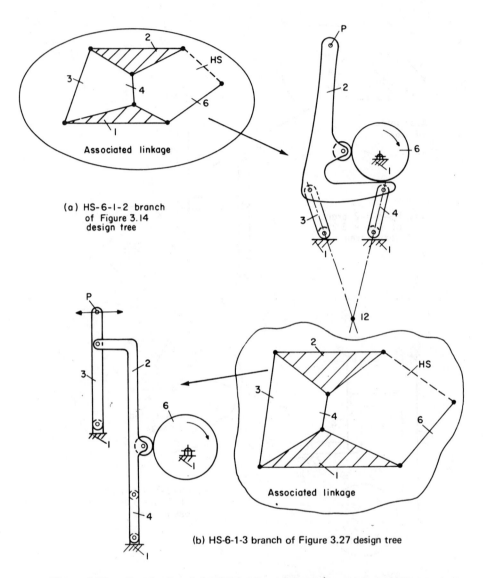

(a) HS-6-1-2 branch
of Figure 3.14
design tree

Associated linkage

(b) HS-6-1-3 branch of Figure 3.27 design tree

Associated linkage

Figure 3.29   Synthesis of feasible drive-mechanism components from the design tree in Figure 3.27 for application to the setting in Figure 3.24. (Optimum dimensions are not shown in sketches.)

Figure 3.30   Improved cam-controlled slider drive mechanism with larger mechanical advantage than the initial configuration of Figure 3.23 (drive-mechanism component of Figure 3.29(a) in setting of Figure 3.24, with notation of the latter).

Figure 3.31   Epicyclic Gear Linkage and Its Associated Linkage.

Incidentally, we recognize the associated linkage of Figure 3.31 as the acceptable isomer previously derived in Figure 3.12(b). Hence, our reference catalog will be Appendix B for acceptable isomers of $F = +1$ with higher joints.

The *basic steps in the synthesis of new epicyclic gear linkages* (EGL)'s are summarized as follows:

1. Refer to Appendix B and select an associated linkage which is compatible with the preceding general requirements.

2. Sketch the linkage chosen in (1), and with curved arrow notation indicate the four-jointed closed loops which you intend to use.

3. For each of the loops in (2) designate the rotation points by R and the base points by B, in the necessary *RRBB* sequence.

4. Designate the HS imaginary binary links, and number all of the other real links. Compound gears will be links having more than one base point in more than one loop. Number these links with multiple digits, using the same number of digits as base points on the link.

Figure 3.32   Synthesis of an Epicyclic Gear Linkage from (B8) of Appendix B.

5. Introduce corresponding numerical subscripts to the R and B joints. All base points B will have single digit subscripts, whereas the subscripts for rotation points R will designate the relative connection between the real links at the joint.

6. Using the associated linkage now developed, as an aid for the work of synthesis, sketch the epicyclic gear linkage. It is generally easier to start with

Figure 3.33   Synthesis of an Epicyclic Gear Linkage from $\textcircled{B4}$ of Appendix B.

the sketching of the most complex compound gear, shown as a set of concentric circles with single link connection designations as illustrated in Figure 3.31. Then sketch the meshing gears as tangent circles, being careful for the sake of clarity that they do not overlap. Introduce all rotation point notation and number all gears, according to the associated linkage. Finally, sketch the remaining frame links if they exist showing their numerical designations, and

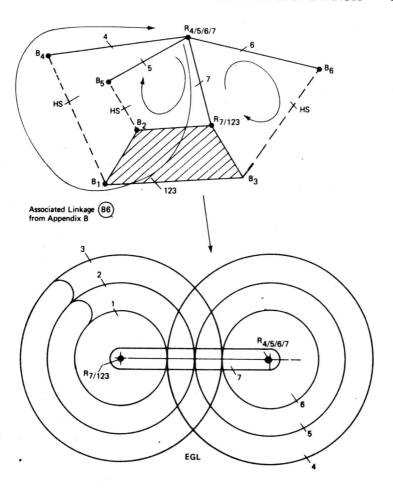

Figure 3.34   Synthesis of an Epicyclic Gear Linkage from $\boxed{B6}$ of Appendix B.

designate all rotation point connections for all of the links, based on the associated linkage.

To facilitate step (6) in particular, the use of colored pencil can be a great help in avoiding confusion with respect to meshing gears, compound gears (same color for the two or more gears), and the connections involved between links. Hence, at the start of step (6), shade all real links in the associated linkage with a different color. Then, as the epicyclic gear linkage is sketched use the appropriate color for each of the links sketched.

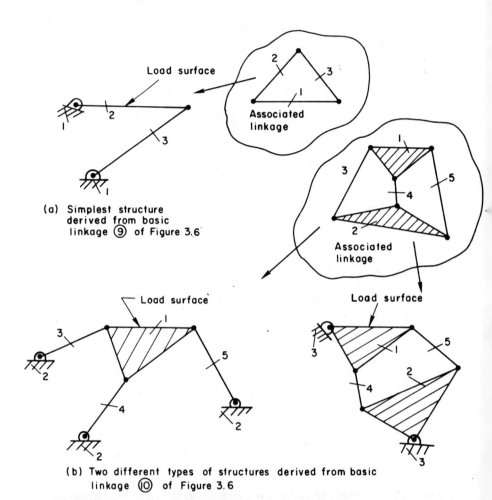

(a) Simplest structure
derived from basic
linkage ⑨ of Figure 3.6

(b) Two different types of structures derived from basic
linkage ⑩ of Figure 3.6

Figure 3.35 Some different types of structures having a horizontal load surface with all components below the surface.

After the completion of step (6) above, the difficult part of creative design is complete. Deciding which link to fix, and showing the stacking of gears and frames in a side view, are relatively easy to carry through. These steps are illustrated in reference 3-3, and they will not be covered further here. However, steps (1) through (6) previously explained will be illustrated with some examples presented in Figures 3.32 through 3.34. The interested reader can obtain some exercise in the technique by continuing in Appendix B for the selection of other associated linkages and execution of steps (1) through (6).

It is interesting to note that a second type of EGL can be derived from the associated linkages in each of Figures 3.32 and 3.33. This is done by choosing the other binary link in the left loop as the HS imaginary link, for each case. The interested reader should sketch the two additional EGL's obtained in this way. Also, as an exercise two EGL's can be derived in a similar manner from the acceptable isomer (4) shown in Figure 3.13(d). Also, it should be observed that only one EGL can be derived from acceptable isomer (B6) of Appendix B, and this is the one shown in Figure 3.34. The interested reader might wish to check his exercise EGL's with reference 3-3.

3-5 SYNTHESIS OF SOME STRUCTURES FOR SUPPORTING EXTERNAL LOADS

***Example* 3-5**   We wish to design a structure having a horizontal load surface with all supporting components located beneath. A linkage of $F = 0$ characteristic is to be used in the synthesis work. Incidentally, such a linkage will automatically accommodate small changes in ground-point locations due to earth movement and temperature variations, without developing additional internal forces. The simplest design of fewest parts is desired, and from Figure 3.6 we first try basic linkage (9) as the associated linkage, thereby giving the structure in Figure 3.35(a). If this configuration is not acceptable for the particular application at hand, we proceed to the next degree of complexity and try basic linkage (10) of Figure 3.6 as the associated linkage. From this we synthesize the two structure types shown in Figure 3.35(b) for consideration. If we extend the requirements for inclusion of a hand adjustment feature for tilting the load surface, we synthesize the configuration of Figure 3.36. This is derived from basic linkage (10) of Figure 3.6 using the clamping of the pivot for a binary link as the adjustment method. Thus, in releasing the clamp the number of links is reduced by one and the number of joints is reduced by two, thereby increasing F by one for the linkage by Grübler's equation. Therefore, we obtain $F = +1$ upon release of the clamp in Figure 3.36, which is what we desire for the adjustment operation, whereas we have $F = 0$ for the structural function of the device with the clamp engaged. If we extend the requirements further by desiring a higher mechanical advantage through the introduction of gears and cams for the adjustment process, we refer to Appendix A for a more complicated associated linkage. Many could be used, but the idea easily synthesized from acceptable isomer (A25) is shown in Figure 3.37. As an exercise, the interested reader should synthesize other structural configurations for solution of the stated problem, using Figure 3.6 and Appendix A.

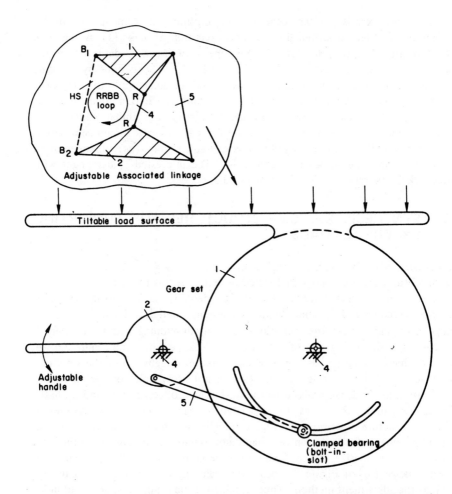

Figure 3.36   Adjustable structure with tiltable load surface synthesized from basic linkage ⑩ of Figure 3.6.

3-6 SYNTHESIS OF SOME INTERNAL-FORCE-EXERTING DEVICES

An internal force exerting device generally operates in a self-adjusting capacity to accomodate small differences in dimensions because of tolerances, wear, etc. for the workpiece on which it acts. Examples are many, including jacks, clamping devices, shears, yoke riveters, differential brakes, clutches, etc. Generally, the operation of such devices can be resolved into two stages as follows.

Stage 1: Initial engagement of the device to the workpiece, during which the internal forces are not developed. In this first stage, the small dimensional variations of the workpiece n_{dv} in number are automatically accomodated by the device in preparation for the next stage.

Stage 2: Internal forces are developed in the device for its intended operation acting on the workpiece. In this second stage, n_{if} internal forces in number are introduced, including those acting on the workpiece and perhaps an actuating force such as from a piston-cylinder.

In applying the systematics of linkages to the creative design of internal force exerting devices, we must consider *both stages* of operation previously described. For synthesis use, our associated linkage will be sketched for the Stage 2 operation, with n_{if} internal forces shown as hypothetical binary links, based on the two-force principle of elementary mechanics. However, for the Stage 1 operation these n_{if} internal forces do not act, but the device must accommodate n_{dv} dimensional variations of the workpiece as it comes into engagement. These considerations of both stages will be taken into account by the proper choice for number-of-degrees of freedom, F_{al}, of the associated linkage. Hence, we can apply equation (3.8) directly, going from the device of Stage 1 to the device of Stage 2 where n_{if} hypothetical binary links have been added. Thus, $F_{ba} = F - n_{ba}$ gives by direct substitution

$$F_{al} = n_{dv} - n_{if} \tag{3.9}$$

In equation (3.9), F_{al} is the number of degrees of freedom for the associated linkage in the Stage 2 operation, n_{dv} is the number of dimensional variations on the workpiece which must be accommodated by the device in the Stage 1 operation, and n_{if} is the number of internal forces which will be added as hypothetical binary links in sketching the associated linkage.

Equation (3.9) will be applied to many practical examples which follow. However, to illustrate its application in a very simple introductory example, consider the nut cracker device of Figure 3.1(e) and the associated linkage shown. For the nut cracker we must accommodate one dimensional variation because of different sizes of nuts, and so we use $n_{dv} = 1$. Also, forces N and P will be represented by hypothetical binary links in the associated linkage, so we use $n_{if} = 2$. Hence, by equation (3.9), the number of degrees of freedom for the associated linkage should be $F_{al} = 1 - 2 = -1$, and this is the value for the associated linkage shown in Figure 3.1(e) as calculated by Grübler's equation (3.1).

In general, there are two types of actuation for developing internal forces in a device, as follows.

1. *Force actuation*, such as from a pneumatic or hydraulic piston-cylinder, an electric solenoid, a helical spring, or a simple hand force. The other internal forces depicted by hypothetical binary links, in number, are the same as the dimensional variations accommodated by the device. Hence, $n_{if} = n_{dv} + 1$ is generally true for this case. Thus, by equation (3.9), for the associated linkage we require

$$F_{al} = n_{dv} - n_{if} = n_{dv} - (n_{dv} + 1)$$

$$= -1$$

Therefore, we would choose an appropriate acceptable isomer from the $F = -1$ catalog for the associated linkage, if we are to use *force actuation* in the internal force exerting device. This will be illustrated specifically in the practical Examples 3-6 through 3-10 which follow.

2. *Torque actuation*, such as obtained from a rotary hydraulic actuator, rotary solenoid, or torsion spring. In general, the torque will be applied between any two real links at a pin joint connection. Hence, in such cases, the actuation will not be depicted by a hypothetical binary link representing an internal force in the associated linkage. Instead, the actuating torque will be introduced between two pin connected real links in the associated linkage and the device. Thus, the internal forces depicted by hypothetical binary links, in number, are the same as the dimensional variations accommodated by the device, so we use $n_{if} = n_{dv}$ in equation (3.9) giving for this case

$$F_{al} = n_{dv} - n_{if} = n_{dv} - n_{dv} = 0$$

Therefore, we would choose an appropriate acceptable isomer from the $F = 0$ catalog for the associated linkage, if we are to use *torque actuation* in the internal force exerting device. This will be illustrated specifically in practical Examples 3-8 and 3-9 which follow.

As a note of caution for the case of torque actuation, you will find that the $F = 0$ catalog of acceptable isomers contains linkages with two binary links in general series, such as linkage (11) of Figure 3.6, and these are capable of supporting external loads. However, for internal force exerting devices two binary links in general series cannot transmit internal forces alone, based on elementary mechanics. Hence, such linkages will generally be discarded as an associated linkage for this type of problem. The exception, of course, is when the actuating torque acts as an external transverse load on one of the binary links. In this case, the binary links in general series can function in the internal force exerting device.

Figure 3.37 **Adjustable Structure of High Mechanical Advantage Synthesized from** (A25) **of Appendix A.**

Let us now consider several examples of creative design by the systematics of linkages, for several types of internal force exerting devices.

***Example* 3-6** We wish to design a compound lever jack or pry for overcoming large resisting force F_r with hand force P. Hence, since this may be classified as an internal-force-exerting device with force-actuation, our associated linkage should have the characteristic of $F = -1$. Also, for obtainment

Figure 3.38   Synthesis of a compound lever jack from basic linkage (15) of Figure 3.6.

of a good force multiplication we should have at least two moving links and the frame link in addition to the imaginary binary links for the forces F_r and P, in the associated linkage. Thus, from the catalog of Figure 3.6 we derive the compound lever jacks shown in Figures 3.38 and 3.39 as relatively simple configurations to consider. The interested reader should derive other configurations of compound lever jacks from the basic linkages of Figure 3.6 and Appendix A.

(a) Compound lever jack derived from associated linkage with notation shown in inset

Figure 3.39 Synthesis of two different compound lever jacks from basic linkage ⑯ of Figure 3.6.

Associated
linkage

F_r (resisting force)

(b)   Compound lever jack derived from associated linkage
with notation shown in inset

Figure 3.39   (*continued*)

Figure 3.40 Synthesis of Compound Lever Snips from Acceptable Isomer (15) of Figure 3.6.

Example 3-7 Simply constructed compound lever snips are to be designed for cutting through tough materials with a relatively small amount of effort. We will designate the actuating force by P and the resisting force by F_r. Assume that the compound lever snips should be hand-operated and mobile, so there will not be a ground link in the device. For high mechanical advantage, forces P and F_r must act between links which are different at least at one end; otherwise, a simple lever type of construction will result with relatively low force multiplication. Hence, the basic requirements for a suitable associated linkage may be summarized as follows

1. $F = -1$, from equation (3.9) since $n_{dv} = 1$ and $n_{if} = 2$.
2. Must have at least two binary links, corresponding to forces P and F_r.
3. The two binary links for P and F_r must not connect the same two links; otherwise, the snips derived will be simple instead of compound.

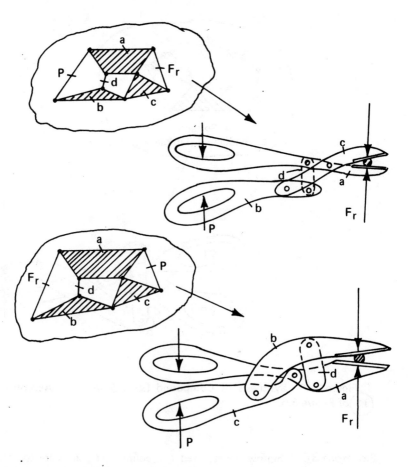

Figure 3.41   Synthesis of Two Compound Lever Snips from Acceptable Isomer ⑯ of Figure 3.6.

The simplest acceptable isomers of $F = -1$ are ⑮ , ⑯ and ⑰ of Figure 3.6, with the above requirements imposed in the selection review. Using these for the associated linkages, by the systematics of linkages we derive the compound lever snip ideas shown in Figures 3.40, 3.41 and 3.42. As an exercise, the interested reader should sketch other compound lever snip ideas derived from other associated linkages selected from the $F = -1$ catalogs of Figure 3.6 and Appendix A.

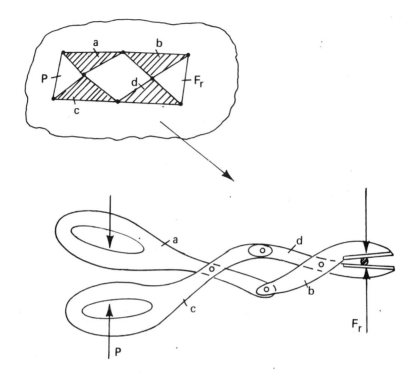

Figure 3.42   Synthesis of Compound Lever Snips from Acceptable Isomer (17) of Figure 3.6.

***Example* 3-8**   A given workpiece is to be clamped by prescribed vertical and horizontal forces V and H respectively, both acting to ground as shown in Figure 3.43. Small dimensional variations δ_v and δ_h are anticipated from one workpiece to another as depicted in the figure, because of manufacturing tolerances. The creative design problem is to synthesize some ideas for self-adjusting clamping devices using the systematics of linkages. Considered as an internal force exerting device, the ground link will be included as part of the associated linkage.

For the case of *force actuation* using a hydraulic piston-cylinder for delivering force P, the basic requirements for a suitable associated linkage are summarized as follows.

1. $F = -1$, from equation (3.9) with $n_{dv} = 2$ (for δ_v and δ_h) and $n_{if} = 3$ (for V, H, and P).

Figure 3.43   Basic problem of Clamping for Given Workpiece (WP).

2. Must have at least three binary links (for V, H, and P), at least two of which must be connected to a common link (since V and H both act on the ground link).

3. The ground link must have at least three joints if P is not acting on it (two for V and H, and at least one more to resist their resultant force).

4. The ground link must have at least four joints if P acts on it (three for V, H, and P, and at least one more to resist their resultant force).

The simplest acceptable isomers satisfying the preceding requirements are ⑭ , ⑮ and ⑯ of Figure 3.6. Hence, using these three as the associated linkages we readily synthesize the four simple ideas for clamping devices shown in Figures 3.44 through 3.46. For a greater mechanical advantage we might wish to synthesize a more complicated design. Thus, in light of our basic requirements, we would review the $F = -1$ category of the catalog of acceptable isomers in Appendix A, and a multitude of ideas for clamping devices could be synthesized. To illustrate specifically, the clamping device of high mechanical advantage sketched in Figure 3.47 is derived from linkage Ⓐ36 of Appendix A as the associated linkage. As an exercise, the interested reader should sketch several other clamping device ideas from appropriately selected linkages of Appendix A. If advantageous, you may wish to introduce gears and/or cams by techniques previously described, for special effects. This will be illustrated in Figures 3.49 and 3.50.

For the case of *torque actuation*, assuming that torque T will be delivered to a pivot link by a rotary hydraulic actuator mounted on ground, the basic

Associated Linkage
(14) from Figure 3.6.

Figure 3.44   Simple Force-Actuated Clamping Device Sketched from Linkage (14) of Figure 3.6.

requirements for a suitable associated linkage are summarized as follows.

1. $F = 0$, from equation (3.9) with $n_{dv} = 2$ (for δ_v and δ_h) and $n_{if} = 2$ (for V and H).

2. Must have at least two binary links connected to a common link (for V and H acting on the ground link).

3. The ground link must have at least three joints (two for V and H, and at least one more to react the resultant force from T, V, and H).

4. Avoid linkages with two binary links in series, unless one of these binary links is to be acted on by T.

Figure 3.45   Simple Force-Actuated Clamping Device Sketched from Linkage (15) of Figure 3.6.

The simplest acceptable isomer satisfying the preceding basic requirements is (10) of Figure 3.6. Hence, with this as the associated linkage, the simple type of clamping device shown in Figure 3.48 is derived. For a greater mechanical advantage we might wish to synthesize a more complicated design. Thus, in light of our basic requirements, we .would review the $F = 0$ category of Appendix A, and a multitude of ideas for clamping devices with torque actuation could be synthesized. Suppose, in addition we would like to illustrate the introduction of cams or gears, which we believe might help in the obtainment of a large mechanical advantage. Hence, using acceptable isomer (A26) of

Associated Linkage
(16) from Figure 3.6

Associated Linkage
(16) from Figure 3.6

Figure 3.46   Two Simple Force-Actuated Clamping Devices Both Sketched from Linkage (16) of Figure 3.6.

Appendix A as the associated linkage we derive the torque actuated clamping device idea shown in Figure 3.49, incorporating a cam and follower which could be designed for obtaining a very large mechanical advantage. On the other hand, using acceptable isomer (A27) of Appendix A as the associated linkage, we derive the torque actuated clamping device idea shown in Figure

Figure 3.47   More Complicated Force-Actuated Clamping Device of Greater Mechanical Advantage, Sketched from Linkage (A36) of Appendix A.

3.50, which incorporates a meshing gearset. Finally, using acceptable isomer (A28) of Appendix A as the associated linkage we derive the clamping device idea shown in Figure 3.51. As a suggested exercise, the interested reader should sketch several other torque actuated clamping device ideas from appropriately selected linkages of Appendix A. For the sake of simplification, start by synthesizing some devices without gears or cams, and expand the work later by inclusion of such refinements in design.

Associated Linkage
(10) from Figure 3.6.

Figure 3.48   Simple Torque-Actuated Clamping Device Sketched from Linkage (10) of Figure 3.6.

***Example* 3-9**   Problem is the creative design of a self-adjusting braking device, which is to act on a given rotating drum through diametrically opposite brake shoes, as shown in Figure 3.52. Drum diameter D is given as well as shoe design with friction angle ϕ known. Specified shoe forces R_L and R_R are to constitute a braking couple $M_b = R_L D \sin \phi$ acting on the drum. The device must accommodate independent wear on left and right shoes, so the number of dimensional variations will be $n_{dv} = 2$, for δ_L and δ_R shown in Figure 3.52. The attack for creative design will be to consider the device as an internal force exerting one, with forces R_L and R_R acting on the ground link, which will be included in the associated linkage.

For the case of *force actuation* using a hydraulic piston-cylinder for delivering force P, the basic requirements for a suitable associated linkage are summarized as follows.

1. $F = -1$, from equation (3.9) with $n_{dv} = 2$ and $n_{if} = 3$ (for R_L, R_R, and P).

Figure 3.49 Torque-Actuated Clamping Device with Cam for High Mechanical Advantage, Sketched from Linkage (A26) of Appendix A.

2. Must have at least three binary links (for R_L, R_R, and P), at least two of which must be connected to a common link (since R_L and R_R both act on the ground link).

3. The ground link must have at least four joints (two for R_L and R_R, and at least two more to resist the braking couple).

4. R_L and R_R binary links must be connected to the ground link at one end and to *different links* at their other ends. (Otherwise, the brake shoes would be on a common rigid body link, and with diametrically opposite shoes the two independent components of wear could not be accommodated. Then the device would not be self-adjusting.)

The simplest acceptable isomers satisfying the preceding requirements are (15) and (16) of Figure 3.6, and using these as the associated linkages we

Figure 3.50   Torque-Actuated Clamping Device with Meshing Gearset for High Mechanical Advantage, Sketched from Linkage (A27) of Appendix A.

readily synthesize the two self-adjusting brake ideas shown in Figures 3.53 and 3.54.

In different situations, we might wish to consider more complicated ideas with the possibility of gaining unique advantages. Thus, in light of our basic requirements, we would review the $F = -1$ category of our available catalog and choose some appropriate acceptable isomers for the associated linkages in this problem. To illustrate specifically, the self-adjusting brake ideas sketched

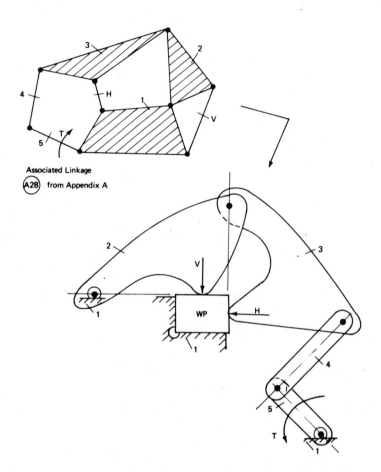

Associated Linkage
(A28) from Appendix A

Figure 3.51   Linkage Type Torque-Actuated Clamping Device with Potential for High Mechanical Advantage, Sketched from Linkage (A28) of Appendix A.

in Figures 3.55 and 3.56 were derived with isomer (2) of Figure 3.11(a) and linkage (A39) of Appendix A as the associated linkages respectively. As an exercise, the interested reader should sketch several other force-actuated self-adjusting brake ideas from appropriately selected linkages of Appendix A, and of course gears and/or cams can be introduced if you wish by the techniques previously described and illustrated.

For the case of *torque actuation*, assuming that torque T will be delivered

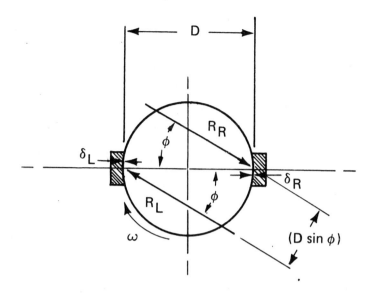

Figure 3.52   Basic Braking Problem Showing Rotating Drum Exerting Forces R_L and R_R on Diametrically Opposite Shoes.

to a pivot link by a rotary hydraulic actuator mounted on ground, the basic requirements for a suitable associated linkage are summarized as follows.

1. $F = 0$, from equation (3.9) with $n_{dv} = 2$ (for δ_L and δ_R) and $n_{if} = 2$ (for R_L and R_R).

2. Must have at least two binary links connected to a common link (for R_L and R_R acting on the ground link).

3. The ground link must have at least four joints (two for R_L and R_R, and at least two more to resist the resultant couple from the braking couple and the actuating torque couple).

4. R_L and R_R binary links must be connected to the ground link at one end and to *different links* at their other ends. (Otherwise, not self-adjusting as previously explained.)

5. Avoid linkages with two binary links in series, unless one of these binary links is to be acted on by T.

The simplest acceptable isomer satisfying the preceding basic requirements is in the seven link category, and the first we encounter in Appendix A is (A24) resulting in the torque-actuated self-adjusting brake design idea shown in Figure

Associated Linkage
(15) of Figure 3.6.

Figure 3.53   Simple Force-Actuated Self-Adjusting Brake Sketched from Linkage (15) of Figure 3.6.

3.57. Be sure to review Figure 3.6 and understand why none of these simpler acceptable isomers can be used for this specific problem. Also, as an exercise the interested reader should sketch several other torque-actuated self-adjusting brake ideas from appropriately selected linkages of Appendix A. Of course, gears and/or cams can be introduced if you wish for special effects, similar to what we did in Example 3-8 for the clamping mechanism problem, using the techniques previously described and illustrated.

Example **3-10** A yoke riveter design is shown in Figure 3,58 with its associated linkage, which we note is acceptable isomer (A35) of Appendix A.

Figure 3.54   Simple Force-Actuated Self-Adjusting Brake Sketched from Linkage ⑯ of Figure 3.6.

We wish to consider some other basic configurations for yoke riveters, and the systematics of linkages will be used for the synthesis of some other ideas. For an acceptable design, force multiplication must be high and side forces on both piston and die sliders should be relatively low; both objectives should be kept in mind when sketching the configurations of new ideas. From acceptable isomer Ⓐ36 of Appendix A we derive the two new ideas shown in Figure 3.59. As an exercise, the interested reader should derive other yoke riveter configurations from basic linkages obtained with the link combinations of Table 3-2 or Appendix A directly.

Associated Linkage
Isomer (2) of Figure 3.11(a)

Figure 3.55   Force-Actuated Self-Adjusting Brake Idea Sketched from Acceptable Isomer (2) of Figure 3.11(a).

3-7 SYNTHESIS OF DIFFERENTIAL MECHANISMS WITH $F \geq +2$

In machinery design it is sometimes necessary to superimpose two or more motions, which can be done by what might be called a differential type of mechanism. In such mechanisms, $F \geq +2$ is required, since two or more input motions must be integrated to obtain a desired output motion characteristic. In general, the number of input motions equals the value of F. The systematics of linkages may be applied in creative design for problems of this type, and

Associated Linkage
(A39) of Appendix A

Figure 3.56   Force-Actuated Self-Adjusting Brake Idea Sketched from Linkage (A39) of Appendix A.

the background for a special technique useful here will be explained next.

A catalog of acceptable isomers for $F = +2$ has been derived and summarized in Figure 3.6 and in Appendix A. Such acceptable isomers obtain their $F = +2$ characteristic by fixing one *link*, and this is the basis upon which Grübler's equation (3.1) was derived. However, we can supplement our derivation of ideas by considering linkages with one *joint* fixed instead of one link.

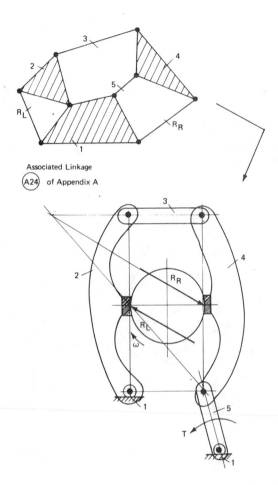

Figure 3.57   Torque-Actuated Self-Adjusting Brake Idea Sketched from Linkage (A24) of Appendix A.

Such a mechanism would have an extra degree-of-freedom in the system, because fixing a joint removes only two degrees-of-freedom whereas fixing a link removes three degrees-of-freedom. Thus, we can write

$$F_{jf} = F_{lf} + 1 \qquad (3.10)$$

where F_{lf} is the number of degrees-of-freedom with one link fixed (normal

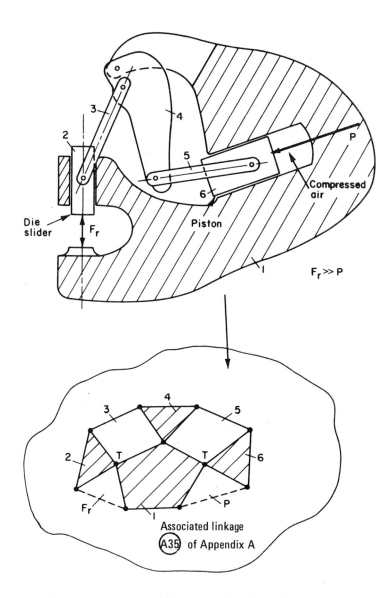

Figure 3.58   Pneumatic yoke riveter with associated linkage.

Associated linkage
(A36) of Appendix A

$F_r \gg P$

(a) Yoke riveter design
derived from associated
linkage of inset

Figure 3.59   Synthesis of two different types of yoke riveters from acceptable isomer (A36) of Appendix A.

Associated linkage

(A36) of Appendix A

(b) Yoke riveter design
    derived from
    associated linkage
    of inset

Figure 3.59 (*continued*)

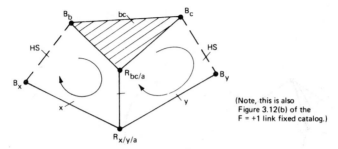

(Note, this is also
Figure 3.12(b) of the
F = +1 link fixed catalog.)

(a)   Associated linkage of existing design, sketched from x-b-c-y-a
heart of mechanism shown in Figure 9.37, showing RRBB gear loops.

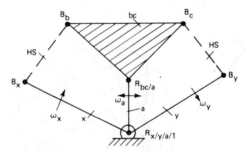

(b)   Mechanism representation from linkage of (a), showing input
motions ω_x and ω_a and output motion ω_y.

Figure 3.60   Associated Linkage and Mechanism Representation for x-b-c-y-a Heart of Present Design Given in Figure 9.37.

case, assumed in Figure 3.6 and Appendices A and B); and F_{jf} is the number of degrees-of-freedom which the same linkage would have with one joint fixed instead.

Thus, for $F = +2$ creative design work as a specific example, we could use an acceptable isomer of $F = +1$ characteristic (link fixed value) from Figure 3.6 or from Appendices A and B, and in application fix one joint instead. This technique would supplement the fixed link catalog of $F = +2$ acceptable isomers shown in Figure 3.6 and Appendix A, and hence more creative design ideas could be derived for differential mechanisms by the systematics of linkages. The value of this technique will be illustrated as follows, in three examples of creative design for differential mechanisms with $F \geq +2$.

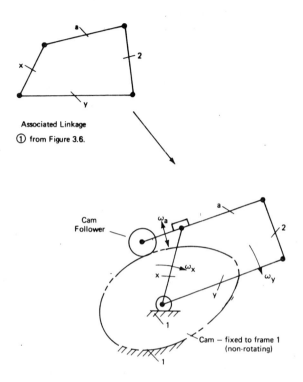

Associated Linkage
① from Figure 3.6.

Figure 3.61   Simplest Type of Differential Mechanism of $F = +2$ Characteristic, Derived from $F = +1$ Catalog with One Joint Fixed Instead of a Link.

***Example* 3-11** In section 9-5 of Chapter 9 we consider optimization of the given design configuration for a differential type mechanism shown in Figure 9.37. In this mechanism, an oscillatory motion is superimposed on a unidirectional motion to give the motion characteristics required for the output rolls. This is done by oscillating the control arm a in an epicyclic gear train mechanism, using the cam as shown in Figure 9.37. Gears d, e and x rotate at constant speed to provide the unidirectional component.

The heart of the present design is the epicyclic gear train x-b-c-y-a shown in Figure 9.37, and for this component we readily sketch the associated linkage shown in Figure 3.60. For it the oscillatory motion is supplied by the cam to link a and the constant speed motion is given to gear x. We note that the associated linkage in Figure 3.60 is the acceptable isomer of Figure 3.12(b) for

Partial shrinkage of link 1 making A and B coincident.

Acceptable Isomer ⑧ of F =+2 from Figure 3.6.

Associated Linkage with RRBB Gear Loops.

Cam Follower

Cam — fixed to frame 1 (non-rotating)

Figure 3.62   A Promising Type of Differential Mechanism of $F = +2$ Characteristic, Derived by Systematics of Linkages as Depicted in Figure.

$F = +1$ with one link fixed. However, the $F = +2$ value needed is obtained by fixing one joint instead of a link. Let us now apply the systematics of linkages to determine whether or not a simpler type of differential mechanism idea can be obtained for driving the rolls with the two components of motion as required in this problem.

Application of the systematics of linkages resulted in 22 different types of differential mechanism configurations, 8 of which showed some promise for the particular requirements of the motion specifications. The simplest idea

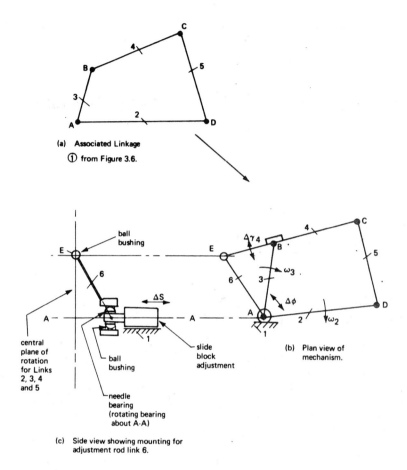

(a) Associated Linkage
① from Figure 3.6.

(b) Plan view of mechanism.

central
plane of
rotation
for Links
2, 3, 4
and 5

ball
bushing

needle
bearing
(rotating bearing
about A-A)

slide
block
adjustment

(c) Side view showing mounting for
adjustment rod link 6.

Figure 3.63   Simple Type of Differential Mechanism with Operationally Adjustable Phase Angle between Links 2 and 3.

derived, from the $F = +1$ catalog by fixing one joint instead of a link, is shown in Figure 3.61. We note that this is an appreciably simpler idea than the present design. A tileboard model of this new idea confirmed validity of its application for the particular motion problem at hand. Also, one of the most promising ideas derived from the catalog of $F = +2$ with one link fixed is shown in Figure 3.62. Note therein that the associated linkage used, of $F = +2$ characteristic with one higher joint, is derived from an acceptable isomer of the $F = +2$ catalog by partial shrinkage of a link. Finally, to save on space we will not present the remaining 20 ideas derived by the systematics of linkages in this problem of creative design.

Example 3-12 In the design of a new machine we wish to adjust the phase angle between two masses 2 and 3 as they rotate about the same axis. The adjustment is to be made while the machine is in operation. After the adjustment is made, the masses will rotate at the same speed $\omega_2 = \omega_3$ during normal operation.

For the adjustment process the mechanism to be designed must have $F = +2$, since rotational speed ω_2 and adjustment motion $\Delta\phi$ (to get the phase angle shift between links 2 and 3) must be independent. The systematics of linkages will be applied in the creative design of some differential mechanisms with an operationally adjustable phase angle feature.

First, we will attempt to obtain the simplest possible mechanism idea for the problem. Hence, using the technique of fixing a joint instead of a link, we choose linkage ① of Figure 3.6 as the simplest starting tool. Thus, referring to the background of the previous example, fixing joint A of Figure 3.63 would give us the simplest basic mechanism with the $F = +2$ characteristic, but to make it complete we must introduce some practical way of adjusting the phase angle between links 2 and 3 while the machine is in full speed operation. Using a cam as we did in Figure 3.61 is not feasible for this application, since we must have $\omega_2 = \omega_3$ constant during normal operation. What we really need is to keep the length EA constant during normal operation, but for the adjustment feature we must be able to vary the length of EA, referring to Figure 3.63(b). This can be done by introducing the adjustment rod 6 mounted in ball bushings and a rotating bearing as shown in Figure 3.63(c).

During normal operation in Figure 3.63, EA is of constant length, and links 2, 3, 4, 5, and 6 rotate in unison about the common axis A-A. Since link 6 rotates on the bearing provided about axis A-A during normal operation, there will be no high speed relative motion within either one of the two ball bushings. This, of course, is a good feature from the standpoint of minimizing wear in the ball bushings.

During the adjustment process, the axis A-A rotating bearing is slowly moved along its axis as shown, and only then will there be some relative motion in the ball bushings, at the slow speed of adjustment. The axial adjustment ΔS causes an effective change in length EA for the planar mechanism, causing a relative angular motion $\Delta\gamma_4$ for link 4 resulting in the phase angle adjustment $\Delta\phi$ between links 2 and 3.

Note that the mechanism as presented in Figure 3.63 is a three dimensional mechanism during the adjustment process, whereas during normal operation the mechanism is planar. Incidentally, creative design variations to consider

(a) Mechanism with Sliding Pivot (Theorem III).

(b) Mechanism with Meshing Gearset (See RRBB four-jointed closed loop).

(c) Mechanism with captured follower cam (Theorem I).

Figure 3.64   Creative Design Variations of Differential Mechanisms with Operationally Adjustable Phase Angle Between Links 2 and 3. (In all cases EA adjustable length by feature similar to Figure 3.63(c)).

for the Figure 3.63 idea are derived by previously presented techniques, as illustrated in Figure 3.64.

As a final thought on the Figure 3.63 design, as a suggested exercise you may wish to consider other ways of making *EA* of variable length during the adjustment process. This is a difficult problem, but perhaps incorporation of a three

(a) Associated Linkage
Figure 3.12(b)
showing RRBB gearset loops.

(b) Mechanism derived from (a),
requires $D_4 = 0$ for $\omega_2 = \omega_3$.

(c) Feasible mechanism derived from (b), can be
designed practically to give $\omega_8 = \omega_9$, with
phase adjustment by gear 4 released.

Figure 3.65   Planar Differential Mechanism Idea with Operationally Adjustable Phase Angle between Gears 8 and 9. (Links must be displaced in parallel planes as necessary to provide clearance in rotation.)

dimensional cam might be possible. There may be other ideas as well to consider.

Let us now go to the next level of complexity and try to obtain an entirely planar mechanism satisfying all of the requirements in the problem at hand. Hence, after reviewing our catalog of acceptable isomers, we choose Figure 3.12(b) and fix its higher joint to obtain the $F = +2$ characteristic needed in the adjustment process. The mechanism derived is shown in Figure 3.65(b), where we see that link 4 is fixed after adjustment is complete thereby giving

ΔEA and ΔFA length adjustments similar to Figure 3.63(c) give independent
phase angle shifts between links 2, 3 and 4. $\omega_2 = \omega_3 = \omega_4$ constant speed
for normal operation.

Figure 3.66   Differential Mechanism of $F = +3$ Characteristic for Operationally Adjustable Phase Angles Between Links 2, 3, and 4.

the required $F = +1$ characteristic in normal operation.

A disappointing revelation comes to light when trying to choose specific dimensions and gear tooth numbers of the Figure 3.65(b) idea, to meet the specific problem requirements at hand. It is found that pitch diameter of gear 4 must be zero to give $\omega_2 = \omega_3$, and this of course is impossible for a practical design. As a general point, carrying through an idea in specific detail sometimes leads to the discouraging conclusion of an infeasible design. However, a cardinal principle of good design is never to give up! Hence, if we choose $\omega_2 \neq \omega_3$ in the basic idea of Figure 3.65(b), and introduce gears 7, 8 and 9 as shown in Figure 3.65(c), we can obtain a feasible design meeting the specific requirements of the problem. The phase adjustment link is still gear 4, but we obtain $\omega_8 = \omega_9$ for coaxial gears 8 and 9 by proper choice of tooth numbers for gears 7, 8 and 9. The planar mechanism derived in Figure 3.65(c) is undoubtedly more complicated than the Figure 3.63 idea, but in certain applications it may have specific advantages.

As a final exercise, the interested reader may wish to sketch other types of mechanism ideas for this very challenging problem of creative design.

Example 3-13   In the design of a new machine, we wish to adjust the phase angles between three masses 2, 3 and 4 as they rotate about the same axis. The adjustment is to be made while the machine is in operation. After adjustment is complete, the masses will rotate at the same speed $\omega_2 = \omega_3 = \omega_4$ during normal operation. Ideas of creative design are desired.

The simpler problem of two rotating masses was solved in the previous example, with the simplest idea being the one shown in Figure 3.63. Hence, we can readily derive an idea for the present problem by taking two of the Figure 3.63 mechanisms and connecting them as shown in Figure 3.66. Phase angle adjustment is made by varying the lengths of EA and FA shown in Figure 3.66, by two slide block adjustments similar to Figure 3.63(c). From a practical standpoint, this is a difficult feat to accomplish in the total mechanism, but not impossible with some ingenuity. The interested reader might wish to pursue this further as an exercise in creative design. Also, other creative ideas for solving the basic problem would be a most challenging exercise.

BIBLIOGRAPHY

3-1   K. Hain, *Applied Kinematics*, 2nd ed., McGraw-Hill Book Co., Inc., New York, 1967, 727 pp.

3-2   R. C. Johnson and K. Towfigh, "Application of Number Synthesis to Practical Problems in Creative Design," ASME Paper No. 65-WA/MD-9, Annual Meeting of the A.S.M.E., Chicago, Nov. 1965, 9 pp.

3-3   R. C. Johnson and K. Towfigh, "Creative Design of Epicyclic Gear Trains Using Number Synthesis," ASME Paper No. 66-MD-A, Dec. 1966, 6 pp.; *Trans. ASME, Ser. B, J. Eng. Ind.* 89, 309-314 (May 1967).

3-4   R. C. Johnson, "Optimum Design By Synthesis," presented at the 1970 Design Engineering Conference, May 14, 1970, Chicago, Ill., sponsored by ASME.

3-5   R. T. Hinkle, "Constrained Motion, Number Synthesis," Chap. 14 in *Kinematics of Machines*, 2nd ed., Prentice-Hall, Inc., Englewood Cliffs, N.J., 1960, pp. 318-328.

3-6   F. R. E. Crossley, "The Permutations of Kinematic Chains of Eight Members or Less from the Graph-Theoretic Viewpoint," *Developments in Theoretical and Applied Mechanics*, Vol. 2, Pergamon Press, 1965, pp. 467-486.

3-7   F. R. E. Crossley, "A Contribution to Gruebler's Theory in the Number Synthesis of Plane Mechanisms," *Trans. ASME, Ser. B, J. Eng. Ind.* 86, 1-8 (Feb. 1964).

3-8   F. Freudenstein and L. Dobrjanskyj, "On a Theory for the Type Synthesis of Mechanisms," *Proceedings of the Eleventh International Congress of Applied Mechanics*, Springer-Verlag, Berlin, Germany, 1964, pp. 420-428.

3-9   R. S. Hartenberg and J. Denavit, *Kinematic Synthesis of Linkages*, McGraw-Hill Book Co., Inc., New York, 1964, Chap. 5.

3-10   R. C. Johnson, "Design Synthesis—Aids to Creative Thinking," *Machine Design*, Penton Publishing Co., Cleveland, Nov. 15, 1973, pp. 158-163.

PROBLEMS

Many suggested exercises are distributed throughout the text in this Chapter, which should provide ample practice for the student in the systematics of linkages.

4

Some Other Aids in Creative Effort

4-1 INTRODUCTION

The systematics of linkages described and illustrated in Chapter 3 can serve as a stimulant or aid for creative work in the synthesis of mechanical devices involving linkages.   In this chapter we will briefly describe and illustrate some other stimulant techniques having broader applications for the precipitation of creative activity in the early stages of mechanical design synthesis. They are the techniques of (*1*) the logical building-block approach, (*2*) synthesis by implication, and (*3*) synthesis with circuit diagrams.   In certain problems of creative design these techniques can serve as the stimulants for execution of the general plan in Figure 2.2 of Chapter 2.

4-2 LOGICAL BUILDING-BLOCK APPROACH

General Technique

In the logical building-block approach to creative design an orderly procedure of basic form is diligently carried through.   The general technique is graphically depicted in Figure 4.1, which consists of several steps that can be outlined briefly in the order of execution as follows:

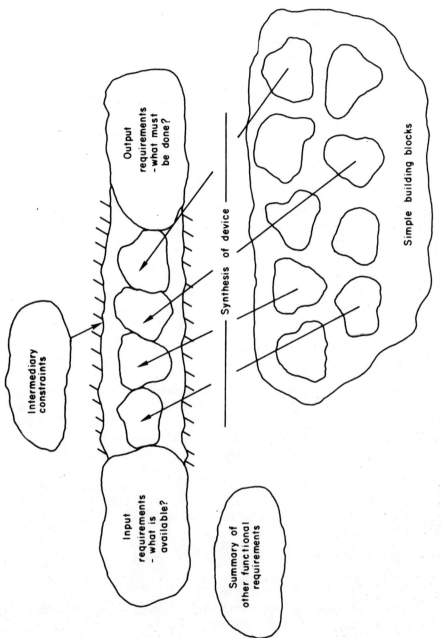

Figure 4.1 Graphical summary of the logical building-block approach to creative design.

1. Establish and summarize all basic boundary conditions of the problem. Resolve these general specifications into steps (2), (3), (4), and (5), which follow; if possible, a graphical free-hand sketch on a single sheet of paper should be made to summarize the specifications.

2. Summarize the *output requirements* of the device to be conceived, based on a consideration of the functions which it must perform.

3. Summarize any *input requirements* of the device to be conceived, based on a consideration of what is available or acceptable for the total system.

4. Summarize all *intermediary constraints* on the device to be conceived, based on a consideration of such factors as space available in the entire system, manufacturing facilities, and cost limitations.

5. Summarize all *other functional requirements* of the device to be conceived, such as low cost, low frictional power loss, high rigidity, low wear rate, etc.

6. Diligently think of simple components which satisfy at least in part some of the boundary conditions and requirements now summarized. Often it is helpful to resolve the specifications of the device into basic steps or functions which must be satisfied in operation, as in example 4-2. Then many simple component building blocks should be conceived if possible, disregarding at least initially any consideration of the total assemblage of components which eventually will be synthesized. Generally, this step will require experience and knowledge, and appreciable time should be allotted to the conception of basic ideas for the building blocks.

7. Finally, in a step of synthesis, use *some* of the basic building blocks now summarized for logically connecting in the simplest conceivable fashion the output and input requirements, abiding by all of the summarized boundary constraints. Often, in this concluding step of synthesis the appropriate simple building blocks will naturally fall into place for the logical derivation of a suitable mechanical device.

Application of the logical building-block approach to the creative design of mechanical devices generally requires a great deal of perseverance for the orderly execution of the outlined steps. Generally, many free-hand sketches must be made in the process, with order-of-magnitude feasibility calculations made where appropriate. Also, one must have the ability to integrate knowledge and experience with ingenuity in a consideration of the total system requirements for the synthesis cf a satisfactory device. However, if diligently pursued, the orderly execution of the logical steps now outlined can lead to the synthesis of a useful and sometimes surprising device, which otherwise would not be conceived. Let us briefly consider some industrial results where this aid to creative activity played an important part in mechanical design synthesis.

Example **4-1**   The detachable tensile socket configuration shown in Figure 1.13 of Chapter 1 was conceived by the logical building-block approach. The input and output requirements of steel rope attachment were logically connected by functional simple building blocks of ears, tensile flanges, basket, and spring-loaded wedges for the synthesis of the design.   The patent drawing for the device is shown in Figure 1.11 of Chapter 1.

Example **4-2**   Let us briefly consider a problem in creative design which was given by the Singer Company to the author in a consulting capacity. In a sewing machine the cloth must be moved with a stepping motion between the presser foot and the throat plate during the sewing operation.   This is accomplished by means of a ratchet-shaped feed dog.   The motion of the feed dog must be adjustable from the standpoints of both height above the throat plate and the stepping distance, the latter of which determines the stitch length in sewing of the cloth.   It must be possible to adjust the stitch length during operation of the machine, and it must be possible to feed the cloth in both the forward and reverse directions.   Path of the feed dog had to satisfy certain specified characteristics, the details of which will not be disclosed because of proprietary reasons.

It was desired to synthesize a feed-dog drive mechanism of *fewest possible parts,* for use in a family-type sewing machine.   This objective of optimum design by synthesis was introduced for purposes of simplification and cost

Figure 4.2   Model 720 tubular-bed household sewing machine showing feed-dog region.   *Courtesy The Singer Company.*

Figure 4.3 Summary of basic requirements and constraints for creative synthesis of a feed-dog drive mechanism. The feed-dog drive mechanism must connect input to output with the fewest possible parts.

(a) "Ideal" mechanism ①

(b) "Ideal" mechanism ②

Figure 4.4   Ideal feed-dog drive mechanisms (two of the four conceived).

reduction in the machine.   We will briefly describe the creative design of the *few-parts feed mechanism* which was synthesized by the logical building-block approach.

Space constraints were quite stringent on the feed-dog mechanism to be designed because of the required tubular-bed configuration typically shown in the sewing machine of Figure 4.2.   A simplified summary of the basic requirements and constraints for the feed-dog drive mechanism to be conceived are presented in Figure 4.3.   Hence, at this point we have essentially covered steps (1) through (5) of the logical building-block approach, pre-viously described for general application.   Next, in carrying through step (6) of the procedure, we conceived of the simplest possible basic mechanisms for the generation of the required feed path, disregarding temporarily the requirements of stitch-length adjustment and reversed feed.   Four such *ideal mechanisms* were conceived by the systematics-of-linkages technique of Chapter 3, and to save on space only two of these are presented in Figure 4.4.

In the total feed-dog drive mechanism we will require features of stitch-length adjustment and reversed feed.   The simplest mechanism component conceived for accomplishing these functions is the one shown in Figure 4.5. Changing ground pivot 14 of the feed-adjustment swivel link 4 as indicated

Figure 4.5   Simplest amplitude-adjustment and reversed-feed mechanism component conceived.

Figure 4.6   Creative synthesis of the feed-dog drive mechanism of few parts—initial sketch.   The design contains only three moving links besides the drive cam shaft.

in the figure changes instant center 13 of the feed-transfer link 3 relative to the frame 1. Thereby, varying amounts of transferred motion can be obtained from the given motion of cam drive 5 in Figure 4.5.

Let us now follow step (7) in the logical building-block approach previously described. Hence, after some trial in sketching, we combine ideal mechanism ② of Figure 4.4 with the amplitude-adjustment and reversed-feed component of Figure 4.5 to derive the associated linkage shown in the inset of Figure 4.6. Then by the systematics-of-linkages technique explained in Chapter 3 we synthesize the basic feed-dog drive mechanism configuration as sketched in Figure 4.6.

From a consideration of motion requirements, pivot locations are determined by working within the space constraints of the tubular bed, resulting in the initial layout shown in Figure 4.7. A tileboard model was then constructed for generating the actual feed-path characteristics of the mechanism for the various feed adjustments. A photograph of this model with the feed-path characteristics generated is shown in Figure 4.8 for a forward-feed

Figure 4.7   Initial layout of the few-parts feed-dog drive mechanism.

adjustment of eight stitches per inch.   A more inclusive layout of the design
was made incorporating practical mountings for all links within the three-
dimensional space constraints of the total tubular bed, resulting in a feasible
design for the mechanism concept.   Further aspects of design pertaining to
sewing machines are being described by the author in Figure 4.9.

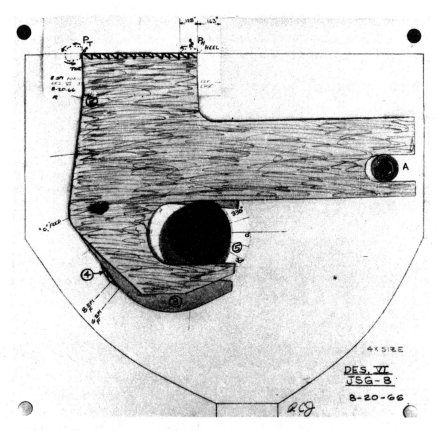

**Figure 4.8   Tileboard model of the layout in Figure 4.7 showing feed path
generated for a forward feed adjustment of eight stitches per inch.**

Figure 4.9   Explaining some further aspects of design. (Sketch by Mr. Tony Piano of The Singer Company during seminar presentation.)

4-3 SYNTHESIS BY IMPLICATION

General Technique

Experience, knowledge, and ingenuity are traits characteristic of the naturally creative person. Underlying objectives for optimization have served as the necessary motivation for inspiring many inventions of the past. In fact, the author has found from personal experience that new configurations are sometimes implied from knowledge of optimization goals on traditional designs. When goals of optimization serve as the impetus for precipitating creative design, we are in effect being motivated by a basic stimulant technique in accordance with the general plan of Figure 2.2 of Chapter 2. This basic stimulant technique we will call *synthesis by implication*. Let us consider some very simple illustrations before presenting some industrial examples.

The desire for a shorter internal-combustion engine resulted in the development of the V-8 engine and the radial engine, instead of the longer inline engine. The desire for minimizing travel time resulted in the development of jet aircraft. The desire for reducing cost and increasing speed of manufacturing led to the invention of the cotton gin. Finally, suppose a design engineer were familiar with the analysis and design of spur, helical, and straight bevel gears, but not with spiral bevel gears. Suppose he knew from experience, or from an optimum design study such as in Chapter 12 of reference 4.1, that for a given size the helical gearset is stronger than the spur gearset. Hence, for a problem requiring increased strength in bevel gears, the configuration change from straight bevel gears as implied from past experience would be to incline the gear teeth. It is well known that this is the thing to do, since for a given size, spiral bevel gears are stronger than straight bevel gears. Let us now consider the basic technique more specifically with two very simple introductory examples. First, suppose the problem is to design a shaft for maximization of torsional rigidity k_θ, in a situation where excessive weight is also of concern. The traditional configuration first coming to mind might be the solid circular shape, shown in Figure 4.10(a). For it we write equation (4.1) from elementary strength of materials, for torsional rigidity k_0.

$$k_\theta = M_t/\theta = J_\theta \, G/L \tag{4.1}$$

From this equation, we readily draw the following goals of optimization for maximization of k_θ.

1. Have polar moment of inertia J_θ as large as practical.
2. Choose a feasible shaft material of large modulus of rigidity, G.
3. Have length L of shaft as small as practical.

Let us look more basically at the goal of maximizing J_θ, by considering the amount contributed by any differential element. Thus, from Figure 4.10(b), we write

$$d J_\theta = r^2 \, (r \, d\theta \, dr) = r^2 \, dA$$

and we see that any differential area dA would contribute more effectively

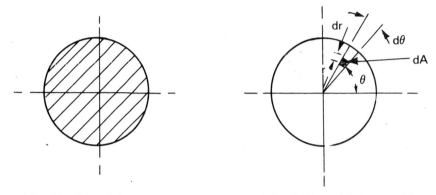

(a)   Traditional shape —
      first conceived.

(b)   Differential element dA
      in traditional shape.

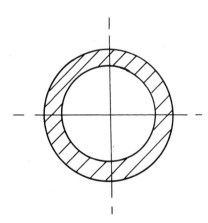

(c)   Improved configuration implied
      from goals of optimization on
      traditional shape.

Figure 4.10   Cross-sectional Shapes of Shaft in Torsion.

to J_θ if it were located at a large radius r from the shaft center. Hence, for a given cross-sectional area (or shaft weight), the hollow shape is implied as the optimum one for maximization of k_θ. Therefore, the basic "goals of design" served as the stimulant for synthesis by implication the hollow shape shown in Figure 4.10(c).

Let us now consider the second introductory example illustrating more specifically the basic technique. Suppose that a rectangular tank configuration is initially chosen for storing heated fluid, with volume V and height H specified. Howevei, we are free to choose the base dimensions x and y shown in Figure 4.11(a). Therefore, after some thought we decide to choose the particular values for x and y which will minimize heat loss from convection and radiation, this being a worthwhile optimization objective for the specific application. Since base area is indirectly specified, $A = V/H$, the heat loss component affected by our choice of x and y comes from the four sides, and it is expressed by equation (4.2),

$$q_{cr} = C(x + y) \tag{4.2}$$

where C is merely a constant for this study.

Since V and H are specified values, we must also satisfy equation (4.3).

$$V = x\,y\,H \tag{4.3}$$

Hence, for determination of the optimum design, we will combine the two equations eliminating y giving the developed primary design equation (4.4)

$$q_{cr} = C\left[x + \frac{V}{xH}\right] \tag{4.4}$$

With this equation we readily sketch q_{cr} versus independent variable x, and point 1 is the optimum design in Figure 4.11(a). This point of minimum q_{cr} is determined from elementary calculus by differentiating equation (4.4) and setting it equal to zero, as follows.

$$\frac{d\,q_{cr}}{dx} = C\left[1 - \frac{V}{x^2 H}\right] = 0$$

$$\therefore x_1 = \sqrt{V/H}$$

$$\therefore y_1 = V/(x_1 H) = \sqrt{V/H} = x_1$$

Therefore, the optimum cross-section for the base of rectangular shape is a square with $x = y$.

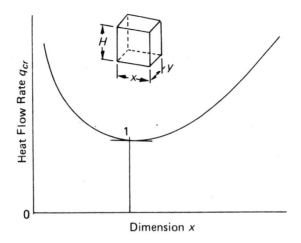

(a)   Initial tank configuration and variation diagram for optimum design.

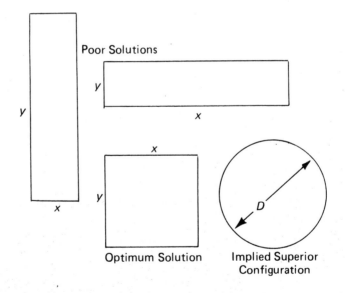

(b)   Implication of an improved configuration from "goals of design" for the initial shape.

Figure 4.11   Tank Design Example for Minimization of Heat Loss.

Looking more closely at the results of this study, we see that the poorest designs are those where x is either much smaller or much larger than y. Hence, improvement of design is achieved by making lateral dimensions x and y more equal, with the best design for this shape being the square. Thus, based on the "goals of design" derived for the traditional rectangular shape, an improved configuration is implied in Figure 4.11(b)—the circular shape for the base cross-section. In fact, from simple analysis with V and H specified, we find that the cylindrical tank would have 11 percent less lateral surface area than the design with a square base, with correspondingly less heat flow. Incidentally, although this simple introductory example served to illustrate the basic technique of creative design by implication, we could apply the calculus of variations described in Chapter 6 to prove explicitly that the circular base is the optimum shape, thus confirming our implied conclusions.

Let us now consider some industrial examples illustrating further the stimulant technique of *synthesis by implication*.

Example 4-3   The problem presented to the author in a consulting capacity by the United States Steel Corporation was to design a bearing-type socket of minimum weight, for use in the attachment of steel rope. The traditional configuration in use at the time is shown in Figure 4.12, for which an optimum design variation study was made as presented in section 8.4 in Chapter 8 of reference 4-1. Some of the determined goals of optimum design for minimization of weight are as follows, referring to Figure 4.12 for notation:

1. Choose h small, as governed by adequate bonding strength of the steel wire rope in the zinc core.
2. Choose R_i small, as governed by adequate clearance with the steel wire rope.
3. Choose D large (as governed by type A stress combination in Figure 4.12) *and* choose D relatively small (as governed by type B stress combination in Figure 4.12).

The last of these optimization goals obviously requires a compromise decision in design if the configuration in Figure 4.12 is retained, since we cannot have D both large and small at the same time. However, the goals in design *imply* a basic change in configuration, if we think in terms of the corresponding goals for angle α of Figure 4.12. Angle α would have to be large if governed by the type-A stress combination of most significance near the middle and top end of the socket basket; α would have to be relatively small if governed by the type-B stress combination of most significance near the bottom or nose end of the socket basket. These goals of optimum design after some thought implied a curved shape for the socket wall as shown in the patent drawing of Figure 4.13.

Figure 4.12   Bearing-type socket for attachment of steel wire rope, showing significant stress combinations.

***Example* 4-4**   In the design of a high-speed machine, conventional practice first led to the spring-loaded cam mechanism of Figure 4.14.   The best design within space constraints for minimization of spring surging resulted in the layout drawing of Figure 4.14, but the lowest natural frequency for internal longitudinal vibrations of the load spring was still too low for safe design. From previous optimum design variation studies made on the conventional configuration of Figure 4.14, we knew that it was desirable to have x small and D large for maximization of lowest natural frequency.   These goals for optimum design served by implication as the stimulant for creative design of the new configuration shown in Figure 4.15. In the new configuration we could inherently obtain a larger D and a smaller x than in the traditional

Oct. 20, 1964   R. C. JOHNSON   3,153,268

WIRE ROPE SOCKET

Filed May 28, 1962

FIG. 1.   FIG. 2.

Figure 4.13   Patent drawing for a new configuration of the bearing-type steel socket.   *Courtesy United States Steel Corporation.*

configuration of Figure 4.14. The result was a 50% increase in lowest natural frequency, thereby giving a safe design for this critical high-speed application.

Example **4-5**   In the development of the detachable tensile socket design shown in Figures 1.13 and 1.14 of Chapter 1 a problem was encountered in removal of the wedges from the socket basket after high load application. Many configurations for removal fixtures were being conceived, most of which pulled on the wedges from the large or top end of the basket. For the sake of simplicity in field operation a hammer-actuated removal device was desired. The one which was successfully developed was initially conceived by implication from a general optimum design study.

Projection shown is
required on follower arm
for other functions not
described

Figure 4.14   **Conventional spring-loaded cam mechanism.**

A given hand-actuated hammer can have only a limited amount of kinetic energy, E_k, at the time of impact with another mass, as depicted in Figure 4.16. A simplified model of the removal operation is sketched in Figure 4.17, showing a general removal fixture acting upon wedges of mass M_w entrenched in a basket of mass M_b. We wish to synthesize a specific removal fixture which will maximize removal force F_r shown in the figure.

Assuming that removal-fixture mass and hammer mass are small compared with the sum of wedge mass and basket mass, given kinetic energy, E_k, of the hammer will be almost entirely transformed to strain energy, W_i, in the elastic region of the removal fixture upon impact.   Thus, referring to Figure

Figure 4.15 New configuration for spring-loaded cam mechanism as implied from knowledge in optimum design for very-high-speed applications.

4.17, strain energy, W_i, is related to hammer force, F_h, and removal-fixture elongation, Δ, as follows:

$$E_k \approx W_i \approx \tfrac{1}{2} F_h \Delta \tag{4.5}$$

However, from elementary strength of materials, we express

$$\Delta \approx F_h L_r / (A_r E) \tag{4.6}$$

where L_r and A_r are effective length and cross-sectional area, respectively, of the removal fixture to be designed, and E is its modulus of elasticity, all of

Figure 4.16   **Hand-actuation of a hammer for impact with a given mass.**

which are depicted in Figure 4.17. Thus, peak hammer force, F_h, will be

$$F_h \approx 2E_k/\Delta \approx \frac{2(E_k)A_r E}{F_h L_r}$$

therefore

$$F_h{}^2 \approx 2(E_k)\, A_r E/L_r \tag{4.7}$$

However, from elementary dynamics, the effective peak removal force, F_r, that would tend to separate the wedges from the basket in Figure 4.17 would be

$$F_r \approx \left(\frac{M_b}{M_b + M_w}\right)F_h \tag{4.8}$$

Figure 4.17 Generalized model of a hammer-actuated removal fixture on the verge of separating wedges from a socket basket.

Thus, combining equations (4.7) and (4.8) by eliminating F_h we obtain primary design equation (4.9), where terms are considered independent of each other.

$$F_r \approx \frac{M_b}{(M_b + M_w)}\left[\frac{2(E_k)A_r E}{L_r}\right]^{1/2} \tag{4.9}$$

Hence, for maximization of removal force F_r, we must satisfy the following goals of optimum design for the removal fixture. (*1*) have A_r and E as large as practical and (*2*) have L_r as small as feasible. Thus, steel will be chosen as the material for the removal fixture, and the design should be as short and stiff as we can conceive for a practical configuration.

After thinking about the goals of optimum design now summarized, the removal-fixture configuration sketched in Figure 4.18 was conceived. The

Figure 4.18 Optimum removal-fixture configuration on a detachable tensile socket assemblage.

Figure 4.19   The new removal-fixture design synthesized by implication from goals of optimum design for the general model of Figure 4.17.   *Courtesy United States Steel Corporation.*

short, stiff removal-fixture design made possible by location at the nose end of the socket was synthesized by implication from the goals of optimum design drawn from the analysis of the generalized model of Figure 4.17. The new removal-fixture design worked very well, and a photograph of the model is shown in Figure 4.19.

4-4   SYNTHESIS WITH CIRCUIT DIAGRAMS

General Technique

The use of circuit diagrams can serve as an aid or stimulant in the creative synthesis of both mechanical and fluidic systems. This is particularly true in problems involving dynamics, if one has a background in electrical-circuit theory and electrical-mechanical-fluidic analogies. There are many sources which cover the basic fundamentals in these fields, and a brief summary of electrical-mechanical analogies is presented in section 4-7 of Chapter 4 in reference 4-1. Also, no attempt will be made to cover the well-known field of analog computers in this book.

The circuit-diagram technique provides an additional vehicle for creative thinking which is uninhibited by the imaginary barriers which one might have associated with the actual physical system. Hence, one can readily manipulate and synthesize system elements, uncovering the many possibilities of circuit configurations which are dynamically acceptable. These can then be used for synthesizing the analagous physical configuration, rejecting at once the ones which are not feasible because of realistic physical barriers.

The technique can be very effective as a stimulant for freeing one's thinking .in the creative synthesis of physical configurations. In using the circuit diagram the engineer will automatically exercise his creativity on something which is unaffected by the traditions of past practices. Thus, only realistic barriers are imposed in transposing to the physical system.

In some challenging problems of design for high-speed machinery we wish to synthesize a mechanical system so it will react in the best possible way with another system for maximizing energy absorbed or transmitted. Thus, techniques of impedance matching, so well-known in electrical engineering, can be directly applied to the creative synthesis of an appropriate mechanical system, by applying the circuit-diagram approach. As is true for any of the creative synthesis techniques, generally some feasibility calculations will be necessary together with the creative design work of the logical building-block approach before a feasible system configuration is decided on. Also, the analysis of the circuit diagram will readily reveal appropriate parameter values for the physical system in order to obtain the desired impedance matching.

The main value of a mechanical circuit diagram in creative design is the ease with which circuit parameters can be manipulated to determine the possible arrangements of the physical parameters. After a number of dynamically feasible circuits are developed, the corresponding physical systems are sketched. Hence, physical constraints previously ignored are introduced at this point, and some of the possibilities are eliminated from further consideration because of practical reasons. However, the circuit diagrams will have revealed all of the possible configurations so that none are overlooked.

The sketching of a circuit diagram from a mechanical system is based upon a thorough understanding of the electro-mechanical analogies. As a brief review of this background, consider the mechanical and electrical analogous systems shown in Figure 4.20. Application of basic fundamentals to the four systems shown results in the similar equations (4.10) through (4.13), with a one-to-one correspondence between the terms involved.

$$F - k \int_{0}^{t} v \, dt - \psi v - m \, \frac{dv}{dt} = 0 \qquad (4$$

$$T - k_\theta \int_{0}^{t} \Omega \, dt - \psi_\theta \, \Omega - I \, \frac{d\Omega}{dt} = 0 \qquad (4$$

(a) Translatory System

(b) Rotational System

(c) Force-Voltage Analogy

(d) Force-Current Analogy

Figure 4.20   **Mechanical and Electrical Systems with Analogous Equ**
of Dynamics.

Figure 4.21   Introductory Example of Circuit Diagrams Showing Manipulation of System Parameters in Creative Design.

Equation (4.10) is for the translatory system of Figure 4.20(a), whereas equation (4.11) is for the rotational system of Figure 4.20(b), and both are derived from the laws of elementary dynamics. Similarly, Kirchhoff's voltage and current laws when applied to the electrical circuits of Figure 4.20(c) and (d) give equations (4.12) and (4.13) respectively.

$$E - \frac{1}{C} \int_0^t i\, dt - R\, i - L\, \frac{di}{dt} = 0 \qquad (4.12)$$

$$i - \frac{1}{L} \int_0^t E\, dt - \frac{1}{R}\, E\, dt - C\, \frac{dE}{dt} = 0 \qquad (4.13)$$

In equations (4.10) through (4.13), t designates time in each case. $F, k, v, \psi,$ and m designate force, spring constant, velocity, damping coefficient, and mass respectively for the translatory system in Figure 4.20(a). $T, k_\theta, \Omega, \psi_\theta,$ and I designate torque, torsional spring constant, angular velocity, torsional damping constant, and mass moment of inertia respectively for the torsional system in Figure 4.20(b). Finally, in equations (4.12) and (4.13), $E, i, C, L,$ and R designate voltage, current, capacitance, inductance, and resistance respectively for the electrical circuits in Figure 4.20(c) and (d).

A comparison of corresponding terms in equations (4.10), (4.11) and (4.12) leads to what is known as the force-voltage analogy. On the other hand, a comparison of corresponding terms in equations (4.10), (4.11) and (4.13) leads to what is known as the force-current analogy. Either one of the two analogies can be used to represent a physical system in a corresponding circuit diagram. However, we will choose the force-voltage analogy for what follows in the remaining part of the chapter.

In the force-voltage analogy, the mechanical terms $m,\ 1/k,\ \psi,\ F$ and v are analogous to the electrical terms of L, C, R, E and i respectively. These analogous physical and electrical terms can be used interchangeably on a circuit diagram.

Again, a simple introductory example will be presented first to illustrate the use of circuit diagrams in the creative design of physical systems. Suppose we want to determine the possible physical arrangements of a mass m, spring k, and damping device ψ. First, consider the circuit possibilities for combining m and k alone, which are circuits A and B in Figure 4.21. Next, for each introduce damping ψ in all of the possible ways, and the results of this manipulation are shown in the center column of Figure 4.21. Finally, by our thorough understanding of the force-voltage analogy, we can use each of these circuit diagrams as a guide for sketching the corresponding physical configurations. The eight possibilities are shown in the right column of Figure 4.21. In application, the last step would be to discard the impractical configurations, with the

remaining ones constituting a catalog of feasible ideas. Of course, choice of
the optimum configuration would be desired in a specific application, and this
will be considered in Chapter 5.

Let us now consider two industrial examples illustrating the use of circuit
diagrams as an aid in the synthesis of mechanical configurations.

Example **4-6**   A load drum is to be indexed with intermittent motion, and
the initial configuration is conceived as sketched in Figure 4.22. We see
that it is necessary to connect the Geneva drive to the driven load of inertia
J_L with a long connecting shaft having torsional stiffness k_θ. With a large
drive motor being used, we will assume a displacement-type command for

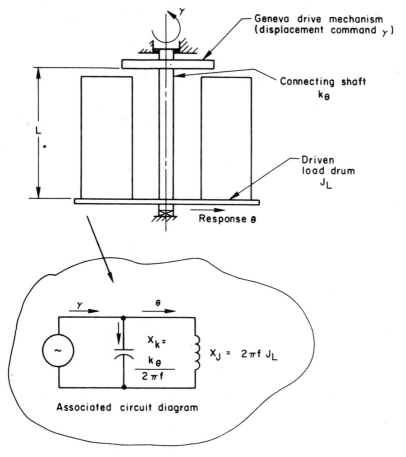

Figure 4.22   Sketch of the initial configuration for intermittent drive of a
load drum by the Geneva mechanism.

excitation of the driven system.   Hence, the associated circuit diagram for the initial configuration is shown in Figure 4.22.

From our knowledge of electrical circuits, we at once recognize the possibility of "circulating currents" in the "tank circuit" of Figure 4.22. This would correspond to torsional vibrations of the load drum in the physical system, which we would like to mitigate by proper design.   In fact,

Figure 4.23   Synthesis of a damping device for the drive system of Figure 4.22.

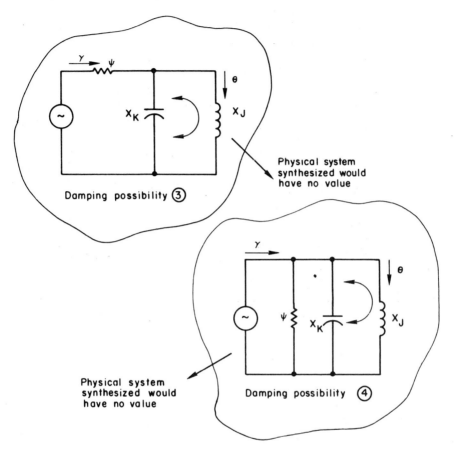

Figure 4.23 (*continued*)

for proper control of the system we would like to have response displacement θ of the driven load as close as practical to command displacement γ of the Geneva drive.   Hence, from the circuit diagram of Figure 4.22 we reach the following goals of design. (Also, see section 9-7 of Chap. 9.)

1. Spring reactance, X_k, of connecting shaft should be as large as practical; i.e., we want shaft stiffness as large as feasible.
2. Mass moment of inertia reactance, X_J, should be as small as possible; i.e., if a choice exists, we want load inertia as small as feasible.
3. If feasibility calculations indicate the necessity, some damping should be present in the tank circuit to mitigate the "circulating currents" or vibrations to a tolerable level.

The first two goals will automatically inspire some thinking along the lines of increasing stiffness or rigidity and decreasing mass. With respect to the introduction of damping in the tank circuit, from the circuit diagram we at once recognize that there are two worthwhile possibilities to consider; these are shown in Figure 4.23, from which the physical configurations in the figure are sketched. The third and fourth damping possibilities shown in the circuit diagrams of Figure 4.23 would not mitigate the circulating currents in the tank circuit, and thus are abandoned at once without sketching the physical configuration.

From a review of the physical configurations of damping possibilities synthesized in Figure 4.23 with the aid of the circuit-diagram technique, we recognize that the two configurations have both advantages and disadvan-

$$X_m = j \ 2\pi f \ m$$

$$X_k = -j \ k/(2\pi f)$$

Associated circuit diagram

Figure 4.24 The design initially conceived for stack weight, with associated circuit diagram.

tages.   For instance, in possibility ① we have the advantage of a simple damping-brake construction, but the disadvantage of energy loss during the "direct current" component of motion in indexing.   On the other hand, in possibility ② we have less energy loss in the damping device but its physical construction is more complicated and difficult to achieve.   Because of the latter, we choose possibility ① for the configuration to develop.

***Example* 4-7**   In the development of a high-speed film handling machine for the Eastman Kodak Company a film stack weight had to be designed to perform several functions in operation of the machine system.   The initially conceived design (Figure 4.24) resulted in unacceptable surging in the film stack, which was excited at fundamental frequency f by machine components at the other end.   The difficulty was due to improper impedance matching between the terminal weight and the film stack.   A new stackweight configuration was conceived, to a great extent by the logical building-block approach generally described in the first part of this chapter.   A patent

Figure 4.25   Patent drawing of the improved design eventually conceived and synthesized by the circuit-diagram technique and the logical building-block approach.   *Courtesy Eastman Kodak Company.*

drawing for this improved design is shown in Figure 4.25, some of the features being required from a consideration of the total system operation.

Final synthesis of the improved stack-weight design was made using the associated circuit diagram shown in Figure 4.26(a). The approach chosen

Figure 4.26 Associated circuit diagram (top) and model (bottom) of the improved design of the film stack weight. *Courtesy Eastman Kodak Company.*

was to make mass reactances X_{m1} and X_{m23} relatively low, by designing for small masses of shoe, body, and arm since we know that $X_m = 2\pi f m$ from Table 4-3 of reference 4-1. On the other hand, spring reactances X_{k1} and X_{k2} were made relatively high, by designing for sufficient stiffness in these springs since we know that $X_k = k/(2\pi f)$ from Table 4-4 of reference 4-1. Finally, after much calculation work, a proper choice was made for the value ψ of cork-pad damping in Figure 4.26(a), resulting in an acceptable degree of impedance matching between the film stack and weight. Problems of film-stack surging were no longer encountered in the machine and the improved stack-weight design is shown in Figure 4.26(b). Incidentally, calculations with the circuit-diagram approach also revealed that an alternate "tuned tank" type of configuration would be unacceptable because of physical size limitations and frequency sensitivity effects from required variations in operational speed of the machine. Thus, the synthesized configuration of Figures 4.25 and 4.26 was used for the film stack weight in the high-speed film-handling machine.

BIBLIOGRAPHY

4-1 R. C. Johnson, *Optimum Design of Mechanical Elements*, John Wiley & Sons Inc., New York, 1961, 535 pp.

4-2 R. C. Johnson, "Optimum Design By Synthesis," presented at the 1970 Design Engineering Conference on May 14, 1970, Chicago, Ill., sponsored by ASME.

4-3 R. C. Johnson, "Impact Forces In Mechanisms," *Transactions of the Fourth Conference on Mechanisms*, Penton Publishing Company, Cleveland, Ohio, 1958, pp. 16-24.

4-4 R. C. Johnson, "Design Synthesis—Aids to Creative Thinking," *Machine Design*, Penton Publishing Co., Cleveland, Ohio, Nov. 15, 1973, pp. 158-163.

5

Selection of
Optimum Configuration

5-1 INTRODUCTION

Decision making plays an important part in the total process of mechanical design synthesis. For problems of design that are relatively routine in nature, such as the synthesis of standard components to form a system, explicit mathematical techniques may be of help in the reaching of decisions along the way. In such problems of decision making we must have on hand specific information, generally statistical in nature, for execution of the techniques. The interested reader might consult references 5-2, 5-4, 5-5, and 5-6 for presentation of some of these techniques.

Unfortunately, for challenging problems of synthesis that are original in nature, generally relatively little quantitative information is available in the early stages of mechanical design synthesis. In such situations the designer most often must rely on his developed talent in the exercising of good judgement for integrating appropriate aspects of engineering science, experience, art, and ingenuity in the reaching of design decisions. At present, in such cases, he has no other choice but to follow that quasi-scientific course of action, particularly in view of economic and time limitations existing in most realistic environments of design. The job of design must be carried through in the best possible way, but this must be done within the confines of such practical limitations.

In Chapters 2 through 4 we described and illustrated some explicit techniques for helping in the creative synthesis of configurations.   Hence, at the start of stage 4 of the morphology of the mechanical design process presented in Figure 1.3 of Chapter 1, generally we would have several basic design concepts to choose from. The task of selecting the optimum configuration is not simple in most cases, since all significant aspects of a total system should be considered as well as possible. Also, the choice of optimum configuration generally must be made with relatively little expenditure of time and money.   In most realistic situations of design it is not feasible to carry through each design concept to the point of accurate evaluation.

In many cases an experienced and competent engineer can select the optimum configuration from a brief review of several design concepts. Although he might very well be right, the decision reached may be misinterpreted as being too subjective, particularly if several designers have contributed the several design concepts under review. With a relatively small amount of additional effort, a more objective approach can often be applied even though it may require the making of comparative approximations in some of the details of execution. Nevertheless, the decision reached in selection of optimum configuration should be more readily accepted by all concerned, and in some problems such an approach will be necessary even to the experienced engineer to reach a decision.   The basic technique is generally described in many sources, such as references 5-1, 5-2, and 5-3. In the remaining part of this chapter we will present and illustrate such a technique for helping in the selection of optimum configuration from several design concepts, particularly applicable to challenging problems of mechanical design synthesis that are original in nature.

5-2   GENERAL TECHNIQUE

A comparative rating technique has been developed for application to the problem of selecting the optimum configuration from several alternative designs. This decision matrix technique has been applied successfully in realistic design situations. With some practice this systematic and logical technique can be applied expediently to original problems, where more explicit techniques would fail because of insufficient information obtainable in the time allowed for design. Generally a good background in engineering science and practical experience is required together with the exercising of good engineering judgement, keeping in mind that only calculations of the comparative approximation type may be possible with the information at hand. The technique is resolved into various basic steps which we will now outline briefly before presentation of some examples. The six steps are summarized as follows.

1. *Decide on attributes of importance.*   Good judgement must be exercised in considering the operation of the total system in deciding on the attributes of significance.   The attributes selected depend on the particular product being designed, its intended use, the product line in which it will be sold, and basic policies of the Company.   Examples of attributes which might be significant for a product are cost, reliability, weight, size, backlash, wear rate, appearance, etc.   In what follows we will designate an attribute by the subscript j, for $j = 1, \ldots, m$, assuming that there are m attributes of importance for the product being designed.

2. *Estimate attribute factors for each of m attributes in each of n design alternatives.*   This step is generally made by simple approximation-type calculations for each of m attributes in each of n design alternatives.   If possible, base such calculations on fundamental laws of engineering science. For example, friction power loss at a region of sliding contact may be considered as proportional to the product of normal force N, coefficient of friction f, and sliding velocity V.   Thus, the friction power-loss factor (FF), as a negative attribute for a design alternative of interest, would be determined by simple calculations from the equation

$$(FF) = \sum_{p=1}^{q} (NfV)_p$$

This equation assumes that there are q friction power-loss regions of sliding contact in the particular design alternative under consideration.

In what follows we will designate a design alternative by the subscript k, for $k = 1, \ldots, n$, assuming that there are n design alternatives being considered.   Hence, there are $m \times n$ attribute factors to be calculated, which must be done as simply as possible; the calculations can often be based on fundamental proportionalities for the sake of expediency.   With the subscript notation now defined, we will designate an attribute factor of interest by $(AF)_{jk}$.

3. *Select a reference design alternative and convert all attribute factors* $(AF)_{jk}$ *to attribute comparison ratings* $(ACR)_{jk}$.   If possible, select as the reference design alternative the one on which you have the most experience; this can often be obtained from your work on similar designs of the past.   The selected reference design alternative will be designated in what follows by the particular subscript $k = r$.

The attribute comparison ratings $(ACR)_{jk}$ are calculated from the attribute factors by equation (5.1) which follows, using the subscript notation now defined.

$$(ACR)_{jk} = (AF)_{jk}/(AF)_{jr} \tag{5.1}$$

Hence, $(AF)_{jk}$ is the jth attribute factor previously estimated for the kth design alternative, and $(AF)_{jr}$ is the corresponding value for the reference design alternative. The normalization step for attribute factors from application of equation (5.1) results in a common base value of unity for each of the attributes of the reference design alternative.

The attribute-comparison-rating values, $(ACR)_{jk}$, as determined by equation (5.1), can be summarized in a decision matrix of order (m, n). However, for what follows, we must recognize that there are in general both positive and negative types of attributes. For example, cost, wear rate, weight, backlash, and size might be considered of the negative type in a certain application, whereas reliability and appearance might be classified of the positive type. The difference can be accounted for by the assignment of positive and negative signs to the $(ACR)_{jk}$ values. However, if not carefully done, this might be misleading in the relative comparison of design alternatives at the end, because of significance of error when taking the difference between two numbers. Instead, it is sometimes more appropriate to reciprocate the $(ACR)_{jk}$ values of the less predominant type to account for the difference in attribute types; all attributes are then considered as algebraically positive values in the decision matrix. This reciprocation technique (to be illustrated later) is most appropriate if the $(ACR)_{jk}$ value is not too far from unity; the technique can result in a relative comparison of design alternatives at the end which is more meaningful, particularly if there is a balance in distribution of positive and negative attribute types in the decision matrix.

4. *Decide on the appropriate relative weighting* $(RW)_j$ *for each attribute of importance.* This step will require establishing a relative order of significance for the various attributes, depending on their relative importance in the total picture for success of the product. To do this, a scale of relative importance (e.g., Table 5-1) can be established. Obviously, the relative weighting value assigned to any one attribute requires the exercising of good engineering judgement backed by much experience in the proper viewing of a total system. However, working cautiously, in most realistic problems the experienced

TABLE 5-1   TYPICAL RELATIVE WEIGHTING SCALE FOR ATTRIBUTES AS RELATED TO SUCCESS OF PRODUCT

9–10	of extremely high importance
8–9	of very high importance
6–7	of above average importance
5	of average or moderate importance
3–4	of minor importance
1–2	of almost insignificant importance

design engineer can select a meaningful value for the relative weighting term, $(RW)_j$, of each attribute. As before, subscript j designates the particular attribute under consideration.

5. *Calculate the overall rating factor* $(ORF)_k$ *for each design alternative.* This step is carried through by multiplying each of the attribute comparison ratings, $(ACR)_{jk}$, by the corresponding relative weighting value $(RW)_j$ and summing.   Stated mathematically, using the notation now defined, we would calculate for each design alternative

$$(ORF)_k = \sum_{j=1}^{m} [(ACR)_{jk}(RW)_j] \qquad (5.2)$$

The value of the overall rating factor, $(ORF)_k$, as calculated by equation (5.2), is an index of relative merit for the design alternative in the particular application under consideration, and it will be used in the next step for selection of the optimum configuration.

6. *List the recommended order of preference for the design alternatives.* This step is determined by a comparison of the values of the overall rating factor, $(ORF)_k$, thereby establishing an order of preference for the various configurations being considered.   The best choice is the design alternative of most favorable $(ORF)_k$ value.   Also, from a comparison of the $(ORF)_k$ values we can readily recognize whether the choice of optimum configuration is an unequivocal decision or a marginal decision.   If it is the latter, a more refined analysis may be in order, or other factors previously considered insignificant might be the deciding influence.

The procedural steps now outlined are suitable for use with a tabular summary.   A typical decision-table format is shown in Table 5-2, depicting

TABLE 5-2 TYPICAL DECISION-TABLE FORMAT

	Design Alternative #1[a]				Design Alternative #2		
	ACR	×	RW	=	ACR	×	RW	=
Attribute #1				→				→
Attribute #2				→				→
Attribute #3				→				→
Overall Rating Factor — — — — — — → (basis for order of preference)			↓		— — — — — — →		↓	

a Selected reference.

in particular the determination of the $(ORF)_k$ values explained in step 5 by equation (5.2).

Let us now consider some specific examples illustrating application of the decision table to the selection of optimum configuration.

5-3   ILLUSTRATION OF APPLICATION

Example 5-1   As an introductory simple example let us apply the decision-table technique to the selection of the optimum configuration in Figure 4.23 of Example 4-6 from Chapter 4.   Suppose that attributes of importance in this particular case are (*1*) short-range manufacturing cost, (*2*) long-range service cost, including replacement of parts, and (*3*) energy loss in operation of the system.

Using our experience and engineering judgement, together with some order-of-magnitude calculations, we reach the following estimates for the damping configurations of Figure 4.23.   Short-range manufacturing cost of design ② would be three times as much as the cost of design ①.   Long-range service cost of design ① would be two times as much as that of design ②. Energy dissipated in design ① would be four times as much as that dissipated in design ②.   Thus, the attribute comparison ratings are calculated and entered under the ACR columns of Table 5-3, normalized for base unity in the selected reference design ①.

Next we must estimate the relative weightings for the attributes, which depend on the relative importance of each in the total picture of system operations for this product line of our Company.   Hence, suppose we decide in this situation that short-range manufacturing cost is twice as important as long-range service cost and five times as important as energy dissipated during

TABLE 5-3   DECISION TABLE FOR SELECTION OF OPTIMUM DAMPING CONFIGURATION

	DESIGN ① of Figure 4.23 (Reference Design Alternative)			DESIGN ② of Figure 4.23		
	ACR ×	RW	=	ACR ×	RW	=
Short-range manufac-turing cost	1	5	5	3	5	15
Long-range service cost	1	2.5	2.5	0.5	2.5	1.25
Energy loss in operation	1	1	1	0.25	1	0.25
Overall rating factor	– – – – – – – – →		8.5	– – – – – – – – →		16.5

operation. Thus, we enter the appropriate relative weighting values under the RW columns in the preceding decision table. Finally, we calculate the overall rating factor for each design alternative by application of equation (5.2); the factor is 8.5 for design ① and 16.5 for design ②. Since the attribute ratings were all of the negative type, the smallest overall rating factor is for the optimum configuration. Therefore, we select design ① of Figure 4.23 as the optimum choice of the two alternatives for damping configuration in this situation. It should be emphasized that in other cases of different total system requirements with different attributes and/or different relative weightings, the optimum choice might not be the same from the Figure 4.23 configurations.

Example **5-2**   In example 4-2 of Chapter 4, an industrial problem of creative design was presented. Suppose four mechanism configurations, which we will designate as Ⓐ, Ⓑ, Ⓒ, and Ⓓ, had been conceived and they were under consideration for the feeding of cloth in a new machine being designed. From a consideration of the total picture of system operations as related to success of the machine, it was decided that attributes of importance for the mechanism were as follows, with a brief explanation given for each:

1. *Initial backlash* for a new machine was important since lost motion of the output link was undesirable for proper functioning of the process operation which the mechanism must perform. Hence, for each mechanism configuration it was necessary to estimate the required backlash at each joint and its effect on lost motion, referred to the output link. In this way, by summation of component terms, *backlash factors* were estimated for each mechanism, proportional to lost motion on the output link, where units are unimportant as long as we are consistent in our calculations. The values estimated for backlash factor are summarized in Table 5-4; the values are normalized to unity for design Ⓑ as the reference, giving the attribute comparison ratings in the second row of the table (see equation (5.1)).

2. *Wear rate* is an important attribute, again of the negative type, since it is related to the *development* of lost motion of the output link as the machine

TABLE 5-4

	Design Alternative			
	Ⓐ	Ⓑ	Ⓒ	Ⓓ
Backlash factor	11.5	7.5	6	5
Comparison rating	1.53	1.0	0.8	0.67

is being used.   In review of the mechanism configurations being considered, it was believed that lost motion producing wear would be primarily of the adhesive type, with large amounts of sliding at the critical regions.   Thus, applying the law of adhesive wear from reference 5-8, we concluded that depth h of wear for contacting surfaces as developed after a certain period of time would be proportional to load L times distance slid x divided by contact area A_c.

$$h \sim L\,x/A_c$$

Orders of magnitude for values proportional to L, x, and A_c could be estimated in a consistent way at each critical wear region of the mechanism configurations.   In this way by summation of component terms *wear rate factors* were estimated for each mechanism, proportional to lost motion developed on the output link after a certain period of time.   Again, units are unimportant as long as consistency is maintained throughout for the proportionality estimates of the component terms.   The values obtained are summarized in Table 5-5; these values are normalized to unity for design Ⓑ as the reference, giving the attribute comparison ratings (see equation (5.1)) in the second row of the table.

TABLE 5-5

	Design Alternative			
	Ⓐ	Ⓑ	Ⓒ	Ⓓ
Wear rate factor	337	257	234	364
Comparison rating	1.31	1.0	0.91	1.42

3. *Friction power loss* was an important attribute of the negative type, since use of a small motor was desired for the machine for economic reasons and surplus power was desired for the process operation external to the mechanism.   As previously discussed, a friction power-loss factor could be estimated at each critical region of a mechanism, proportional to the product of normal load, coefficient of friction, and sliding velocity.   In this way, by summation of component terms *friction power-loss factors* were estimated for each mechanism, where again units are unimportant as long as consistency is maintained in the calculations.   The values obtained are summarized in Table 5-6; the values are normalized to unity for design Ⓑ as the reference, giving the attribute comparison ratings in the second row of the table (from equation (5.1)).

TABLE 5-6

	Design Alternative			
	(A)	(B)	(C)	(D)
Friction power-loss factor	113	117	101	145
Comparison rating	0.97	1.0	0.86	1.24

4. *Cost and complexity* of the mechanism was an important attribute of the negative type, since the machine was to be manufactured in large quantities in a highly competitive field.   Hence, relative order-of-magnitude estimates were made for each part of each mechanism, after having first established a relative cost scale for rapid application.   In this way, by the summation of component terms *cost and complexity factors* were estimated for each mechanism, where again units are unimportant as long as consistency is maintained in the calculations.   The values obtained are summarized in Table 5-7; the

TABLE 5-7

	Design Alternative			
	(A)	(B)	(C)	(D)
Cost and complexity factor	45	45	36	31
Comparison rating	1.0	1.0	0.8	0.69

values are normalized to unity for design (B) as the reference, giving the attribute comparison ratings (from equation (5.1)) in the second row.

5. *Motion characteristics* of the output link were important for proper functioning of the process operation which the mechanism must perform. This attribute was difficult to evaluate quantitatively, but after careful review of requirements the relative ratings in Table 5-8 were decided upon.

TABLE 5-8

	Design Alternative			
	(A)	(B)	(C)	(D)
Motion characteristics of output link	Excellent	Good	Good	Fair
Comparison rating	1.25	1.0	1.0	0.75

In accordance with the comparison ratings of Table 5-8, we see that this attribute has been considered of the positive type whereas the preceding four attributes were all of the negative type.   Also, note that the motion characteristic comparison ratings are already normalized for unity of the reference design Ⓑ.

6. *Shaking forces* in high-speed operation were important attributes of the negative type, since smoothness is a desired characteristic of the machine when placed on a supporting structure that is not highly rigid.   Hence, each mechanism was reviewed and from basic dynamics·order-of-magnitude estimates were made for the peak inertia forces of significance.   Comparative results are summarized in Table 5-9; the values are again normalized to unity

TABLE 5-9

	Design Alternative			
	Ⓐ	Ⓑ	Ⓒ	Ⓓ
Peak shaking force	0.705	1.53	0.52	0.49
Comparison rating	0.46	1.0	0.34	0.32

for design Ⓑ as the reference, giving the attribute comparison ratings (from equation (5.1)) in the second row of the table.

At this stage in our work we have completed steps 1 through 3 of the general technique, having selected design Ⓑ as the reference for the calculation of attribute comparison ratings already presented.   Next, in step 4 of the general technique, we must decide on the relative weighting for each attribute of importance, based on a consideration of its relative significance in the total picture of system operations for success of the product.   Using the scale of relative importance given in Table 5-1, the relative weighting values in Table 5-10 were decided on for the various attributes.

TABLE 5-10   RELATIVE WEIGHTING OF ATTRIBUTES FOR EXAMPLE 5-2

Attribute	Relative Weighting
1. Initial backlash	3
2. Wear rate	8
3. Friction power loss	5
4. Cost and complexity	10
5. Motion characteristics	3
6. Peak shaking force	5

Our next step will be 5 of the general procedure, and overall rating factors will be calculated for each design alternative by equation (5.2). The summary of these calculations is presented in Table 5-11. Attributes 1 through 4 and

TABLE 5-11   DECISION TABLE FOR EXAMPLE 5-2

Attribute	Design Alternative											
	(A)			(B)			(C)			(D)		
	$(ACR) \times (RW) =$			$(ACR) \times (RW) =$			$(ACR) \times (RW) =$			$(ACR) \times (RW) =$		
1. Initial backlash	1.53	3	4.59	1.0	3	3.00	0.8	3	2.40	0.67	3	2.00
2. Wear rate	1.31	8	10.48	1.0	8	8.00	0.91	8	7.28	1.42	8	11.36
3. Friction power loss	0.97	5	4.85	1.0	5	5.00	0.86	5	4.30	1.24	5	6.20
4. Cost and complexity	1.0	10	10.00	1.0	10	10.00	0.8	10	8.00	0.69	10	6.90
5. Motion characteristic	0.8	3	2.40	1.0	3	3.00	1.0	3	3.00	1.33	3	3.99
6. Peak shaking force	0.46	5	2.30	1.0	5	5.00	0.34	5	1.70	0.32	5	1.60
Overall rating factor (ORF) — — — →	34.62			— — — → 34.00			— — — → 26.68			— — — → 32.05		

6 are all of the negative type and these (ACR) values are taken directly from the calculations already made and previously summarized in the short tables. However, attribute 5 is of the positive type, and as previously discussed we enter the reciprocal of this (ACR) value in Decision Table 5-11, which is then based on the negative type. Hence, by execution of equation (5.2) in Decision Table 5-11, we determine the overall rating factors for each mechanism and recommend the order of preference listed in Table 5-12. From this listing we select design (C) as the optimum configuration, and relatively little expenditure of time and effort has been required in the calculations made to reach this decision objectively. Note that the design alternative of lowest (ORF) value was selected as the optimum choice because of the negative base for the

TABLE 5-12   ORDER OF PREFERENCE FOR DESIGN ALTERNATIVES

Order of Preference	Overall Rating Factor (ORF)	Design Alternative	Merit Rating (MR)	Comments
1	26.68	(C)	1.27	Optimum choice
2	32.05	(D)	1.06	
3	34.00	(B)	1.00	Reference design
4	34.62	(A)	0.98	

(ACR) values in Table 5-11.   In such cases of a negative base in the decision table, it is sometimes of value to calculate a *merit rating*, $(MR)_k = (ORF)_r/(ORF)_k$, for each design alternative, giving unity for the reference design with the optimum configuration then having highest value in the listing.   Also, if this is done it is generally easier to make comparisons between design alternatives on a percentage basis, which is often desired from a quick glance of the results.

BIBLIOGRAPHY

5-1   J. R. M. Alger and C. V. Hays, *Creative Synthesis In Design*, Prentice-Hall, Inc., Englewood Cliffs, N.J., 1964, pp. 17–22.

5-2   J. P. Vidosic, *Elements of Design Engineering*, The Ronald Press Company, New York, 1969, Chap. 4.

5-3   J. R. Dixon, *Design Engineering: Inventiveness, Analysis, and Decision Making*, McGraw-Hill Book Co., Inc., New York, 1966, Chap. 11.

5-4   M. Asimow, *Introduction To Design*, Prentice-Hall, Inc., Englewood Cliffs, N.J., 1962, Chap. 9.

5-5   W. H. Middendorf, *Engineering Design*, Allyn and Bacon, Inc., Boston, 1969, Chap. 6.

5-6   R. C. Jeffrey, *The Logic of Decision*, McGraw-Hill Book Co., Inc., New York, 1965.

5-7   R. C. Johnson, "Optimum Design By Synthesis," presented at the 1970 Design Engineering Conference on May 14, 1970, Chicago, Ill., sponsored by ASME.

5-8   E. Rabinowicz, *Friction and Wear of Materials*, John Wiley & Sons, Inc., New York, 1965, 244 pp.

PART II

SELECTION OF MATERIALS AND DIMENSIONS WITH DESIGN OPTIMIZATION

6

Introduction to Advanced Design

6-1 SOME ASPECTS OF ADVANCED DESIGN

In using the word *advanced* we mean to designate something as being in the forefront, ahead of the times, very progressive, or unconventional. Thus, with the word design treated as a noun, we presented some examples of advanced design from various industrial activities in Chapter 1. With the word design treated as a verb, some techniques have been published in recent years which should be placed in the category of advanced design. For example, much literature has been published in the mechanisms field of kinematic synthesis, such as typically listed in references 6-1, 6-2, and 6-3 of the bibliography. Techniques well covered in such sources will not be repeated in this book, although this omission does not minimize their value in some problems of design. In Part I of this book we have presented some techniques of advanced design with respect to the creative synthesis of configurations, emphasizing applications of this relatively unknown field. In Part II, which follows, we will present some advanced design techniques which are relatively unknown for the selection of materials and dimensions, again emphasizing applications in practical problems of mechanical design synthesis.

6-2 CRITICAL REGIONS OF DESIGN

In reference 6-4 Oliver Wendell Holmes calls to our attention some important philosophy to keep in mind when applying techniques of advanced design. With respect to the optimum design of a chaise in days long past, from the reference better known as "The Wonderful 'One-Hoss Shay'," we quote the following excerpts:

> Now in building of chaises, I tell you what,
> There is always *somewhere* a weakest spot,—
> In hub, tire, felloe, in spring or thill,
> In panel, or crossbar, or floor, or sill,
> In screw, bolt, thoroughbrace,—lurking still,
> Find it somewhere you must and will,—
> Above or below, or within or without,—

and

> It should be so built that it *couldn'* break daown,
> —"Fur," said the Deacon, "'t's mighty plain
> Thut the weakes' place mus' stan' the strain;
> 'n' the way t' fix it, uz I maintain,
>   Is only jest
> T' make that place uz strong uz the rest.

The preceding philosophy is in agreement with what is well-known to the experienced design engineer.   Designs of appreciable complexity inherently have weak spots or critical regions upon which the techniques of advanced design should be applied.   In this way the overall system performance will be optimized, and valuable engineering effort will be spent where the results pay off.

Being able to recognize critical regions of design is often an engineering talent which is developed with experience.   This skill can be of great value in the work of creative design, where the avoidance of too many critical design regions is a worthwhile goal in the synthesis of configurations.   On the other hand, once an idea is conceived it is a valuable talent for knowing where the advanced design emphasis should be placed.

An explicit technique, which the author has found of value in the recognition of critical design regions for high-speed machinery, is what we will call the *threshold-speed chart*.   By means of this technique critical design regions are placed in their proper order of significance.   Based on calculations of analysis for an existing or initial design, the speed at which a failure phenomenon impends is determined as governed by each suspected critical design region.   The lowest governing threshold speed indicates the most critical region and advanced design techniques might then be applied in an effort to maximize the governing threshold speed.   A typical threshold-speed chart for a high-speed machine, giving the first five items, is presented in Table 6-1.

TABLE 6-1   TYPICAL THRESHOLD-SPEED CHART FOR A HIGH-SPEED
MACHINE

Drive Shaft Speed (rpm)	Phenomenon Reached
1186	Excessive tooth wear for gearset "ef"
1230	In danger of fatigue breakage for drive shaft
1238	30,000-hour B-10 life for ball bearing at joint "C"
1300	Excitation frequency becomes too close to lowest natural frequency, f_{n1} of helical spring
1329	Dynamic deflection of link "b" reaches unacceptable level of 0.005 in.

Hence, if we wished to maximize the safe speed of operation for the machine, the threshold-speed chart indicates the critical regions which should be alleviated by special design effort and perhaps the application of optimization techniques.

6-3   MATERIALS, DIMENSIONS, AND SYSTEM REQUIREMENTS

In Part II, which follows, we will present and illustrate advanced design techniques for the selection of materials and dimensions which meet total system requirements.   Optimization techniques will be of central importance when applied to critical regions of design, thereby maximizing total system performance.   Emphasis will be on not-so-well-known but nevertheless powerful optimization techniques having broad application in realistic problem situations of mechanical design synthesis.   Where appropriate, use of the digital computer will be illustrated as an important tool for executing the determined procedures of optimum design in some problems.

On the selection of materials, long-range research and development of new and improved materials will not be considered.   This very important aspect in the total long-range picture is of greatest concern to the materials specialist. On the other hand, most mechanical-design engineers, for whom this book is written, must select materials of certain properties which are available to them at the time of design.   Hence, in what follows we will assume that a finite list of materials is available for use in design, the specific content of which depends on both economic and manufacturing considerations.

On the selection of dimensions, there are two types of problems encountered.   First, in Chapter 7, we will consider the selection of what could be called *continuous* dimensions, which determine the *shape* of a surface.   Then, of much broader concern, we will consider the selection of *singular* dimensions in Chapters 8 and 9.   Both types are illustrated in the prismatical bars of Figure 6.1.   In (a) of that figure, the cylindrical shape has been assumed and

(a) Cylindrical Bar

(b) Rectangular
parallelepiped

(c) Special shape for
cross section

Figure 6-1   Various types of prismatical bars.

singular dimensions must be selected for L and D. In (b) of the figure, the rectangular shape has been assumed and singular dimensions must be selected for L, B, and H. In (c) of the figure, a shape defined by function f in $r = f(\theta)$ must be determined and the singular dimension L must be selected. Generally, for the selection of continuous dimensions it is necessary to determine a defining mathematical function, such as $r = f(\theta)$ of the example.

For completion of design of all three of the simple prismatical-bar examples in Figure 6.1, the materials would also have to be selected; the materials are related to the dimensions generally in a complex fashion for satisfaction of the functional requirements both internal and external to an assemblage of elements in the total system. In the following Chapters we will consider some explicit techniques of advanced design which are applicable in realistic situations of mechanical design synthesis requiring the selection of materials and dimensions for complex problems.

In realistic problems of design, seldom can we uniquely select the optimum dimensions and materials independently, as is possible in introductory example 8-1 of Chapter 8. More often the material properties, dimensions, and functional requirements are related by a system of equations, as is very well illustrated in the examples of sections 9-2, 9-3, and 9-4 of Chapter 9. For such cases, we determine the best dimensions for each material combination of interest, with the overall optimum design revealed in the end by a simple comparison of optimization quantity values. Hence, as specifically illustrated in the forementioned sections, selection of the optimum materials and dimensions is often facilitated by use of the digital computer for execution of a simple program explicitly derived by the *method of optimum design* which is introduced in Chapter 8.

BIBLIOGRAPHY

6-1 K. Hain, *Applied Kinematics*, 2nd ed., McGraw-Hill Book Co., Inc., New York, 1967, 727 pp.

6-2 R. Beyer, *The Kinematic Synthesis of Mechanisms*, McGraw-Hill Book Co., Inc., New York, 1963, 353 pp.

6-3 R. S. Hartenberg and J. Denavit, *Kinematic Synthesis of Linkages*, McGraw-Hill Book Co., Inc., New York, 1964, 435 pp.

6-4 O. W. Holmes, "The Deacon's Masterpiece," *Atlantic Monthly*, Sept. 1858. (Also, O. W. Holmes, *The Autocrat of The Breakfast-Table*, Houghton, Mifflin and Co., Boston, 1858, pp. 252–256.)

7

Shape Design

7-1 INTRODUCTION

The selection of shape for an element is often an important problem of mechanical design synthesis. In practice, the experienced design engineer most often must rely on knowledge, good judgement, and ingenuity in deciding on the basic shape for an element. Problems are often too complicated for explicit solution by advanced mathematics in the time available for design. Also, in selecting a basic shape for an element, the functional requirements of a total system, both practical and theoretical in nature, must be considered.

There are many opportunities for applying advanced mathematics in the determination of optimum shape for critical mechanical elements. The basic theory is well developed and available to the interested reader in references such as those listed in the bibliography. This list is not meant to be complete in any sense, but it does at least provide some readable sources for the derivation of basic theory as an important background to the design engineer. Space is not available in the book to repeat derivations of basic theory available in many other sources, and it is assumed that the reader will have an understanding of this important work from such references. Instead, in what follows we will emphasize *applications* of the basic techniques to some problems of mechanical design synthesis, since such examples are not too common in the literature.

7-2   SHAPE DESIGN BY SIMPLE INTEGRATION

In some critical problems of design, the optimum shape of a surface or element can be determined directly by elementary integration from calculus. An example of application in high-speed machinery will now be presented.

Example 7-1   In a high-speed machine a cam-actuated driver is to reciprocate with specified motion s of known maximum acceleration value a_{max}. Given mass M is to be driven with the same motion, and we wish to design the connecting arm between the two as shown in Figure 7.1. From manufacturing considerations, connecting arm thickness is to be of specified value t, and the problem is to determine the optimum shape function $y(x)$ which will minimize dynamic forces.

In Figure 7.1(a), subscript 0 designates a section of interest having coordinates (x_0, y_0). Outer fiber bending stress $(\sigma_b)_0$ will govern how small y_0 can be rather than centroidal axis shear stress, the assumption of which was verified by analysis in the end for the particular numerical problem. Peak bending moment at 0 from dynamics is in terms of "inertia forces" F_i as follows:

$$(M_b)_0 = (F_i)_M x_0 + \int_0^{x_0} (x_0 - x)\, dF_i$$

where

$$(F_i)_M = M\, a_{max}$$

and

$$dF_i = \left(\frac{w}{g}\, yt\, dx\right) a_{max}$$

Thus, we obtain

$$(M_b)_0 = \left[Mx_0 + \frac{wt}{g} \int_0^{x_0} (x_0 - x) y\, dx \right] a_{max} \tag{7.1}$$

Terms in equation (7.1) have been defined except w and g, which are the weight density of the arm material and the gravitational constant, respectively.

From strength of materials, outer fiber bending stress at section 0 is

$$(\sigma_b)_0 = \frac{(M_b)_0(y_0/2)}{(1/12)ty_0^3} = \frac{6(M_b)_0}{ty_0^2}$$

(a) Notation for shape design of a connecting arm

(b) Layout of optimum arm shape as
determined for numerical problem

Figure 7.1   Shape design for the connecting arm in example 7-1.

Hence, by combination with equation (7.1) we obtain

$$(\sigma_b)_0 = \frac{6a_{max}}{ty_0{}^2}\left[Mx_0 + \frac{wt}{g}\int_0^{x_0}(x_0 - x)y\,dx\right] \qquad (7.2)$$

For fully reversed stress variation from the given reciprocating motion, we have as the *allowable* limit

$$(\sigma_b)_0 = S_e/N_e \qquad (7.3)$$

where S_e is the fatigue strength of the connecting arm material, and N_e is an appropriate factor of safety now selected. Thus, by combining equation (7.2) and (7.3) and eliminating $(\sigma_b)_0$ we obtain

$$\left(\frac{S_e t}{6a_{max}N_e}\right)y_0{}^2 = Mx_0 + \frac{wt}{g}\int_0^{x_0}(x_0 - x)y\,dx$$

which can be written as

$$C_1 y_0{}^2 = C_2 x_0 + \int_0^{x_0}(x_0 - x)y\,dx \qquad (7.4)$$

where

$$C_1 = g\,S_e/(6\,a_{max}\,N_e\,w) \qquad (7.5)$$

and

$$C_2 = g\,M/(wt) \qquad (7.6)$$

Solution of the integral equation (7.4) will give the optimum shape for the connecting arm. This was obtained quite readily by a finite difference approach. First we differentiate twice both sides of equation (7.4) with respect to x_0. If necessary for review, see typically equation (9.3) on page 320 of reference 7-2 for differentiation of the right side integral. Thus,

$$2C_1 y_0\left(\frac{dy}{dx}\right)_0 = C_2 + \int_0^{x_0} y\,dx$$

$$2C_1\left[y_0\left(\frac{d^2y}{dx^2}\right)_0 + \left(\frac{dy}{dx}\right)_0^2\right] = y_0 \qquad (7.7)$$

Applying finite difference approximations (2-4) and (2-5) from reference 7-6, for $(dy/dx)_0$ and $(d^2y/dx^2)_0$, respectively, equation (7.7) becomes as follows:

$$\left(\frac{d^2y}{dx^2}\right)_0 + \frac{1}{y_0}\left(\frac{dy}{dx}\right)_0^2 = \frac{1}{2C_1}$$

therefore

$$\frac{y_1 + y_2 - 2y_0}{(\Delta x)^2} + \frac{1}{y_0}\left(\frac{y_2 - y_1}{2\Delta x}\right)^2 \approx \frac{1}{2C_1}$$

$$y_2 \approx 2y_0 - y_1 + \frac{(\Delta x)^2}{2C_1} - \frac{(y_2 - y_1)^2}{4y_0} \qquad (7.8)$$

In finite difference approximation (7.8) subscripts 1 and 2 designate the sections adjacent to 0 in Figure 7.1(a), algebraically preceding and following section 0 at the increment (Δx) away as shown in the figure. Successive application of equation (7.8), solving for y_2 in an iterative approach at each step, and working from left to right in Figure 7.1(a), gives the optimum shape of the connecting arm. Incidentally, for the very first calculation we use $y_1 = 0$ for $x = 0$, and it is necessary to estimate the appropriate value of y_0 at $x = \Delta x$, which is done assuming that the bending-moment effect of mass M predominates in that vicinity since connecting-arm mass is insignificant for $0 \leq x \leq \Delta x$.

By the technique now described, optimum shape was determined for a connecting arm in a specific numerical problem. A layout drawing for the determined shape is shown to scale in Figure 7.1(b). As we see, a minor modification was necessary in the end by introducing fillets for joining the connecting arm to both driven mass M and the driver body. Finishing touches of this type are generally required for making a theoretical design practical.

7-3   SHAPE DESIGN BY THE CALCULUS OF VARIATIONS

General Technique

Occasionally in mechanical design synthesis it is desired to find a particular function $y(x)$ defining a shape which minimizes (or maximizes) some phenomenon I of interest, which is expressed in terms of a definite integral having the general form

$$I = \int_{x_1}^{x_2} F(x, y, \dot{y}) \, dx \qquad (7.9)$$

In this equation F designates the integrand function of x, y, and $\dot{y} = dy/dx$. The problem is to find the particular function $y(x)$ which minimizes (or maximizes) the value of I.

A necessary condition for $y(x)$ is satisfaction of the Euler-Lagrange differential equation (7.10).

$$\frac{\partial F}{\partial y} - \frac{d}{dx}\left(\frac{\partial F}{\partial \dot{y}}\right) = 0 \qquad (7.10)$$

The interested reader should refer to the derivation of equation (7.10), such as typically found on pp. 326–329 of reference 7-2.

Solution of equation (7.10) will give the function $y(x)$ sought, and in addition it *may* give other curves which do *not* minimize (or maximize) integral value I. From a practical standpoint it is generally relatively easy to select the sought function $y(x)$ if a choice exists. However, if greater explicitness is desired, we might apply the Legendre and the Jacobi necessary conditions as sufficient together with satisfaction of the Euler-Lagrange equation (7.10) to assure that the function $y(x)$ is the one sought. These techniques are covered in the literature, such as in reference 7-4.

Example 7-2   We wish to find explicitly the path of shortest distance between two given points P_1 and P_2 in Figure 7.2(a). In reference to the figure, length of a general path shown is

$$S = \int ds = \int_{x_1}^{x_2} \sqrt{1 + (\dot{y})^2} \, dx$$

Hence, comparing the preceding with equation (7.9), we see that the integrand function is

$$F = \sqrt{1 + (\dot{y})^2}$$

Thus, applying Euler-Lagrange equation (7.10) we obtain

$$\frac{\partial F}{\partial y} - \frac{d}{dx}\left(\frac{\partial F}{\partial \dot{y}}\right) = 0$$

therefore

$$0 - \frac{d}{dx}\left(\frac{\dot{y}}{\sqrt{1 + (\dot{y})^2}}\right) = 0$$

$$\frac{\dot{y}}{\sqrt{1 + (\dot{y})^2}} = C, \text{ constant}$$

$$(\dot{y})^2 = \frac{1}{(1/C)^2 - 1} = C_1, \text{ constant}$$

Therefore, $\dot{y} = dy/dx$ is a constant, and we have explicit proof that the optimum shape of path for shortest distance between two points is a straight line.

Example 7-3   We wish to determine the ideal shape for a toboggan run which will minimize the time of descent between given points A and B in Figure 7.2(b). Ideally, we will assume negligible retarding forces from ice

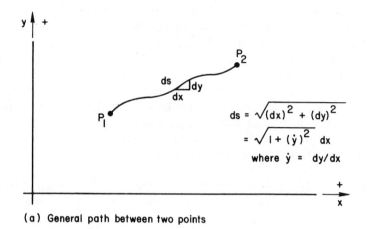

(a) General path between two points

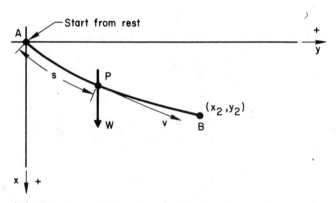

(b) Toboggan path between two points in vertical xy plane

Figure 7.2   Introductory examples 7-2 and 7-3 for the calculus of variations.

friction and windage. Thus, starting from rest at origin point A of the figure, from which s is measured, we have, from elementary mechanics

$$v = ds/dt = \sqrt{2gx}$$

As for the previous example we use

$$ds = \sqrt{1 + (\dot{y})^2}\, dx$$

where again $\dot{y} = dy/dx$. Thus, for time T of descent, we obtain

$$T = \int_0^T dt = \int \frac{ds}{v} = \int_0^{x_2} \frac{\sqrt{1+(\dot{y})^2}\, dx}{\sqrt{2gx}}$$

therefore

$$T = \frac{1}{\sqrt{2g}} \int_0^{x_2} \frac{\sqrt{1+(\dot{y})^2}}{\sqrt{x}}\, dx \qquad (7.11)$$

The problem is to find the particular function $y(x)$ which minimizes descent time T.

Comparing equations (7.9) and 7.11), we see that the integrand function is

$$F = \sqrt{1+(\dot{y})^2}/\sqrt{x}$$

Thus, applying the Euler-Lagrange differential equation (7.10), we obtain

$$\frac{\partial F}{\partial y} - \frac{d}{dx}\left(\frac{\partial F}{\partial \dot{y}}\right) = 0$$

$$0 - \frac{d}{dx}\left(\frac{\dot{y}}{\sqrt{x}\sqrt{1+(\dot{y})^2}}\right) = 0$$

Hence,

$$\frac{\dot{y}}{\sqrt{x(1+\dot{y}^2)}} = C, \text{ constant}$$

Solution of this differential equation gives the $y(x)$ sought, and it is

$$y = -\sqrt{xC_1 - x^2} + C_1 \sin^{-1}\sqrt{x/C_1} \qquad (7.12)$$

Equation (7.12) defines the ideal brachistochrone shape of the toboggan run, and constant C_1 is determined by the condition that the curve starting at origin A must pass through point B of specified coordinates (x_2, y_2) in Figure 7.2(b).

Variational Problems with Accessory Conditions

Let us now consider a basic extension to the type of problem so far discussed and illustrated. Suppose that we wish to find the particular function $y(x)$ which minimizes (or maximizes) the definite integral I of the equation (7.9) form, while also satisfying a specified constant value I_1 for another integral of form

$$I_1 = \int_{x_1}^{x_2} F_1(x, y, \dot{y})\, dx \qquad (7.13)$$

$$= \text{specified value}$$

The type of problem now described is known as an *isoperimetric problem*. It can be handled quite nicely by defining another definite integral I_0 as follows:

$$I_0 = \int_{x_1}^{x_2} F_0(x, y, \dot{y})\, dx \qquad (7.14)$$

where the integrand function is specifically taken as

$$F_0 = F + \mu F_1 \qquad (7.15)$$

where F and F_1 are the integrand functions from equations (7.9) and (7.13), respectively, and μ is a parameter to be determined near the end of the derivation from satisfaction of the accessory condition. We readily see that the particular function $y(x)$ which makes integral value I a minimum (or maximum) in equation (7.9) also makes integral value I_0 a minimum (or maximum) in equation (7.14), since integral value I_1 of equation (7.13) must be a specified constant. Hence, solution to the isoperimetric problem is obtained by applying the Euler-Lagrange differential equation (7.10) to integrand function (7.15).

$$\frac{\partial F_0}{\partial y} - \frac{d}{dx}\left(\frac{\partial F_0}{\partial \dot{y}}\right) = 0 \qquad (7.16)$$

Let us now illustrate application of the technique with some examples of isoperimetric problems.

Example 7-4 The problem is to determine the boundary curve shape which minimizes perimeter S while enclosing a planar area of specified value A. Specifically, this might be a problem of minimizing the required length of fence for enclosing a certain amount of land area. Or, it might be one of minimizing surface area SL of a prismatical bar of length L, meeting specified cross-sectional area A for satisfaction of strength requirements.

In reference to Figure 7.3, polar coordinates will be used, with perimeter S to be minimized expressed as

$$S = \int ds = \sqrt{(dr)^2 + (rd\theta)^2}$$

$$S = \int_0^{2\pi} \sqrt{(\dot{r})^2 + r^2}\, d\theta \qquad (7.17)$$

where

$$\dot{r} = dr/d\theta$$

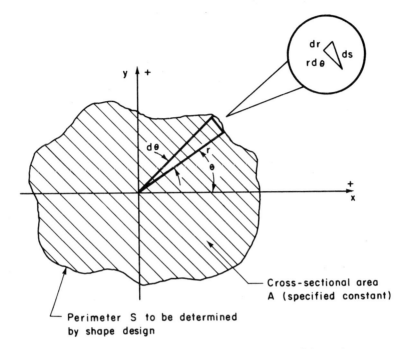

- Cross-sectional area
  A (specified constant)

- Perimeter S to be determined
  by shape design

Figure 7.3 **Specified area enclosed by the general boundary curve of example 7-4.**

Also referring to Figure 7.3, specified enclosed area A is expressed as

$$A = \int \tfrac{1}{2}(rd\theta)r$$

therefore

$$A = \tfrac{1}{2} \int_0^{2\pi} r^2 \, d\theta \qquad (7.18)$$

The problem is to find the function $r(\theta)$ which minimizes integral S while satisfying specified A for area. Equation (7.17) is analogous to (7.9), equation (7.18) is analogous to (7.13), with coordinates (θ, r) analogous to (x,y) respectively. Hence, using the previous notation we have the following integrand functions:

$$F = \sqrt{(\dot{r})^2 + r^2}$$
$$F_1 = r^2/2$$

Thus, as previously described for equations (7.14) and (7.15), we define a composite integrand function F_0 as

$$F_0 = F + \mu F_1$$

therefore

$$F_0 = \sqrt{(\dot{r})^2 + r^2} + \mu r^2/2 \qquad (7.19)$$

where μ is a parameter to be determined near the end of derivation.

The Euler-Lagrange differential equation (7.16) expressed in polar coordinates becomes

$$\frac{\partial F_0}{\partial r} - \frac{d}{d\theta}\left(\frac{\partial F_0}{\partial \dot{r}}\right) = 0 \qquad (7.20)$$

Applying equation (7.20) to integrand function (7.19) we obtain

$$\left[\frac{r}{\sqrt{(\dot{r})^2 + r^2}} + \mu r\right] - \frac{d}{d\theta}\left[\frac{\dot{r}}{\sqrt{(\dot{r})^2 + r^2}}\right] = 0$$

Introducing the notation $\ddot{r} = d^2r/d\theta^2$ and proceeding,

$$\left[\frac{r}{\sqrt{(\dot{r})^2 + r^2}} + \mu r\right] - \frac{\ddot{r}\sqrt{(\dot{r})^2 + r^2} - \dfrac{\dot{r}(\dot{r}\ddot{r} + r\dot{r})}{\sqrt{(\dot{r})^2 + r^2}}}{[(\dot{r})^2 + r^2]} = 0$$

Thus, by clearing fractions and collecting terms, we obtain

$$r\left\{\mu + \frac{r^2 + 2(\dot{r})^2 - r\ddot{r}}{[(\dot{r})^2 + r^2]^{3/2}}\right\} = 0$$

Therefore,

$$\frac{r\ddot{r} - 2(\dot{r})^2 - r^2}{[(\dot{r})^2 + r^2]^{3/2}} = \mu \qquad (7.21)$$

The absolute value of the left side of equation (7.21) we recognize as the curvature, $1/\rho$, expressed in polar coordinates.   Hence, radius of curvature ρ of the boundary for the optimum shape is

$$\rho = |1/\mu| = \text{constant}$$

Therefore, we have explicitly determined that the boundary curve shape which minimizes perimeter S while enclosing specified area A is a circle. The value for radius of curvature, ρ, is determined from the specified enclosed area A, which also gives the value for μ.

Example 7-5   In reference to Figure 7.4, a high-speed mechanism was designed for reciprocating the slider of given mass M through given stroke S with known acceleration characteristics.   For the critical position shown in the figure, "inertia forces" $(F_i)_M$ and $(F_i)_b$ are to be cancelled from $(F_i)_h$ of

Figure 7.4   High-speed mechanism with counterbalance to be designed for optimum shape in example 7-5.

a counterbalance h to be designed.   At this stage of design, required "inertia force" $(F_i)_h$ of the counterbalance has been determined from the force polygon of Figure 7.4.   Also, at this stage we have determined the desired values of R_h, L_h, M_h, and $\ddot{\theta}_h$ for the counterbalance h.   Referring to Figure 7.4, R_h and L_h are dimensions from pivot O_h to mass center G_h and to link connection joint H, respectively. M_h is the necessary counterbalance mass and $\ddot{\theta}_h$ is the angular acceleration of the counterbalance for the critical position shown. We wish to minimize the inertial load of the counterbalance as reflected to the main mechanism, by choosing the optimum shape for the counterbalance. In effect this will be achieved by minimizing Δ of Figure 7.4, where Δ is the displacement of "inertia force" $(F_i)_h$ from mass center G_h.   From dynamics of machinery, such as typically covered in reference 7.7, we derive the expression for Δ of Figure 7.4 as follows.

$$\Delta = \frac{\bar{I}_h \ddot{O}_h}{M_h (R_h \ddot{\theta}_h)} = \frac{\bar{I}_h}{M_h R_h}$$

Hence, with M_h and R_h specified the objective for optimum shape design can be considered as minimization of counterbalance mass moment of inertia, $\bar{I}_h$, about the mass center.   Thus, in reference to Figure 7.5(a), from basic mechanics we derive

$$\bar{I}_h = \iint \frac{w}{g} t\rho^2 (\rho \, d\theta \, d\rho)$$

$$= \frac{wt}{g} \int_0^{2\pi} \int_0^r \rho^3 \, d\rho \, d\theta$$

therefore

$$\bar{I}_h = \frac{wt}{4g} \int_0^{2\pi} r^4 \, d\theta \tag{7.22}$$

In equation (7.22), w is the weight density of the counterbalance material, t is its thickness perpendicular to the plan view of Figure 7.5(a), g is the gravitational constant, and r and θ are defined in the figure.   Also, since required mass M_h is known at this stage of design, we derive the following equation referring to Figure 7.5(a).

$$M_h = \iint \frac{w}{g} t(\rho \, d\theta \, d\rho)$$

$$= \frac{wt}{g} \int_0^{2\pi} \int_0^r \rho \, d\rho \, d\theta$$

therefore

$$M_h = \frac{wt}{2g} \int_0^{2\pi} r^2 \, d\theta \tag{7.23}$$

All terms have been defined in equation (7.23).

(a) General shape of counterbalance for Figure 7.4 mechanism, with differential elements shown

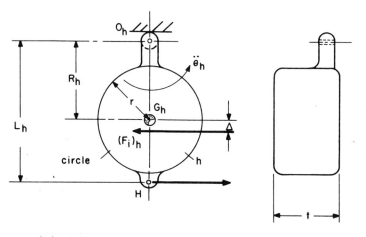

(b) Optimum shape for counterbalance of example 7.5

Figure 7.5 Counterbalance design for minimization of reflected inertial load in the high-speed mechanism of example 7-5.

The isoperimetric problem which we now face is to determine the function $r(\theta)$ which minimizes the value of $\bar{I}_h$ from integral equation (7.22) *and* which also satisfies the specified value of M_h from the integral equation (7.23). Hence, using the notation previously defined, our integrand functions are taken as

$$F = r^4/2$$

and

$$F_1 = r^2$$

since constant $wt/(2g)$ common to both equations will have no effect on the basic shape being determined. Thus, as before we extend equations (7.14) and (7.15) to polar coordinates, and define a composite integrand function F as follows:

$$F_0 = F + \mu F_1$$
$$F_0 = (r^4/2) + \mu r^2 \tag{7.24}$$

where μ is a parameter to be determined near the end of derivation.

Applying the Euler-Lagrange differential equation (7.20) to integrand function (7.24), we obtain

$$\frac{\partial F_0}{\partial r} - \frac{d}{d\theta}\left(\frac{\partial F_0}{\partial \dot{r}}\right) = 0$$

therefore

$$2r^3 + 2\mu r = 0$$

therefore

$$\mu = -r^2 = \text{constant}$$

Hence, the optimum shape for the counterbalance is circular, as shown in Figure 7.5(b). Incidentally, these findings were in agreement with several random checks made on other shapes for the counterbalance. Thus, the cylindrical counterbalance was the shape selected for cancelling the "inertia" or "shaking forces" of the mechanism and yet minimizing the inertial load reflected from the counterbalance through the drive mechanism.

Example **7-6** The calculus of variations has been applied extensively to practical problems of shape design in aerodynamics, and the interested reader should consult reference 7.3 for many excellent studies. As a specific example, consider the problem of optimum shape design for a stabilizing airfoil to give minimum pressure drag in linearized supersonic flow. The airfoil is not to function in a lift capacity. From equation (4.93) on page 121 of reference 7-3 we obtain the following pressure coefficient equation.

$$C_p = 2\dot{y}/\sqrt{M^2 - 1} \tag{7.25}$$

Equation (7.25) assumes simple linearized theory, and it is applicable to a two-dimensional symmetric airfoil design such as in Figure 7.6(a), moving at supersonic speed with zero lift in an inviscid, nonconducting perfect gas.

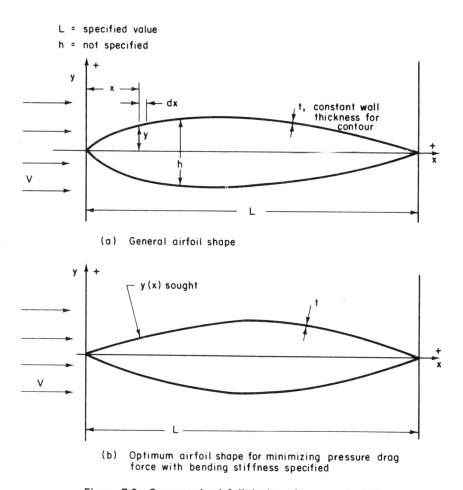

(a)   General airfoil shape

(b)   Optimum airfoil shape for minimizing pressure drag
      force with bending stiffness specified

Figure 7.6   Supersonic airfoil designs for example 7-6.

In the equation (7.25), C_p is the pressure coefficient, $\dot{y} = dy/dx$ is for the upper airfoil surface in Figure 7.6(a), and M is the free stream Mach number of value $M = V/c$, where V is the free stream velocity of Figure 7.6(a) and c is the speed of sound in the gas.   From equation (4.94) on page 121 of reference 7-3 we obtain the following aerodynamic drag force equation for a unit length of span for the Figure 7.6(a) symmetric general shape.

$$D_p = 2q \int_0^L C_p \dot{y} \, dx$$

To this is applied equation (7.25), giving

$$D_p = [4q/\sqrt{M^2 - 1}] \int_0^L (\ddot{y})^2 \, dx \qquad (7.26)$$

In equation (7.26), D_p is the pressure drag force acting on the airfoil of the Figure 7.6(a) symmetric general shape, q is the free stream dynamic pressure, $\rho V^2/2$, where ρ is the density of the gas, and all other terms have been previously described.

For adequate stiffness against bending as anticipated in some modes of system operation we will assume that moment of inertia of the contour must have a specified value I_x. Hence, in reference to Figure 7.6(a) we derive

$$I_x \approx 2 \int_0^L y^2 t \, dx$$

which assumes an airfoil of slender shape. Thus,

$$I_x = 2t \int_0^L y^2 \, dx \qquad (7.27)$$

In equation (7.27) all terms have been described or are shown in Figure 7.6(a).

The isoperimetric problem which we now face is to determine the particular function $y(x)$ which minimizes the drag force integral D_p of equation (7.26) while satisfying the specified value I_x from integral equation (7.27). Hence, using the previous notation we have integrand functions

$$F = (\ddot{y})^2$$

and

$$F_1 = y^2$$

Thus, as previously described for equations (7.14) and (7.15) we define a composite function F_0 as

$$F_0 = F + \mu F_1$$

$$F_0 = (\ddot{y})^2 + \mu y^2 \qquad (7.28)$$

where μ is a parameter to be determined near the end of derivation.

Applying Euler-Lagrange differential equation (7.16) to integrand function (7.28), we obtain

$$\frac{\partial F_0}{\partial y} - \frac{d}{dx}\left(\frac{\partial F_0}{\partial \dot{y}}\right) = 0$$

$$2\mu y - \frac{d}{dx}(2\ddot{y}) = 0$$

$$\mu y - \ddddot{y} = 0 \qquad (7.29)$$

where μ is a constant to be determined. Hence, this last equation we can write in the well-known form

$$\frac{d^2y}{dx^2} + ky = 0 \qquad (7.30)$$

whose solution is

$$y = C_1 \sin [\sqrt{k}\, x + C_2] \qquad (7.31)$$

The boundary conditions referring to Figure 7.6(a) are:
for $x = 0$, $y = 0$

$$C_2 = 0$$

and for $x = L$, $y = 0$

$$\sqrt{k}\, L = \pi$$

Thus, the solution is

$$y = C_1 \sin [(\pi/L)\, x] \qquad (7.31a)$$

where constant C_1 is determined to satisfy the specified value of I_x, using equation (7.31a) and (7.27) to give

$$I_x = 2t \int_0^L C_1{}^2 \sin^2[(\pi/L)x] \, dx$$

$$= tC_1{}^2 L$$

therefore

$$C_1 = [I_x/(t\,L)]^{1/2}$$

which is used in equation (7.31a), giving

$$y = \left[\frac{I_x}{tL}\right]^{1/2} \sin[(\pi/L)x] \qquad (7.31b)$$

Hence, the function $y(x)$ which we sought is equation (7.31b), giving minimum pressure drag force D_p and meeting specified bending stiffness I_x of the contour. The optimum sinusoidal-shaped airfoil is drawn to scale in Figure 7.6(b). Incidentally, the solution obtained is in agreement with that obtained on pages 124–125 of reference 7-3.

Example 7-7 An industrial example for the calculus of variations will be considered briefly, with respect to the ideal shape of a tensile socket basket designed for the United States Steel Corporation. The objective for optimum design was minimization of basket weight, W_{SB}. Referring to Figure 7.7,

Figure 7.7   Ideal tensile socket basket configuration, showing notation and critical stress combinations.

we derive from basic calculus,

$$W_{SB} = w \int 2\pi r t \, ds$$

where

$$ds = \sqrt{(dr)^2 + (dz)^2}$$
$$= \sqrt{(\dot{r})^2 + 1} \, dz$$

and

$$\dot{r} = \frac{dr}{dz}$$

Hence,

$$W_{SB} = 2\pi w \int_0^h tr\sqrt{(\dot{r})^2 + 1}\, dz \qquad (7.32)$$

From a complicated analysis of combined stresses in the ideal basket of Figure 7.7, we derive an equation for allowable stress which is combined with equation (7.32), eliminating wall thickness t. Thus, we obtain the following equation for W_{SB}:

$$W_{SB} \approx 2\pi w C_s \left\{ \int_0^h \frac{\sqrt{(\dot{r})^4 + (\dot{r})^2}\, dz}{r^{(a-0.34)}} + \int_0^h \frac{C_s\sqrt{(\dot{r})^6 + (\dot{r})^4}}{r^{(2a+1.32)}}\, dz \right\} \qquad (7.33)$$

In equation (7.33), r, z, and h are defined in Figure 7.7, as before $\dot{r} = dr/dz$, w is the weight density of the socket steel, and C_s and a are constants determined by the boundary conditions.

The problem now at hand is to determine the function $r(z)$ which minimizes basket weight W_{SB} of equation (7.33). The Euler-Lagrange differential equation (7.10) was applied, which takes the following form:

$$\frac{\partial F}{\partial r} - \frac{d\left(\dfrac{\partial F}{\partial \dot{r}}\right)}{dz} = 0 \qquad (7.34)$$

where from equation (7.33) we have the integrand function

$$F = \frac{\sqrt{(\dot{r})^4 + (\dot{r})^2}}{r^{(a-0.34)}} + \frac{C_s\sqrt{(\dot{r})^6 + (\dot{r})^4}}{r^{(2a+1.32)}} \qquad (7.35)$$

Thus, application of equation (7.34) to (7.35) gives differential equation (7.36),

$$\ddot{r}r[A(\dot{r})^2 + B] + (\dot{r})^2[C(\dot{r})^2 + D] = 0 \qquad (7.36)$$

where A, B, C and D are merely constants determined by C_s and a, and as before we designate $\dot{r} = dr/dz$ and $\ddot{r} = d^2r/dz^2$.

Solution to differential equation (7.36) is explicit but lengthy, giving function $r(z)$ as follows:

$$r = [C_6 z + C_7]^{B/(D+B)} \qquad (7.37)$$

In equation (7.37), C_6 and C_7 are constants of integration which we will not take the space to describe, determined by the boundary conditions.

Hence, in the outlined way we have now derived equation (7.37), which defines the optimum shape of the ideal socket basket for minimization of W_{SB} with stress limited to an allowable value.   The solution was applied to numerical problems, and the shape so determined is presented graphically to scale in Figure 7.8.

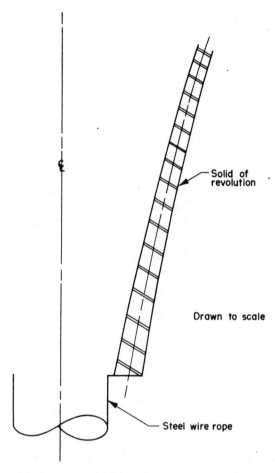

Figure 7.8   Layout of an ideal tensile socket basket shape as determined by the calculus of variations for minimum weight with limited stress.

BIBLIOGRAPHY

7-1   M. J. Forray, *Variational Calculus In Science and Engineering*, McGraw-Hill Book Co., Inc., New York, 1968, 221 pp.

7-2 L. A. Pipes, *Applied Mathematics for Engineers and Physicists*, 2nd ed., McGraw-Hill Book Co., Inc., New York, 1958, 723 pp.

7-3 G. Leitmann, *Optimization Techniques With Applications to Aerospace Systems*, Academic Press, New York, 1962, 453 pp.

7-4 S. E. Dreyfus, *Dynamic Programming and The Calculus of Variations*, Academic Press, New York, 1965, 248 pp.

7-5 E. F. Beckenbach, *Modern Mathematics for the Engineer*, McGraw-Hill Book Co., Inc., New York, 1956, Chap. 4.

7-6 R. C. Johnson, *Optimum Design of Mechanical Elements*, John Wiley & Sons, Inc., New York, 1961, 535 pp.

7-7 R. L. Maxwell, *Kinematics and Dynamics of Machinery*, Prentice-Hall, Inc., Englewood Cliffs, N.J., 1960, 477 pp.

7-8 D. R. Smith, *Variational Methods in Optimization*, Prentice-Hall, Inc., Englewood Cliffs, N.J., 1974, 378 pp.

8

Introduction to Optimum Design

8-1 THE BASIC PROBLEM OF OPTIMIZATION

Having selected the general configuration for our mechanical device and the basic shapes of elements, by techniques described in previous chapters, we are now at the stage in the design process of Figure 1.3 where decisions must be reached for the selection of discrete value dimensions and materials. For routine problems that are noncritical in nature this can be done by cut-and-try design in conjunction perhaps with some simple calculations of analysis. For such cases this is the optimum approach to follow since design cost is minimized, providing that the product achieved is satisfactory from all standpoints of significance throughout its expected period of performance.

On the other hand, in challenging problems of design there are critical areas where the decisions cannot be reached satisfactorily by the cut-and-try approach. At such places there are phenomena of significance which should be either amplified or mitigated as much as feasible by optimum design. For instance, in some cases it may be appropriate to maximize a quantity such as energy-absorption capability, power-transmission capability, factor of safety, speed of reliable operation, or force-transmission capability. In other situations it may be appropriate to minimize some quantity such as cost, weight, a critical dimension, deflection, wear rate, backlash, or dynamic shaking force. Regardless of the objective for optimization at a critical design region of interest, all constraints of significance must be adhered to in the decisions reached.

From the preceding discussion we readily recognize that practical problems of optimum design are not simple in nature. First, the engineer must recognize the critical areas of design where optimization techniques are likely to pay off in the results achieved. Secondly, he must be able to formulate the optimization problem, not only from the standpoint of the optimization quantity itself but also from the standpoint of significant subsidiary quantities, specifications, and various constraints. Finally, he must be able to handle expeditiously the *system of equations* thus formulated, drawing design conclusions of value from the variation study made, which can be applied to the design problem at hand. Hence, we see that application of optimum design in a realistic decision-making process requires an understanding of optimization theory integrated with engineering science, practical experience, design talent, and good engineering judgement.

The system of equations as first derived in an optimization problem will be referred to as the *initial formulation*. In general, it will consist of a criterion function or primary design equation (P.D.E.), subsidiary design equations (S.D.E.), and limit equations (L.E.). A *typical initial formulation* expressed in general mathematical terms would be as follows:

Objective: maximize Q.

$$Q = f_1 (u, v, w) \qquad\qquad\qquad \text{(P.D.E.) (8.1)}$$

$$y = f_2 (u, v, w) \qquad\qquad\qquad \text{(S.D.E.) (8.2)}$$

$$A = f_3 (u, v, w) \qquad\qquad\qquad \text{(S.D.E.) (8.3)}$$

$$u \geq u_{min} \qquad\qquad\qquad\qquad\quad \text{(L.E.) (8.4)}$$

$$v = v_s, \text{ for } s = 1, \ldots N \qquad\qquad \text{(L.E.) (8.5)}$$

$$y_{min} \leq y \leq y_{max} \qquad\qquad\qquad \text{(L.E.) (8.6)}$$

Specified values: A, u_{min}, v_s for $s = 1, \ldots N$; y_{min}, y_{max}.

In this typical *initial formulation*, Q is the optimization quantity to be maximized for this case, u, v, and y are constrained variables, and w is an unconstrained or free variable. The constraints on u and y are of the regional type, whereas the constraint on v is of the discrete-value type such as would be imposed by a standard size limitation. We assume that boundary conditions have been established so that the specified values are known at this time. Also, at this time we are not restricting the functions $f_1, f_2,$ or f_3 to any particular mathematical type.

For the *typical* initial formulation now summarized the problem of optimum design is to determine the values of u, v, w, and y which will maximize Q while adhering to all of the specifications and constraints as described. In the remaining part of this chapter we will present and illustrate several techniques for solving problems of optimum design. For each technique we will also summarize the types of problem application, and the particular limitations which each has. As will become evident in what follows, emphasis will be placed on the *method of optimum design* technique because of the relatively high degree of success which the author has experienced with it in both academic teaching and industrial applications for realistic problems of mechanical design synthesis. In Table 8-4 of this chapter we will compare the various application features and limitations of the various optimization techniques. Before our study of optimization techniques, some items of general interest will be discussed briefly.

First, one might wonder how he should handle a problem where there is more than one optimization quantity. For instance, in a device being designed we might wish to minimize wear rate and also to minimize deflection. The problem can be handled in one of two ways as follows.

1. Select one of the criteria as the optimization quantity Q, and establish a constraint of acceptability on the other quantity which will be included as a subsidiary design equation and limit equation in the initial formulation.
2. Decide on a relative weighting based on the degree of relative importance for the optimization quantities, thereby establishing a single criterion function, $Q = f_4 + f_5$, as the primary design equation. This is then included with the subsidiary design equations and limit equations in the initial formulation, with the objective of maximizing or minimizing Q within the boundaries of all constraints and specifications.

It has been the author's experience that the first approach is the easier to apply in realistic problem situations. This is true since it is generally easier to establish a constraint of acceptability on a parameter rather than to decide on the relative weighting or relative importance between two parameters in a problem situation. Often the latter is more elusive in nature.

A second point of general importance should be mentioned with respect to the derivation of equations as related to accuracy. In optimization problems in realistic situations occasionally it is necessary to derive equations which are only approximations having appreciable error. This might lead one to conclude that the design decisions reached will also have appreciable error. However, such a conclusion is not necessarily true. As an example, suppose we wish to maximize Q by the appropriate choice of a design variable x constrained to the acceptability range $x_{min} \le x \le x_{max}$. Suppose we consider both the exact function and an approximate function of appreciable error, as

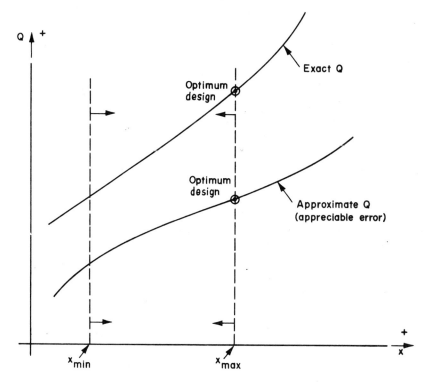

Figure 8.1 Exact and approximate functions for Q leading to the same design decision: place $x = x_{max}$ for maximizing Q.

shown in Figure 8.1.   Regardless of which criterion function we use the same design decision is reached, which is to place $x = x_{max}$.   Thus, in problems of optimum design, do not automatically assume that error in function expression is synonymous with inaccuracy in the design decisions reached.   As in the illustrative example the two might be totally independent, and often exact design decisions are reached from variation studies using an approximate criterion function.

8-2   THE METHOD OF OPTIMUM DESIGN

A very general technique was presented and illustrated in reference 8-1 for solving optimization problems in realistic design situations.   This *method of optimum design* is virtually unlimited in scope of application, as long as the various equations can be manipulated and variations can be studied as required in the procedural steps to be outlined as follows.   The author has applied the technique successfully for more than sixteen years in professional practice in the work of mechanical design synthesis. He has also taught the subject to both senior and graduate-level mechanical engineering students and practicing engineers for more than twelve years. It is not an easy subject either to teach or to master, but it is extremely powerful for the solution of practical optimization problems of mechanical design synthesis involving complex equation systems such as in the examples of Chapter 9. Also, it has been found that the better students with perseverance have no difficulty in mastering the systematic technique of handling complex equation systems for the making of variation studies leading to the derivation of computational flow charts in the method of optimum design. Favorable feedback from such former students who have gone on toward their masters or doctoral degrees, as well as from those who have gone on to work in industry, have made this pioneering effort of teaching the method of optimum design worthwhile.

The method of optimum design is described in sections 7-12 and 7-14 of reference 8-1, and we will expound on this background in what follows. From a very broad viewpoint, the technique consists of the following general steps:

1. Derivation of the *initial formulation* equation system, previously described and illustrated.
2. Reduction and transformation of the initial formulation equation system to a workable *final formulation* arranged in standard format.
3. Derivation of an *optimum design variation study* from the final formulation equation system.
4. *Application of conclusions* from the variation study *to decisions of optimum design*, in the form of
   a. general "goals" of design, indicating the directions in which to head and serving as possible guides for creative change, and/or
   b. specific procedures of design in outline form or in flow chart form, for numerical application in the explicit determination of the optimum design.
5. *Summary and evaluation of the optimum design* so determined, making sure that all phenomena of interest are satisfactory and calculating what has been achieved for the criterion function value.

In the specific execution of the preceding steps there are three types of problems encountered. They are the *case of normal specifications*, the *case of redundant specifications*, and the *case of incompatible specifications*, as described in sections 7-12, 7-14, and 7-16, respectively, in reference 8-1. Before explaining more specifically the details of execution in each case, let us tie together in flow-chart form what has been generally presented so far. This flow chart is given in Figure 8.2, which we will further explain in what

follows.   Some of the general nomenclature used in the method of optimum
design is given below.

Nomenclature Used in the Method of Optimum Design

A_t = number of approaches theoretically possible for derivation of
final formulations from the initial formulation

D_{vs} = number of dimensions for variation study required

n_c = number of constrained variables

n_e = number of eliminated parameters in final formulation

n_f = number of free variables in (S.D.E.)s of initial formulation,
not counting optimization quantity Q

n_i = number of inseparable variables in a developed (P.D.E.)

n_r = number of related parameters in final formulation

n_v = number of variables in initial formulation not counting Q

N_{ff} = number of equations in final formulation system of equations

N_{if} = number of equations in initial formulation system of
equations, not counting (L.E.)s

N_r = number of relating equations in final formulation

N_s = number of (S.D.E.)s in initial formulation

(P.D.E.) = primary design equation, or criterion function

(S.D.E.) = subsidiary design equation

(L.E.) = limit equation, expressing acceptable values for a design
variable

(I) = developed (P.D.E.) of ideal problem, in final formulation

(II), (III),
(IV), etc. = relating equations, in final formulation

Before presentation of some illustrative examples of application we will
describe in further detail the Figure 8.2 summary of the *method of optimum
design*.   In that figure the numbers in parentheses correspond to the general
steps previously outlined.   To start, we assume that the basic configuration
has been selected at this stage of design and that the experienced engineer
recognizes an appropriate place for application of the method of optimum
design.   Hence, a free-hand sketch is made of the configuration in its particular setting summarizing the input, output, and intermediary boundary
conditions and specifications in parametric form, such as for functional
requirements, space constraints, and other limitations of importance.   Also,
the independent geometrical parameters are recognized and labeled on this
sketch.   Next we must derive the criterion function expressing the optimization quantity Q, and often it is necessary to resort to reasonably accurate
approximations for this *primary design equation*.   Now we must derive the
subsidiary design equations which express the functional requirements and undesirable effects of significance, tying together mathematically the functional

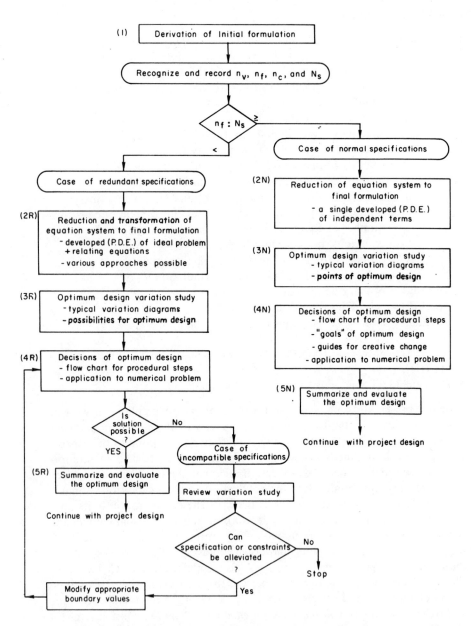

Figure 8.2   Summary of the method of optimum design.

relationships that are internal and external to the critical part. These *subsidiary design equations* are followed by a summary of *limit equations*, which express the domains of acceptability for various parameters in the form of regional and discrete value constraints. At this time we have derived and summarized a system of equations which we call the *initial formulation*, typically illustrated in equations (8.1) through (8.6) previously discussed, and step (1) in Figure 8.2 is complete. Incidentally, for a supplementary and more detailed description of the initial formulation specifically see steps 1 through 4 in section 7-12 of reference 8-1.

If upon inspection of the initial formulation we see that there is a design variable which exists in only the (P.D.E.) and not in any of the (S.D.E.)s, it is a *truly independent parameter*. If possible we should make a decision on its optimum value at this time. In the variation study procedure which follows, we would consider such a truly independent parameter to be held constant at this optimum value.

In the initial formulation at this time we can readily recognize the number of variables involved, designated as n_v in what follows, not counting the optimization quantity Q. The number of variables involved can be resolved into constrained variables of number n_c and free variables of number n_f, again not counting the optimization quantity Q. Thus, we have equation (8.7).

$$n_v = n_c + n_f \qquad (8.7)$$

The constrained variables are the *variables* at this stage which have either discrete value or regional constraints *directly* imposed by limit equations in the initial formulation. On the other hand, a free variable at this stage is one which does *not* have either a discrete value or regional constraint directly imposed by a limit equation on the variable in the initial formulation.

Let us designate the number of subsidiary design equations in the initial formulation by N_s. Generally the *case of normal specifications* exists if $n_f \geq N_s$, and the decisions of optimum design can be explicitly extracted in a relatively simple fashion for such a situation, as we will explain in what follows. On the other hand, the *case of redundant specifications* exists if $n_f < N_s$, and in such situations the decisions of optimum design are explicitly determined in a more complicated fashion which we will also describe in conjunction with Figure 8.2. A special type of problem encountered in numerical application of the latter case is that of *incompatible specifications*, where it is actually impossible to find any solution which satisfies all of the specifications and constraints of the initial formulation. Application of the method of optimum design explicitly reveals such a situation and it indicates the necessary changes in boundary values for alleviating the barriers to design.

As a specific example of application for what was described in the preceding paragraph, consider the typical initial formulation expressed by the system of equations (8.1) through (8.6). The constrained variables are u, v, and y, so $n_c = 3$. The only unconstrained or free variable not counting Q is w, so we have $n_f = 1$. Thus, by equation (8.7) the number of variables not counting Q is

$$n_v = n_c + n_f$$
$$= 3 + 1 = 4$$

In this system of equations we have two subsidiary design equations, so $N_s = 2$. We see that the situation exists where $n_f = 1 < N_s = 2$, and referring to Figure 8.2 we would follow the path for the case of redundant specifications. This will be continued further under our discussion of that case in what follows.

Case of Normal Specifications

Assuming that we have an initial formulation where $n_f \geq N_s$, we can reduce the initial system of equations to a *final formulation* consisting of merely a single *developed primary design equation* of independent terms and from which decisions of optimum design can be concluded explicitly. Hence, an equation-reduction step is executed by combining the subsidiary design equations with the primary design equation and eliminating a free variable *from the system* for each subsidiary design equation combined in the procedure. Thus, the number of equations in the initial formulation system is reduced by one for each free variable so combined in deriving the developed primary design equation.

If a variable other than the optimization quantity Q does *not* exist in *all* of the equations, i.e., (P.D.E.) *and* (S.D.E.)s, of the initial formulation, it is sometimes helpful for a better understanding of the equation-combination procedure to introduce such variables raised to the zero power; these are known as *phantom variables* and are used where necessary in equations of the initial formulation. They are sometimes helpful for those who are not experienced with execution of the method of optimum design. Incidentally, for a further more detailed description of the equation-reduction step (2N) of Figure 8.2, see step 5 in section 7-12 of reference 8-1.

From the developed primary design equation now derived, we have optimization quantity Q expressed in terms of independent variables. For example, suppose the initial formulation consists of the system of equations (8.1) through (8.4) and (8.6), with both w and v as free variables since (L.E.) (8.5) has been deleted. In such a case we would have $n_f = 2$ and $N_s = 2$, and from Figure 8.2 we recognize that with $n_f = N_s$ we have a case of normal specifications. Thus, in the equation-reduction step we combine subsidiary design equations (8.2) and (8.3) with primary design equation (8.1), eliminating free variables v and w in the process, thereby obtaining developed primary design equation (8.8).

$$Q = f_4(u,y) \qquad \text{dev. (P.D.E.) (8.8)}$$

In this equation u and y are independent terms, by nature of the equation-combination procedure, and we could arbitrarily assign any values to u and y within the domains of acceptability expressed by limit equations (8.4) and (8.6). Hence, the next step (3N) in Figure 8.2 is one of *variation study* which will enable us to make the best choices of u and y for optimization of Q. In general two-dimensional variation studies are required for the case of normal specifications, such as for Q as a function of u and for Q as a function of y, assuming that u and y are separable in function f_4 of equation (8.8). On the other hand, if function f_4 is such that u and y are inseparable, a three-dimensional variation study of Q versus u and y would be necessary. In general, if D_{vs} designates the number of dimensions required for the variation study, for the case of normal specifications we have

$$D_{vs} = 2$$

· for each separable variable in the developed (P.D.E.), and

$$D_{vs} = n_i + 1 \qquad (8.9)$$

for each group of n_i inseparable variables in the developed (P.D.E.). In the variation-study step it is usually helpful to make a general free-hand sketch (not plotting points) in the form of a *typical* variation diagram, such as for Q versus u with y constant and for Q versus y with u constant, imposing on each diagram the appropriate *typical* constraint, u_{min} and (y_{min}, y_{max}), respectively. For the variation study, often calculus must be applied to the developed (P.D.E.) in order to determine in general how Q varies with respect to an independent variable, although in many cases the variation is simple enough so this is •unnecessary. Incidentally, a further discussion of the variation-study step (3N) of Figure 8.2 is presented specifically as step 6 in section 7-12 of reference 8-1.

The purpose for the variation study is to reach the decisions of optimum design for application to the numerical problem. Hence, in step (4N) of Figure 8.2 we reach the decisions on independent variables in the developed (P.D.E.) which will optimize Q. For instance, depending upon the specific characteristics of function f_4 in equation (8.8), we might conclude for maximization of Q that we should place $u = u_{min}$ and $y = y_{max}$. Or, function f_4 might be such that for maximization of Q we desire u as large as possible.

In that event we recognize the significance of a constraint u_{max} previously overlooked in the initial formulation (L.E.) (8.4), and we would have to estimate the appropriate value for u_{max} at this time from a review of the boundary conditions of the particular problem. Thus, in such a situation for the domain of acceptability $u_{min} \leq u \leq u_{max}$ with boundary values now established, from our variation study we make the design decision to place $u = u_{max}$. The decisions of optimum design, as concluded from the general variation study with boundary conditions imposed, can be summarized in procedural steps in flow-chart form for numerical application, or more loosely as "goals" of design for application in the graphical work of layout drawings.

The decisions of optimum design are made with respect to the independent and constrained variables in the developed (P.D.E.) from the variation study, based on the objective of optimizing Q. Next, values of the free variables which were eliminated in the equation-combination procedure are uniquely determined at this stage by reversal of the procedure. Thus, at this time decisions have been reached on the optimum values for both the constrained and free variables. Hence, we could now program explicitly the procedural steps of optimum design. If the value of a free variable now determined is found to be unacceptable, because of an overlooked constraint in our initial assumptions, an estimate should be made for the appropriate boundary value to be included as a limit equation on that variable in the initial formulation, followed by a second application of the Figure 8.2 procedure. Incidentally, a further discussion of step (4N) of Figure 8.2 is presented specifically as steps 7 through 9 in section 7-12 of reference 8-1.

If the optimum design has been successfully determined at this point, we should summarize the specifications for the design and carry through any further evaluation to be sure that all phenomena of interest are satisfactory. For one, we certainly would want to calculate the value of the optimization quantity to determine what we were able to achieve in the optimum design. Also, we would want to analyze the optimum design for all other phenomena of possible significance previously ignored in our initial formulation. Hence, if unacceptable results are uncovered in this step (5N) of analysis in Figure 8.2, inclusion of the appropriate constraints previously ignored must be made in the initial formulation. With experience and good engineering judgement iterative application of Figure 8.2 is generally not necessary. Thus, it is good practice to work cautiously in the derivation and summary of the initial formulation. The initial inclusion of too many constraints unnecessarily complicates the variation study and the extraction of design decisions, and this is time-consuming and costly from an engineering standpoint. On the other hand, if too many constraints are initially ignored they will become apparent in evaluation step (5N), but this will require an iterative application of Figure 8.2, which also can be time-consuming and costly from an engineering standpoint. With practice and good engineering judgement, the derivation and summary of the correct initial formulation can be made at the start for most practical applications of optimum design, thereby minimizing time consumed and engineering cost in the design process.

Case of Redundant Specifications

If the initial formulation is such that $n_f < N_s$ we cannot reduce the initial formulation by the described equation-combination procedure to a single developed primary design equation. There are too few free variables compared with the number of subsidiary design equations, so we cannot eliminate a free variable for each (S.D.E.) combined with the (P.D.E.). In other words, there are too many constraints for the development of the (P.D.E.) to the point where the optimum design can be directly extracted as explained previously for the case of normal specifications. Hence, for this case of redundant specifications where $n_f < N_s$, at best we can reduce the initial formulation to a lower degree of complexity for a system of fewer equations. This equation reduction and transformation step (2R) in Figure 8.2 will be described further in the next several paragraphs.

In derivation of the *final formulation*, eliminate all free variables in the initial formulation by the combination procedure previously described for the case of normal specifications. Also, if the optimization quantity Q appears in any of the (S.D.E.)s of the initial formulation, eliminate it at the start from each such (S.D.E.) by combination with the (P.D.E.), retaining, however, all of the equations in the system.

For the case of redundant specifications we have $n_f < N_s$, which means that the equation-combination procedure by eliminating free variables for each step is stopped before complete development of the (P.D.E.). Thus, in general our final formulation will consist of N_{ff} equations in number, determined by equation (8-10).

$$N_{ff} = N_s - n_f + 1 \qquad (8.10)$$

Since we have $n_f < N_s$, where both n_f and N_s are positive integers, from equation (8.10) we see that $N_{ff} \geq 2$ is required. Hence, for the case of redundant specifications the final formulation will always be a system of two or more equations. In fact, for an initial formulation having no free variables we have $n_f = 0$ and by equation (8.10)

$$N_{ff} = N_s - 0 + 1$$

$$= N_s + 1 = N_{if}$$

Thus, for $n_f = 0$ the number of equations, N_{ff}, in the final formulation will be the same as the number of equations, N_{if}, in the initial formulation.

In the derivation of the final formulation, we will strive for a certain form for the reduced system of equations that is compatible with the variation study to follow. The first equation will express optimization quantity Q in terms of certain constrained variables such as u and v. The second equation will express another constrained variable such as y in terms of the *same* variables u and v as for the first equation. If other equations exist in the final formulation system, the format will be such that they too will express other constrained variables in terms of the same variables u and v as for the other equations. In this way, we obtain a final formulation system of equations set up in such a form that a variation study can be made by the method of optimum design. Thus, in the case of redundant specifications the *decisions of optimum design are reached based on a consideration of a final formulation system of equations arranged in a suitable format*, rather than from a single equation by itself.

Consider more specifically the initial formulation system of equations (8.1) through (8.6). Since for that example we have $N_s = 2$ and $n_f = 1$, we have $n_f = 1 < N_s = 2$, indicating the case of redundant specifications with $N_{ff} = N_s - n_f + 1 = 2 - 1 + 1 = 2$ for the number of equations in the final formulation, as calculated from equation (8.10). Specifically, we could combine the initial formulation system of equations by eliminating the only free variable, w, thereby giving the following *final formulation system of equations*, transformed to the previously described *format suitable for an optimum design variation study*:

$$Q = f_5(u,v) \qquad \text{(I)} \qquad\qquad\qquad (8.11)$$

$$y = f_6(u,v) \qquad \text{(II)} \qquad\qquad\qquad (8.12)$$

If we did not have the constraints on y expressed by (L.E.) (8.6), we could extract the optimum design directly from a variation study of equation (8.11) in what would be a case of normal specifications. Hence, equation (8.11) would be the *developed (P.D.E.) of an ideal problem*, where we have *temporarily ignored or eliminated* considerations of the effects of constraints on y. Hence, in the final formulation we will designate y as an *eliminated parameter*. However, we realize that the constraints on y must be included in the total picture, and from equation (8.12) we see that y is related to u and v which are in the ideal problem developed (P.D.E.) (8.11). Thus, equation (8.12) will be designated as a *relating equation* and u and v will be called *related parameters*. This general terminology has significance in understanding the final formulation set-up in the case of redundant specifications, and it will be used in the remaining part of the book. Incidentally, a further description of the equation reduction step (2R) of Figure 8.2 together with the definitions of this general terminology used for the final formulation format are presented specifically in steps 2 through 4 in section 7-14 of reference 8-1.

Throughout the remaining part of the book we will use the following *notation* for particular types of equations in the *final formulation format of* the method of optimum design:

Notation	Meaning
(I)	developed (P.D.E.) of ideal problem
(II), (III), (IV), etc.	relating equations

For specific problems, the use of the notation (I), (II), (III), ... will both simplify the presentation and emphasize the basic meaning of the various boundary curves of significance in the typical variation diagram, which follows, as related to the final formulation.

As previously discussed, if a variable other than the optimization quantity Q does *not* exist in *all* of the equations of the initial formulation, it is often helpful to introduce *phantom variables* previously defined where necessary in the equations before starting the equation combination procedure in step (2R) of Figure 8.2. Thus emphasizing the presence of phantom variables in the final formulation, which have not been eliminated in the combination procedure, is helpful in the variation study which follows. If a variable exists in the developed (P.D.E.) of the ideal problem but appears only as a phantom variable in *all* of the relating equations of the final formulation, it is a special type of related parameter known as a *truly independent parameter* and the decision of optimum design can be made for it at this time before proceding further in the variation study. Incidentally, a supplementary explanation of truly independent parameters is presented specifically in step 5 in section 7-14 of reference 8-1.

The *variation study* in step (3R) of Figure 8.2 *involves* a study of the *final formulation system* of equations. First, we study the variation of optimization quantity Q versus the related parameter(s), from the developed (P.D.E.) of the ideal problem, imposing the constraints on the related parameter(s). On the same diagram we superimpose the variation(s) of the eliminated parameter(s) versus the related parameter(s), as expressed by the relating equation(s), imposing the constraints on the eliminated parameter(s). This defines the possible domains of feasible design solution and from the variation of optimization quantity Q we can determine the points therein which are possible for optimum design. Incidentally, the variation-study approach to optimum design as now described is further explained in steps 5 through 8 in section 7-14 of reference 8-1. Upon completion of such a general variation study in step (3R) of Figure 8.2, the procedure of optimum design can be logically programmed in flow-chart form for application to the solution of a numerical problem.

As a more specific example of the optimum design variation-study step (3R), consider the final formulation expressed by equations (8.11) and (8.12). First, we would investigate the variation of Q versus u and v from ideal problem (P.D.E.) (8.11) imposing typical constraints from (L.E.)s (8.4) and (8.5) on the typical three-dimensional variation diagram. A domain of acceptability would thus be defined on the u, v plane. Next, we would superimpose the variation of y versus u and v from relating equation (8.12), with typical constraints on y from (L.E.) (8.6) defining another domain of acceptability on the u, v plane. By superimposing the now defined domains of acceptability, we explicitly determine the possible domains of feasible design solution on the u, v plane. Hence, from the variation of Q versus u and v in such domains of feasibility, we uniquely determine the possibilities for optimum design. The procedure of optimum design is then logically formulated in flow-chart form for application to numerical calculation. Several specific three-dimensional variation studies in optimum design are presented in reference 8-2; a number of such studies will be described in this book.

After derivation and summary of the initial formulation, it is often desired to know how many dimensions will be required for the variation study of the final formulation. This we can readily calculate from equation (8.13) by inspection of the initial formulation. Let n_r designate the number of related parameters and n_e the number of eliminated parameters, both in the final formulation. As before, D_{vs} will be the number of dimensions required in the variation study of the final formulation, with N_s, n_v, and n_c being the number of (S.D.E.)s, the number of variables not counting optimization quantity Q, and the number of constrained variables, respectively, in the initial formulation. Thus, we derive

$$D_{vs} = n_r + 1$$

$$n_c = n_e + n_r$$

therefore

$$n_r = n_c - n_e$$

and

$$D_{vs} = n_c - n_e + 1$$

where from equation (8.7) we have

$$n_c = n_v - n_f$$

But, the number of eliminated parameters, n_e, is the number of equations, N_{ff}, in the final formulation minus one, which by equation (8.10) becomes

$$n_e = N_{ff} - 1 = N_s - n_f + 1 - 1$$
$$= N_s - n_f$$

Thus, we obtain

$$D_{vs} = n_c - n_e + 1$$
$$= (n_v - n_f) - (N_s - n_f) + 1$$
$$D_{vs} = n_v - N_s + 1 \qquad (8\text{-}13)$$

Therefore, as a specific example consider again the initial formulation system of equations (8.1) through (8.6). We at once count the number of variables which are u, v, w, and y not counting optimization quantity Q and obtain $n_v = 4$. Next, from that system we at once count two (S.D.E.)s, so $N_s = 2$. Hence, from equation (8.13) we calculate

$$D_{vs} = n_v - N_s + 1$$
$$= 4 - 2 + 1 = 3$$

which indicates that a three-dimensional variation study would be required, which is in agreement with our previous discussion on the final formulation equations (8.11) and (8.12).

Incidentally, the number of dimensions required for the optimum design variation study can be reduced from the D_{vs} value of equation (8.13) by the number of truly independent parameters in the final formulation. As previously discussed decisions of optimum design can be made at once from the developed (P.D.E.) of the ideal problem for such parameters. Because of this, each truly independent parameter in the final formulation will reduce by one the value of D_{vs} calculated by equation (8.13) for the actual dimensions of the variation study required.

After the final formulation has been derived, one might wonder if other approaches could be derived which perhaps might be simpler for the variation study to follow. For example, from the initial formulation system equations (8.1) through (8.6) we derived the final formulation system equations (8.11) and (8.12). Referring to the latter, we could equally as well have treated either u or v as the eliminated parameter instead of y. With u as the eliminated parameter, our final formulation format would have been as follows:

$$Q = f_7(y,v) \qquad \text{(I)} \qquad\qquad (8.14)$$

$$u = f_8(y,v) \qquad \text{(II)} \qquad\qquad (8.15)$$

where (8.14) would be the developed (P.D.E.) of the ideal problem. (8.15) would be the relating equation, u would be the eliminated parameter, and y and v would be the related parameters.

On the other hand, with v as the eliminated parameter our final formulation format would have been as follows:

$$Q = f_9(u,y) \quad \text{(I)} \tag{8.16}$$

$$v = f_{10}(u,y) \quad \text{(II)} \tag{8.17}$$

where (8.16) would be the developed (P.D.E.) of the ideal problem, (8.17) would be the relating equation, v would be the eliminated parameter, and u and y would be the related parameters.

Hence, from the initial formulation system of equations (8.1) through (8.6) we see that three different approaches could have been taken for the derivation of final formulations, treating either y, u, or v as the eliminated parameter. Of course, any one of these three approaches would have given the same optimum design solution in application for a given numerical problem. However, one of the three approaches might very well have been simpler from the standpoint of execution in the optimum design variation study or in the programming of the optimum design decisions, i.e., in steps (3R) and (4R) of Figure 8.2.

In general, the theoretical number of approaches, A_t, possible for the derivation of final formulations will be the combinations of n_c things taken n_e at a time, disregarding the order in any combination. Thus,

$$A_t = {}_{n_c}C_{n_e} = \frac{n_c!}{n_e!(n_c - n_e)!}$$

From the initial formulation we readily recognize the number of constrained variables n_c. Prior to equation (8.13) we derived the equation for the number of eliminated parameters n_e, which is as follows.

$$n_e = N_s - n_f \tag{8.18}$$

From the initial formulation we readily recognize the number of subsidiary design equations N_s and the number of free variables n_f. Thus, we can readily calculate n_e from equation (8.18), and the number, A_t, of theoretical approaches possible for derivation of the final formulation can be determined from equation (8.19).

$$A_t = \frac{n_c!}{n_e!(n_c - n_e)!} \tag{8.19}$$

In practical applications it is suggested that a final formulation be first derived, which is generally quite easy to do as previously described. From this final formulation recognize the existence of any truly independent parameters, which then can be placed at their optimum values and treated as specific constants from here on.   From this final formulation we next readily recognize the eliminated parameters and the related parameters, so n_e and n_r are known at this time.   Then calculate n_c from

$$n_c = n_e + n_r \qquad (8.20)$$

which can then be used in equation (8.19) for calculation of the theoretical number, A_t, of different approaches which can be taken for derivation of final formulations.   Most of these approaches will be *determinate* for explicit solution of the optimization problem.   However, it is possible that a particular approach will result in an *indeterminate* final formulation from which we cannot extract explicitly solution to the optimization problem.   Such indeterminate approaches are occasionally encountered and can be recognized from an inspection of the final formulation first derived, if the proposed combination of eliminated parameters all appear as phantom variables in any *one* of the relating equations therein.   More specifically, if in the equation-combination procedure it is necessary to solve explicitly for a phantom variable, other variables such as y in an equation might take the form of $y^{1/0}$, which of course makes the final formulation system of equations indeterminate.   However, for the final formulation special case of only one relating equation, all of the A_t theoretical approaches in number as calculated from equation (8.19) will be determinate, since any phantom variable in that relating equation will actually be a truly independent parameter already set at its optimum value.   Nevertheless, indeterminate approaches are possible if two or more relating equations exist in the final formulation.   Fortunately, the problem of indeterminate approaches is rarely encountered in practical problems of optimum design, and generally it is characterized by special situations where phantom variables are nonuniformly distributed in the equations of the initial formulation.   A specific illustration of an optimum design problem having an indeterminate approach will be presented in Example 8-4.   This particular problem can be solved explicitly by five different approaches, whereas only one of the six possible approaches is indeterminate.

Incidentally, it should be mentioned that the simple equations (8.7), (8.10), (8.13), (8.18), (8.19), and (8.20) are primarily for exploratory purposes in the initial formulation stage.   They are useful as an early guide for indicating what probably lies ahead in the equation reduction and variation study steps (2R) and (3R) of the Figure 8.2 procedure.   In this way they often can serve as a help for establishing confidence if properly applied, and this is of value particularly when gaining initial experience with the method of optimum design.   However, use of these equations in the initial formulation stage should not serve as a deterrent for further progress on the problem.   For instance, if from equation (8.13) we calculate $D_{vs} = 4$, we should not become discouraged with the prospect of having to solve a four-dimensional variation-study problem.   The value of D_{vs} actually may be reduced if truly independent parameters exist in the final formulation, or the use of a digital computer may make it possible for us to extract the optimum design explicitly based on only a three-dimensional variation study.   Hence, things may not be as bad as they first appear, and this will be illustrated in some of the examples which follow.

Application of the decisions of optimum design in a numerical problem for the case of redundant specifications hopefully will result in an optimum design solution which can be summarized and evaluated for all phenomena of interest in step (5R) of Figure 8.2.   If the analysis reveals that a previously ignored constraint should be included in the initial formulation, an iterative application of Figure 8.2 may be necessary.   In this respect, the discussion previously presented under the case of normal specifications applies equally as well here.

On the other hand, application of the decisions of optimum design in a numerical problem for the case of redundant specifications may explicitly lead to the startling and disappointing conclusion that no design solution is possible which satisfies all of the constraints and specifications of the problem. Of course, being able to determine such a situation explicitly by the method of optimum design is really invaluable, since then we can concentrate on alleviation by modification of proper boundary conditions as described later in section 8-9.   As a general example of such a situation, consider the initial formulation system of equations (8.1) through (8.6) reduced to the final formulation system of equations (8.11) and (8.12).   If the constraints on u, v, and y are such that there is no domain of acceptability on the (u, v) plane, regardless of how much we searched we could not find a design solution which satisfied all of the constraints and specifications of the problem.   We would have the *case of incompatible specifications* shown in Figure 8.2, and the optimum design variation study would indicate what constraints should be alleviated for obtainment of a solution, and particularly the optimum design solution once the modifications are made.   The possibility of incompatible specifications is described further in section 7-16 of reference 8-1, and such possibilities will be indicated in some of the illustrative examples which follow.

The mathematical foundation for the *method of optimum design* has now been described in general terms.   Specific problems of application are executed in accordance with this general systematic plan, and problems are often of challenge because of differences in detail encountered in the specific procedures necessary for solution.   Hence, application of the basic general technique is best explained by a consideration of several specific illustrative examples.   For this purpose examples 8-1 through 8-6 will follow, presented in order of increasing complexity and bringing out some subtleties of general significance which are exemplary of what can be expected in specific problems. Also, a wide range of subject matter has been selected for these introductory examples, which thereby illustrates the unlimited breadth of application for the *method of optimum design.*

Introductory Examples for the Method of Optimum Design

Example **8-1**   We wish to design a solid body which will freely fall with maximum velocity in a given viscous fluid.   Assume that the spherical shape has been selected in this situation based on practical considerations, rather than the ideally optimum tear drop shape which could be derived using the calculus of variations.   Hence, the problem of optimum design is to select the appropriate diameter d and the material for the solid sphere which will maximize free-fall velocity v in Figure 8.3.   Suppose that three available materials, designated 1, 2, and 3, are feasible for the casting of such a sphere, having weight densities w_1, w_2, and w_3 lb/in.^{3}, respectively.

In the initial formulation of this problem we must derive the equation for free-fall velocity v in terms of the design parameters.   To do this we must make certain assumptions which we believe will be correct and whose validity can be checked in a step of analysis at the end.   First, we will assume that sphere diameter d will be small compared with container diameter D (see Figure 8.3).   Secondly, we will assume the condition of viscous flow around the falling sphere.   Thirdly, we will assume that a steady-state condition is soon reached for all practical purposes, where retarding force R equals gravitational force W for constant velocity v.

In the free-body diagram of Figure 8.3(b), retarding force R consists of two components.   They are (1) what is commonly called the drag force and (2) the buoyant force of the displaced liquid.   The drag force can be expressed by Stoke's law in fluid mechanics, such as from reference 8-5, assuming Reynolds number $N_R < 0.4$.   Thus drag force on the sphere is 3 $\mu v \pi d$, where μ is the dynamic viscosity of the given fluid.   Hence, the total retarding force R is

$$R = 3\mu v \pi d + w_f \frac{\pi d^3}{6}$$

where the second term is the weight of the displaced liquid; w_f is the weight density of the given fluid, and $\pi d^3/6$ is the volume of the displaced liquid.

(a) Falling body in given viscous fluid

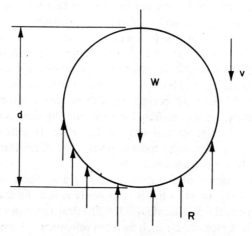

(b) Free body diagram of solid sphere

Figure 8.3   Solid body to be designed for maximization of free-fall velocity in example 8-1.

Gravitational force W on the falling sphere is merely $w\pi d^3/6$, where w is the weight density of the sphere material. Thus, we derive the following equation for free-fall velocity v.

$$W = R$$

$$w\left(\frac{\pi d^3}{6}\right) = 3\mu v \pi d + w_f\left(\frac{\pi d^3}{6}\right)$$

therefore

$$v = \frac{d^2}{18\mu}(w - w_f) \qquad \text{(P.D.E.) (8.21)}$$

In this equation we assume that properties μ and w_f are known for the given viscous fluid. On the other hand, d and w are design variables at this stage associated with the sphere to be designed.

Providing that there are no other constraints of significance, (P.D.E.) (8.21) will be sufficient for the making of the decisions in optimum design, and this equation by itself is the initial formulation for the problem as stated. Hence, for this case of normal specifications there is no equation-reduction step, and the optimum design variation study is made directly from (P.D.E.) (8.21).

The general variations of v with respect to independent parameters d and w are as typically sketched in Figure 8.4, the shapes of which are obvious from an inspection of (P.D.E.) (8.21). For maximization of v we see that diameter d should be as large as possible. Hence, at this time the significance of a constraint d_{max} becomes apparent. Thus, at this stage we must reach a decision on an appropriate boundary value for d_{max}, based on considerations of environment, including handling restrictions on size. The decision of optimum design from Figure 8.4(a) is to place $d = d_{max}$.

Next, from the sketch in Figure 8.4(b) it becomes obvious that we should select the available feasible material of greatest weight density value. In other words, only a finite number of discrete points are accessible on the line in Figure 8.4(b) corresponding to the acceptable materials, and if these materials had the specific relationship $w_2 > w_1 > w_3$ as shown in Figure 8.4(b) the optimum choice would be material 2. In conclusion, the procedure of optimum design for maximization of free fall velocity v is as summarized in the flow chart of Figure 8.5.

In the very simple flow chart of Figure 8.5, note the steps of analysis at the end. Thus we analyze the optimum design to determine the value of optimization quantity v which has been achieved. Also, we analyze the optimum design for all other phenomena of interest, including the checking of validity for the assumptions initially made. If an unacceptable condition is revealed at this time, appropriate revisions would have to be made in the initial formulation before a second application of the Figure 8.2 process.

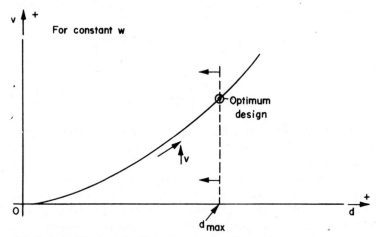

(a) Typical variation diagram of v versus d for constant w

(b) Typical variation diagram of v versus w for constant d

Figure 8.4   Optimum design variation diagrams as sketched from (P.D.E.) (8.21), where w_f and μ are constants for a given viscous fluid.

On the other hand, if all phenomena of interest are acceptable for the optimum design at this stage of analysis, our problem of optimization has been successfully concluded.

As an exercise, the interested reader may wish to extend this problem as

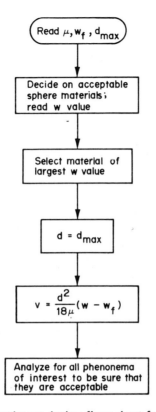

Figure 8.5   Optimum design flow chart for example 8-1.

follows.   Suppose in analysis we find sphere weight W to be unacceptable.
Hence, include in the initial formulation the constraint $W \leq W_{max}$ as well as
$d \leq d_{max}$.   Complete the initial formulation, by also including the (S.D.E.)
$W = w\ (\pi d^3/6)$, and follow the Figure 8.2 procedure to the derivation of the
optimum design flow chart.   Before doing this exercise you may wish to gain
further experience in the method of optimum design by studying examples 8-2
through 8-6 which follow.

***Example* 8-2**   A specified length L of fence is available for enclosing an
area which we wish to maximize.   We could readily prove with the calculus
of variations that the circular shape was ideally optimum.   However, because
of factors associated with this particular application, we have decided to use
the rectangular shape.   The problem of optimum design is to select the
values of x and y which will maximize enclosed area A, referring to Figure 8.6.

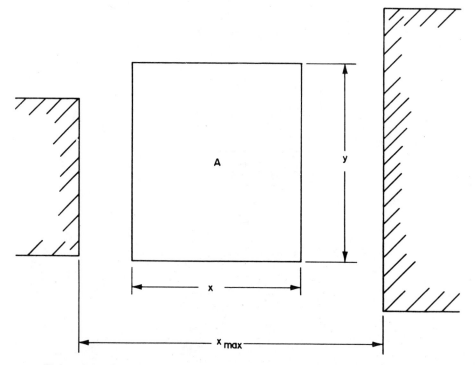

Figure 8.6   Fenced yard of rectangular shape showing space constraints on x.

For our *initial formulation* the primary design equation expresses optimization quantity A in terms of design parameters x and y.

$$A = xy \qquad \text{(P.D.E.) (8.22)}$$

Since length L of fence is specified, we express its relation to x and y by the subsidiary design equation which follows.

$$L = 2(x + y) \qquad \text{(S.D.E.) (8.23)}$$

Also, referring to Figure 8.6 we recognize the possible significance of a space limitation x_{max} on permissible values for x, and we must therefore adhere to the regional constraint 8.24.

$$x \leq x_{max} \qquad \text{(L.E.) (8.24)}$$

In the initial formulation now summarized we have y as a free variable and x as a constrained variable. Therefore, with one (S.D.E.) in the initial formulation, we have

$$n_f = 1 = N_s = 1$$

which indicates the case of normal specifications, referring to Figure 8.2.

For our final formulation we combine (S.D.E.) (8.23) with (P.D.E.) (8.22) by eliminating the free variable y, thus giving the *developed* (*P.D.E.*) (*8.22a*).

$$A = x[(L/2) - x] \qquad (8.22a)$$

For our optimum design variation study we must investigate A as a function of independent parameter x. Thus, we consider

$$\frac{dA}{dx} = x(-1) + [(L/2) - x] \quad (1)$$

$$= (L/2) - 2x$$

and

$$\frac{d^2A}{dx^2} = -2$$

indicating a maximum for A at $x_0 = L/4$, where $dA/dx = 0$. The well-

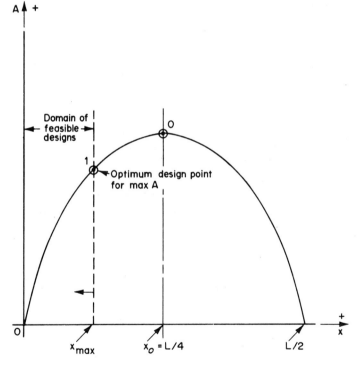

(a) Point 1 is optimum, if $x_{max} \le L/4$

Figure 8.7 Typical variation of A as a function of x, sketched from developed (P.D.E.) (8.22a). (Normally only one variation diagram is sketched, such as (a), with the possibility of (b) understood.)

(b) Point O is optimum, if $x_{max} > L/4$

Figure 8.7 (continued)

known parabolic variation is sketched in Figure 8.7, where we also show the typical space constraint x_{max}. We recognize that x_{max} is independent of $x_0 = L/4$. Hence, we conclude that if $x_{max} \leq x_0$ the optimum design is for $x = x_{max}$. On the other hand, if $x_{max} > x_0$, the optimum design is for $x = x_0$. These possibilities are shown in (a) and (b) of Figure 8.7.

From the general variation study now made we summarize the procedural steps of optimum design in the flow chart of Figure 8.8, which should be studied for an understanding of the logic behind its derivation. Again, the final step is one of analysis to determine the value of A that has been achieved for the optimization quantity.

As an exercise, it is suggested that the reader add limit equation $y \leq y_{max}$ to the initial formulation, carrying through the procedure of Figure 8-2 to the derivation of the optimum design flow chart. You will find that this modified problem is a case of redundant specifications. Hence, before attempting this exercise it might be worthwhile to study first some of the introductory examples which follow.

Figure 8.8   Optimum design flow chart for maximization of A in example 8-2.

***Example* 8-3**   The contact system in Figure 8.9 is to be designed in an application where contact angle θ, force N, and drum speed ω are all specified values.   We wish to minimize the frequency at which worn contact blocks must be replaced.   From our knowledge of wear we believe that the adhesive type will predominate, and from equation (6.7) in Chapter 6 of reference 8-4 we can apply the fundamental law of adhesive wear.

$$V = \frac{kNx}{3p} \qquad (8.25)$$

In equation (8.25), V is the volume of material worn (in.³), k is a dimensionless wear coefficient with typical values from Table 6-6 of reference 8-4, N is the applied load (lb), x is the distance slid (in.), and p is the flow pressure of the material (psi).   Flow pressure p is a material property approximately three times the yield strength S_y, according to references 8-4 and 8-6.

$$p \approx 3S_y$$

Distance slid, x, will be the product of peripheral velocity and time t, giving the relation

$$x = [(D/2)\omega]t$$

Figure 8.9   Basic configuration for the contact system in example 8.3.

where, referring to Figure 8.9, D is drum diameter (in.), ω is drum speed (rad/sec), and t is time of operation (sec). Hence, equation (8.25) becomes

$$V = \frac{kNx}{3p} \approx \frac{kN(D/2)\omega t}{3(3S_y)}$$

$$V \approx \frac{kND\omega t}{18S_y} \qquad (8.25a)$$

However, referring to Figures 8.9 and 8.10, the volume V worn is merely

$$V = bL\Delta \qquad (8.26)$$

where b and L are the contact block dimensions shown in Figures 8.9 and 8.10 (in.), and Δ is the distance worn as depicted in Figure 8.10 (in.).

Volume worn = bL Δ

Figure 8.10   Contact block showing distance worn, Δ.

The objective of optimization will be specifically to minimize wear rate, (Δ/t). Hence, from equations (8.25a) and (8.26) we derive the (P.D.E.) (8.27).

$$V \approx \frac{kND\omega t}{18S_y}$$

$$= bL\Delta$$

therefore

$$(\Delta/t) \approx \frac{kND\omega}{18S_y\,bL} \qquad (P.D.E.)\,(8.27)$$

All terms have been previously defined for this equation.

Suppose the application is one where we could not tolerate excessive frictional moment on the rotating drum from the contact block. Hence, referring to Figure 8.9, we must include in our initial formulation the (S.D.E.) (8.28) which follows.

$$M_f = fND/2 \qquad \text{(S.D.E.) (8.28)}$$

where M_f is the frictional moment on the drum (in.-lb), f is the coefficient of friction which we could estimate from a source such as Figure 6 of reference 8-6, and N and D have already been defined. We must also include the regional constraint on M_f,

$$M_f \le (M_f)_{\max} \qquad \text{(L.E.) (8.29)}$$

where the acceptable boundary value $(M_f)_{\max}$ has been estimated from a consideration of system requirements, such as would be related to available motor size and power loss permitted at the contact block region.

Before proceeding further we should review the boundary constraints of significance for the environment of the Figure 8.9 system. Suppose that space constraints exist on the contact block requiring satisfaction of the limit equations

$$b \le b_{\max} \qquad \text{(L.E.) (8.30)}$$

$$L \le L_{\max} \qquad \text{(L.E.) (8.31)}$$

Also, for the rotating drum suppose for this application that constraint $D_{\max}$ exists because of space considerations, whereas a constraint $D_{\min}$ has been estimated as necessary to assure adequate strength and rigidity of the drum in its mounting. Hence, in selecting our D value we must adhere to the regional constraint which follows.

$$D_{\min} \le D \le D_{\max} \qquad \text{(L.E.) (8.32)}$$

Finally, since contact angle θ of Figure 8.9 is specified, and since width b and drum diameter D are both constrained by (L.E.) (8.30) and (8.32), respectively, we recognize that an entirely geometrical (S.D.E.) must also be included as follows.

$$b = 2 (D/2) \sin (\theta/2)$$
$$b = D \sin (\theta/2) \qquad \text{(S.D.E.) (8.33)}$$

Thus, the initial formulation for this problem consists of (P.D.E.) (8.27), (S.D.E.)s (8.28) and (8.33), and (L.E.)s (8.29), (8.30), (8.31), and (8.32), summarized as follows:

Initial Formulation

$$(\Delta/t) \approx kND\omega/(18S_ybL) \qquad \text{(P.D.E.) (8.27)}$$

$$M_f = fND/2 \qquad \text{(S.D.E.) (8.28)}$$

$$b = D\sin(\theta/2) \qquad \text{(S.D.E.) (8.33)}$$

$$\left.\begin{array}{l} M_f \le (M_f)_{\max}; \, b \le b_{\max}; \\ L \le L_{\max}; \, D_{\min} \le D \le D_{\max} \end{array}\right\} \qquad \text{(L.E.)s}$$

Known or Specified: θ, N, ω, $(M_f)_{\max}$, $b_{\max}$, $L_{\max}$, $D_{\min}$, $D_{\max}$, and estimated values of k, S_y, and f for the available feasible material combinations

Find: Values of D, b, and L which minimize the contact wear rate (Δ/t)

Inspection of the initial formulation summarized above reveals a case of redundant specifications, since $n_f = 0 < N_s = 2$. Also, not counting optimization quantity (Δ/t), the design variables other than material properties are D, b, L, and M_f, so $n_v = 4$. Thus, since $N_s = 2$, from equation (8.13) we calculate

$$D_{vs} = n_v - N_s + 1$$

$$= 4 - 2 + 1 = 3$$

However, from the initial formulation we also recognize that L will be a truly independent parameter since it does not appear in any of the (S.D.E.)s. This will reduce the number of dimensions required for the optimum design variation study by one. Hence, we should be able to extract the optimum design from two-dimensional variation studies. Also, from equation (8.10) we see that the number of equations in the final formulation will be

$$N_{ff} = N_s - n_f + 1$$

$$= 2 - 0 + 1 = 3$$

Finally, from equation (8.18) we see that the number of eliminated parameters in the final formulation will be $n_e = N_s - n_f = 2 - 0 = 2$, and by equation (8.20) the number of related parameters will be $n_r = n_c - n_e = 3 - 2 = 1$. The three constrained variables for the final formulation variation study will be b, D, and M_f, since we will assume that truly independent parameter L is set at its optimum value. Having made the simple calculations of this paragraph based on our inspection of the initial formulation, we have a better understanding of what should follow in derivation of the final formulation and optimum design variation study.

In reference to the initial formulation, at this time we can make a design decision for truly independent parameter L. Thus, for minimization of wear rate (Δ/t), from (P.D.E.) (8.27) we obviously should place $L = L_{max}$. Having now made this decision, we review the initial formulation and must decide on an approach for derivation of the final formulation. Hence, we decide to treat M_f and b as the eliminated parameters and the remaining constrained variable D should be the related parameter. Thus, recognizing that M_f is a phantom variable in (P.D.E.) (8.27), we combine (8.28) and (8.33) with it, eliminating M_f and b in the process, for derivation of the developed (P.D.E.) of the ideal problem which follows.

$$(\Delta/t) = \frac{kND\omega}{18S_y \, bL_{max}} \, M_f{}^0$$

$$= \frac{kND\omega}{18S_y \, L_{max} \, D \sin(\theta/2)}$$

$$(\Delta/t) = \frac{kN\omega D^0}{18S_y \, L_{max} \sin(\theta/2)} \tag{8.27a}$$

In developed (P.D.E.) of the ideal problem (8.27a) we have purposely shown the term D^0 which has not been eliminated by the procedure of combining equations, whereas both M_f and b were so eliminated and thus are not shown as $M_f{}^0$ or b^0 in (8.27a). Also, from inspection of the initial formulation we see that the relating equations will merely be (8.28) and (8.33). Hence, summarizing the final formulation we have the following equation system from which our optimum design variation study will be made. Note the basic format previously described.

Final Formulation

$$(\Delta/t) = \frac{kN\omega D^0}{18S_y \, L_{max} \sin(\theta/2)} \qquad \text{(I)} \tag{8.27a}$$

$$M_f = \frac{fND}{2} \qquad \text{(II)} \tag{8.28}$$

$$b = D \sin(\theta/2) \qquad \text{(III)} \tag{8.33}$$

where

$M_f, b = $ eliminated parameters

$D = $ related parameter

(I) = developed (P.D.E.) of ideal problem

(II) and (III) = relating equations

From the final formulation now summarized we will make a general variation study for optimum design. For the typical two-dimensional variation diagram we will sketch optimization quantity (Δ/t) as a function of related parameter D from equation (I), imposing typical constraints on D in accordance with (L.E.) (8.32). Next, on this typical variation diagram

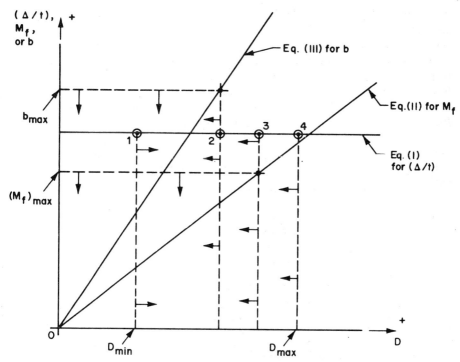

Figure 8.11 Typical variation diagram for a material combination of interest sketched from the final formulation in example 8-3 (for first approach, treating M_f and b as eliminated parameters). (See final formulation for equations I, II, and III.)

we will superimpose the variations of eliminated parameters M_f and b as functions of related parameter D, referring to relating equations (II) and (III) of the final formulation, imposing typical constraints on M_f and b in accordance with (L.E.)s (8.29) and (8.30), respectively. The typical two-dimensional variation diagram so sketched is shown in Figure 8.11, the meaning of which we will briefly discuss next.

From the typical variation diagram of Figure 8.11 we recognize that any accessible point on line (I) would be of equal value for (Δ/t). Hence, for this particular problem we have the situation where *any* acceptable design is an optimum design, of which in general there are an infinite number. If the constraints fell as specifically shown in Figure 8.11, any point on the segment 1–2 of (I) would have the same (Δ/t) value. However, we recognize that constraints D_{min}, D_{max}, b_{max}, and $(M_f)_{max}$ are independent of each other since they are imposed by independent boundary conditions. Thus, the greatest permissible value for D might be governed by either D_{max}, $(M_f)_{max}$, or b_{max}. Also, we recognize that either point 2 or point 3 might fall to the left of point 1 in Figure 8.11, thereby resulting in a case of incompatible specifications.

Occasionally a variation study reveals a situation having an infinite number of points for the optimum design, such as any point on line segment 1-2 of line (I) in Figure 8.11.   In such cases, often it is advisable to select a secondary objective for optimization, such as minimization of frictional moment M_f in Figure 8.11.   Thus, we would choose the particular accessible point on line (I) which has smallest D (or M_f) value, and this would be point 1 in the figure.

In summary, the procedural steps of optimum design as determined in the variation study now made are outlined in the flow chart of Figure 8.12.   The

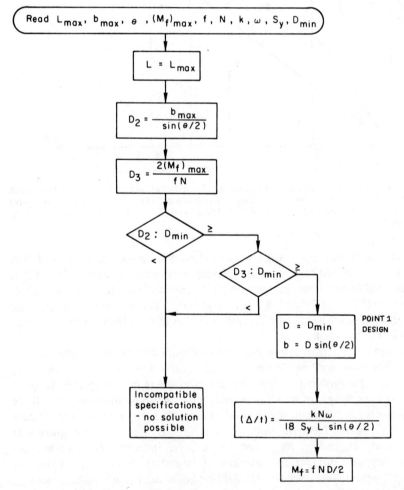

Figure 8.12   Optimum design flow chart for example 8-3 (apply for each material combination of interest).

reader should be sure to understand the logic of its derivation. As in the other examples, the last step is one of analyzing the optimum design to determine what has been achieved for the criterion function value (Δ/t). If several material combinations are of interest, the flow chart would be repeated for each. Then we would select the overall optimum design as the one of smallest (Δ/t) value. Thus, we would explicitly determine in a few steps of calculations and comparisons the material combination and the values of L, b, and D for the overall optimum design which minimizes wear rate (Δ/t), with minimization of frictional moment M_f as a secondary objective of optimization. Incidentally, the value achieved for M_f is calculated in a step of analysis at the end of the Figure 8.12 flow chart.

As an exercise the interested reader should consider other approaches for final formulations. In the presented variation study approach, we chose M_f and b as the eliminated parameters with D as the related parameter. Thus, we have for final formulations $n_e = 2$ and $n_c = 3$. From equation (8.19) the number of approaches possible are

$$A_t = \frac{n_c!}{n_e!(n_c - n_e)!} = \frac{3 \times 2}{2} = 3$$

Hence, the various approaches which are possible have the following eliminated parameter combinations: (1) M_f and b (the presented approach); (2) M_f and D; and (3) b and D.

The *other final formulation possibilities*, which you might wish to derive as an exercise from the initial formulation, are shown in Table 8-1. After deriving the suggested final formulations, sketch a typical two-dimensional variation diagram and derive the optimum design flow chart for each approach. The results determined for each should be in agreement with what we derived in Figures 8.11 and 8.12 by the first approach.

Example 8-4 As another introductory example for the method of optimum design let us consider the simple electrical circuit in Figure 8.13. R_0 represents the resistance of a given lamp, which is to be lit upon closing switch S from discharge of capacitor C. At the instant of closing the switch, let time $t = 0$, and we require a specified level of current I through R_0 at that instant. We have some freedom in design for choosing capacitance C, resistance R, and initial charge voltage E for the capacitor in the figure. Because of space constraints, available stock items, and other environmental considerations for the entire system, we must satisfy the following regional constraints on permissible values for R, C, and E.

$$0 \leq R \leq R_{max} \qquad \text{(L.E.) (8.34)}$$

$$0 \leq C \leq C_{max} \qquad \text{(L.E.) (8.35)}$$

$$E_{min} \leq E \leq E_{max} \qquad \text{(L.E.) (8.36)}$$

Figure 8.13   Electric circuit for example 8-4.

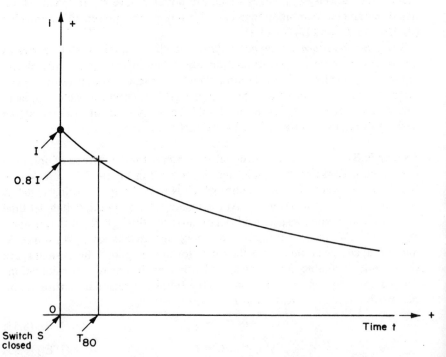

Figure 8.14   Decay of current upon discharge of the capacitor in the electric circuit of Figure 8.13.

TABLE 8-1 OTHER FINAL FORMULATIONS FOR EXAMPLE 8-3

For Approach (2)

M_f, D = eliminated parameters

$$(\Delta/t) = \frac{kN\omega b^0}{18 S_y L \sin(\theta/2)} \quad \text{(I)}$$

$$M_f = \frac{fNb}{2 \sin(\theta/2)} \quad \text{(II)}$$

$$D = \frac{b}{\sin(\theta/2)} \quad \text{(III)}$$

b = related parameter

For Approach (3)

b, D = eliminated parameters

$$(\Delta/t) = \frac{kN\omega M_f^0}{18 S_y L \sin(\theta/2)} \quad \text{(I)}$$

$$D = \frac{2M_f}{fN} \quad \text{(II)}$$

$$b = \frac{2M_f \sin(\theta/2)}{fN} \quad \text{(III)}$$

M_f = related parameter

We assume that boundary values for R_{max}, C_{max}, E_{min}, and E_{max} have been decided upon at this stage of design.

Upon closing the switch in Figure 8.13, the capacitor will start to discharge and current level i through R_0 will drop below the initial value I, as depicted in Figure 8.14. Suppose that the lamp becomes inoperative for all practical purposes when current i through R_0 drops to 80% of its initial value I, and we will designate the time at which this current level is reached by T_{80}. We would like to maximize the operative time for the lamp by the optimum choices for R, C, and E in the electrical circuit. Hence, our primary design equation will express optimization quantity T_{80} in terms of the design parameters, and this (P.D.E.) we derive from basic electrical circuit theory as follows, referring to Figure 8.13 and typically reference 8-7. Thus, application of Kirchhoff's voltage law gives

$$(R + R_0)i + \frac{1}{C}\int_0^t i\, dt = 0$$

$$(R + R_0)\frac{di}{dt} + \frac{i}{C} = 0$$

$$\frac{di}{i} = -\frac{dt}{(R + R_0)C}$$

$$\ln i = -\frac{t}{(R + R_0)C} + \ln K$$

$$i = K\varepsilon^{-t/[(R + R_0)C]}$$

where $\varepsilon = 2.7183$, the base for natural logarithms. Therefore

$$t = (R + R_0)C \ln (K/i)$$

However, at $t = 0$, $i = K = I$; at $t = T_{80}$, $i = 0.8I$; and

$$T_{80} = (R + R_0)C \ln (I/0.8I)$$
$$T_{80} = (R + R_0)C \ln (1.25) \qquad \text{(P.D.E.) (8.37)}$$

Since initial current I is a specified value, and both R and initial voltage E are constrained by limit equations (8.34) and (8.36), respectively, we recognize that (S.D.E.) (8.38), based on satisfaction of Ohm's law, must also be included in our initial formulation.

$$I = E/(R + R_0) \qquad \text{(S.D.E.) (8.38)}$$

Suppose also that after discharge of the capacitor, switch S will be opened and it will be necessary to recharge the capacitor to voltage E, and from a consideration of power limitations in the entire system we wish to limit the energy W_c to be stored in the electrostatic field to a maximum permissible value $(W_c)_{max}$. Hence, we must include (S.D.E.) (8.39) and (L.E.) (8.40) in our initial formulation.

$$W_c = \tfrac{1}{2}CE^2 \qquad \text{(S.D.E.) (8.39)}$$
$$W_c \leq (W_c)_{max} \qquad \text{(L.E.) (8.40)}$$

In (S.D.E.) (8.39) from elementary electrostatics and in (L.E.) (8.40) all terms have been defined, and at this stage of design we will assume that an acceptable limit value for $(W_c)_{max}$ has been estimated.

The initial formulation for this optimization problem is now summarized as follows:

Initial Formulation

$$T_{80} = (R + R_0)C \ln (1.25) \qquad \text{(P.D.E.) (8.37)}$$

$$I = E/(R + R_0) \qquad \text{(S.D.E.) (8.38)}$$

$$W_c = \tfrac{1}{2}CE^2 \qquad \text{(S.D.E.) (8.39)}$$

$$\left. \begin{array}{l} 0 \le R \le R_{\max}; \, 0 \le C \le C_{\max}; \\ E_{\min} \le E \le E_{\max}; 0 < W_c \le (W_c)_{\max} \end{array} \right\} \qquad \text{(L.E.)s}$$

Specified: R_0 and I, $R_{\max}$, $C_{\max}$, $E_{\min}$, $E_{\max}$, $(W_c)_{\max}$

Find: Values of R, C, and E which maximize T_{80}, the time for i to drop from I to $0.8I$ upon closing switch S in Figure 8.13.

Inspection of the initial formulation reveals that we have a case of redundant specifications since $n_f = 0 < N_s = 2$. Also, not counting optimization quantity T_{80}, the design variables are R, C, E, and W_c, so we have $n_v = 4$. Thus, from equation (8.13) we calculate

$$D_{vs} = n_v - N_s + 1$$
$$= 4 - 2 + 1 = 3$$

and we should be able to extract the optimum design from a three-dimensional variation study. From equation (8.10) we see that the number of equations in the final formulation will be

$$N_{ff} = N_s - n_f + 1$$
$$= 2 - 0 + 1 = 3$$

From equation (8.18) we see that the number of eliminated parameters in the final formulation will be

$$n_e = N_s - n_f = 2 - 0 = 2,$$

and by equation (8.20) the number of related parameters will be

$$n_r = n_c - n_e = 4 - 2 = 2.$$

The four constrained variables for the final formulation variation study will be R, C, E, and W_c. Having made the preceding simple calculations, we have a better understanding at this time of what should follow in derivation of the final formulation and optimum design variation study.

At this time we must decide on an approach for derivation of the final formulation. Hence, in review of the initial formulation we choose E and

W_c as the eliminated parameters, leaving R and C for the related parameters. Thus, recognizing in (P.D.E.) (8.37) that E and W_c are phantom variables, the equation-combination procedure results in the same equation for the developed (P.D.E.) of the ideal problem, equation (I). Also, in the final formulation we will have relating equations for E and W_c which must be expressed in terms of related parameters R and C. Therefore, from (S.D.E.)s (8.38) and (8.39), using $E = I(R + R_0)$ in the latter, we directly obtain relating equations (II) and (III). Summary of the final formulation is as follows, presented in the basic format previously described which is suitable for our optimum design variation study.

Final Formulation

$$T_{80} = (R + R_0) \, C \ln (1.25) \qquad \text{(I)}$$

$$E = I (R + R_0) \qquad \text{(II)}$$

$$W_c = \tfrac{1}{2} C I^2 (R + R_0)^2 \qquad \text{(III)}$$

where

E, W_c = eliminated parameters

R, C = related parameters

(I) = developed (P.D.E.) of ideal problem

(II) and (III) = relating equations

From the final formulation now summarized we will make a general variation study for optimum design with the objective of maximizing T_{80}. For the typical three-dimensional variation diagram we will sketch optimization quantity T_{80} as a function of related parameters R and C, imposing typical constraints on R and C in accordance with (L.E.)s (8.34) and (8.35). Next, on this typical variation diagram we will superimpose the variations of relating equations (II) and (III) showing the typical effects of constraints on eliminated parameters E and W_c in accordance with (L.E.)s (8.36) and (8.40), respectively. The possibilities for the domain of feasible designs thus defined on the R, C plane will become evident. Hence, based on an investigation of the variation of optimization quantity T_{80} within the possible domains of feasible design, the points possible for optimum design will be revealed. Such a typical three-dimensional variation diagram is shown in Figure 8.15. In that diagram the necessary variations for T_{80} are obvious, except for the variation of T_{80} along the boundary of equation (III) with $W_c = (W_c)_{max}$. For the variation of T_{80} along that curve we combine equation (III) with (I), eliminating either C or $(R + R_0)$. Specifically, if we

Figure 8.15 Typical three-dimensional variation diagram for example 8-4, by approach 1 of Table 8-2. Points 1, 2, 3, 4, and 5 are possible for optimum design. (For boundary value relationship specifically shown, point 4 would be the optimum design.)

take C, along that boundary we have from (I) and (III)

$$T_{80} \sim (R + R_0)C \sim (R + R_0)\frac{1}{(R + R_0)^2}$$

$$= \frac{1}{(R + R_0)}$$

Hence, we readily conclude that along equation (III) with $W_c = (W_c)_{max}$, T_{80} increases as R decreases, therefore leading us in the directions of points 4, 3, or 5 in Figure 8.15.

From the final formulation and the limit equations of the initial formulation, the boundary lines and acceptable domains in the plan view for the typical variation diagram of Figure 8.15 should be obvious, if one has a basic understanding of simple exponential functions as covered in reference 8-1.   However, we can always apply calculus if questions should arise, such as which side of boundary curve (III) with $W_c = (W_c)_{max}$ is acceptable in Figure 8.15.   For the answer, we determine the differential dW_c in terms of dC and dR from equation (III) of the final formulation as follows.

$$dW_c = \frac{\partial W_c}{\partial C} dC + \frac{\partial W_c}{\partial R} dR$$

$$= \tfrac{1}{2}I^2(R + R_0)^2 dC + CI^2(R + R_0) dR$$

Thus, if we are on the boundary curve (III) with $W_c = (W_c)_{max}$ in Figure 8.15, from the above equation of differentials we see that positive changes dC or dR by themselves would result in a positive change dW_c, which corresponds to an increase in W_c above the limit value $(W_c)_{max}$, and this would be unacceptable.   Therefore we conclude that the acceptable side of this boundary curve (III) with $W_c = (W_c)_{max}$ is down and to the right in Figure 8.15.   The same conclusion is reached by visualizing the surface expressed by equation (III) of the final formulation on the coordinate system of Figure 8.15 (W_c along the same axis as T_{80}), and imposing the ceiling plane, $W_c = (W_c)_{max}$, below which we must stay for satisfaction of the limit equation $W_c \leq (W_c)_{max}$.

Recognizing that constraints E_{min}, E_{max}, R_{max}, C_{max}, and $(W_c)_{max}$ are independent of each other, we realize that the boundaries in Figure 8.15 can fall relative to each other in a number of different ways.   Thus, considering the various possibilities for domains of feasible design together with the variation of T_{80} within each domain we explicitly conclude that the points possible for optimum design are 1, 2, 3, 4, and 5 referring to Figure 8.15 and the final formulation for their definitions.   Incidentally, for a further general discussion of three-dimensional variation diagrams, the reader should see reference 8-2.

In summary, the procedural steps of optimum design as determined in the variation study now made are outlined in the flow chart of Figure 8.16. The logic of its derivation is based on an understanding of the typical variation diagram sketched in Figure 8.15 together with the final formulation previously summarized.   In the flow chart in Figure 8.16, the subscript t merely designates a temporary value for either R or C.   As is typically so in an optimum design flow chart, the last step is one of analyzing the optimum design to determine what has been achieved for the criterion function value, T_{80}. Also, in the optimum design flow chart of Figure 8.16 note the possibilities of incompatible specifications for the cases where $R_2 < 0$ or $R_A > R_{max}$.   For

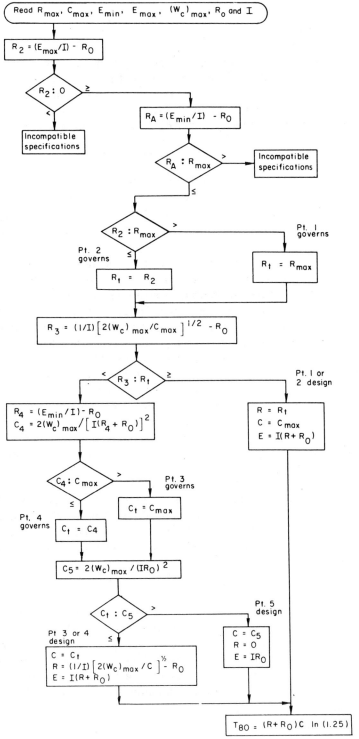

Figure 8.16 Optimum design flow chart for maximization of T_{80} in example 8-4.

either of these two situations, no matter how much we searched it would be impossible to find a combination of R, C, and E values which would satisfy the specifications and constraints of the initial formulation. For alleviation of either situation we would refer to the variation diagram and final formulation and see that boundary values E_{max} and E_{min} would have to satisfy the relations

$$E_{max} \geq IR_0$$

$$E_{min} \leq I(R_0 + R_{max})$$

If these requirements could be met by necessary modifications of the boundary values we no longer would have the case of incompatible specifications, and the optimum design could then be specified by the procedural steps of the flow chart.

A point of general importance should be mentioned at this time with respect to optimization problems. By intuition one might have assumed that time T_{80} would be maximized by having both voltage E and capacitance C as large as allowed, i.e., $E = E_{max}$ and $C = C_{max}$, respectively. This would correspond to the point 2 design in Figure 8.15. However, depending on the particular specifications and boundary values of the problem, from Figure 8.15 we see that points 1, 3, 4, and 5 are also possibilities for optimum design. If points 1 or 3 were optimum we would have $C = C_{max}$ but $E < E_{max}$. On the other hand, if points 4 or 5 were optimum we would have $C < C_{max}$ and $E < E_{max}$. Thus, from this specific example we see the fallacy of drawing general conclusions of optimization based on really only a partial consideration of the specifications and constraints. The entire initial formulation system of equations must be included in the derivation of valid conclusions for optimum design.

As an exercise the interested reader should consider other approaches for final formulations. In the presented variation-study approach, we chose E and W_c as the eliminated parameters with R and C as the related parameters. Thus, for final formulations in this problem we have $n_e = 2$ and $n_c = 4$. Therefore, from equation (8.19) the number of approaches possible are

$$A_t = \frac{n_c!}{n_e!(n_c - n_e)!} = \frac{4 \times 3 \times 2}{2(2)} = 6$$

Hence, the six approaches which are theoretically possible for derivation of final formulations in this problem are summarized in Table 8-2. Note that the eliminated parameter combination of W_c and C is an indeterminate approach. This we could at once spot from the presented approach (1) final formulation as previously discussed in general, since both W_c and C are phantom variables in relating equation (II), $E = I(R + R_0)$, of that approach.

TABLE 8-2 APPROACHES POSSIBLE WITH FINAL FORMULATIONS FOR EXAMPLE 8-4

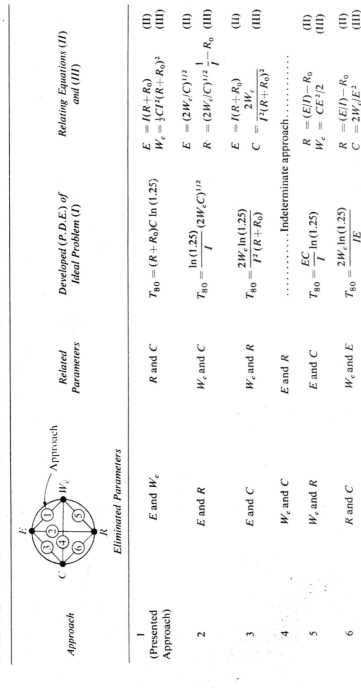

Approach	Eliminated Parameters	Related Parameters	Developed (P.D.E.) of Ideal Problem (I)	Relating Equations (II) and (III)	
1 (Presented Approach)	E and W_c	R and C	$T_{80} = (R+R_0)C \ln(1.25)$	$E = I(R+R_0)$ $\qquad W_c = \frac{1}{2}CI^2(R+R_0)^2$	(II) (III)
2	E and R	W_c and C	$T_{80} = \dfrac{\ln(1.25)}{I}(2W_cC)^{1/2}$	$E = (2W_c/C)^{1/2}$ $\qquad R = (2W_c/C)^{1/2}\dfrac{1}{I} - R_0$	(II) (III)
3	E and C	W_c and R	$T_{80} = \dfrac{2W_c \ln(1.25)}{I^2(R+R_0)}$	$E = I(R+R_0)$ $\qquad C = \dfrac{2W_c}{I^2(R+R_0)^2}$	(II) (III)
4	W_c and C	E and R	.............Indeterminate approach.............		
5	W_c and R	E and C	$T_{80} = \dfrac{EC}{I}\ln(1.25)$	$R = (E/I) - R_0$ $\qquad W_c = CE^2/2$	(II) (III)
6	R and C	W_c and E	$T_{80} = \dfrac{2W_c \ln(1.25)}{IE}$	$R = (E/I) - R_0$ $\qquad C = 2W_c/E^2$	(II) (III)

I and R_0 are specified values.

It is suggested that the interested reader obtain some practice in the equation-transformation process of the method of optimum design by deriving the final formulations of Table 8-2 starting from the previously summarized initial formulation of the problem for each approach. Also, for each approach the typical three-dimensional variation diagram should be sketched showing the points possible for optimum design. The results of course should be in complete agreement with those of Figure 8.15. For a specific illustration, the typical three-dimensional variation diagram for the second approach of Table 8-2 is sketched in Figure 8-17, with points 1, 2, 3, 4, and 5 determined thereby as being possible for optimum design. They correspond to points 1, 2, 3, 4, and 5, respectively, of Figure 8.15. For some practice in the variation-study technique, sketch typical variation diagrams for each of the other approaches which are determinate in Table 8-2, and from the final formulation determine the points possible for optimum design. They should be compatible with points 1, 2, 3, 4, and 5 of Figures 8.15 and 8.17. Also, as an exercise derive optimum design flow charts from the variation studies made, which should be compatible with what we derived in Figure 8.16 from Figure 8.15.

Example **8-5**   The solid cylinder casting in Figure 8.18 is to be designed for an application where we wish to transmit its stored heat to the surrounding environment. The ambient conditions are given, including temperature T_a. Elevated temperature T_c of the cylinder surface is a specified value, and the amount of stored heat Q_s in the cylinder is limited by the available heating source to an allowable value $(Q_s)_{max}$ (in Btu). The cylinder is to stand on the surface of a given heat insulator as shown in Figure 8.18, and for acceptable structural stability of the cylinder the height to diameter *ratio* (H/D) must not exceed an allowable value R_{max}. Space constraints in the environment limit permissible values of D to the range $0 < D \leq D_{max}$, where the value of D_{max} has been estimated at this stage of design. Properties of the environment and of the few feasible casting materials available for the cylinder are known at this stage of design, as estimated from various sources at hand.

The cylinder is to be designed for maximization of initial heat flow rate, q, when placed in the given environment. From our knowledge of heat transfer, we recognize that heat flow from the cylinder surface will be by convection and radiation.

$$q = q_c + q_r$$

Both of these components of heat transfer can be expressed mathematically from the theory of heat transfer, from a reference such as 8-8. Assume that thermal conductivities of the feasible casting materials are sufficiently high to

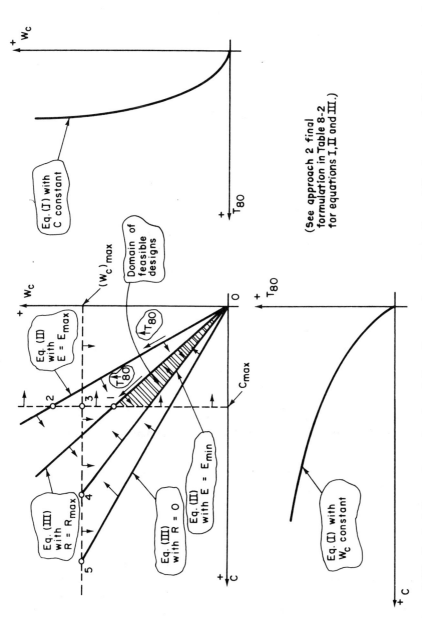

Figure 8.17 Typical three-dimensional variation diagram for maximization of T_{80} in example 8-4, by approach 2 of Table 8-2. Points possible for optimum design are 1, 2, 3, 4, and 5. These correspond to the numbers in Figure 8.15. (For boundary value relationship specifically shown, point 1 would be the optimum design.)

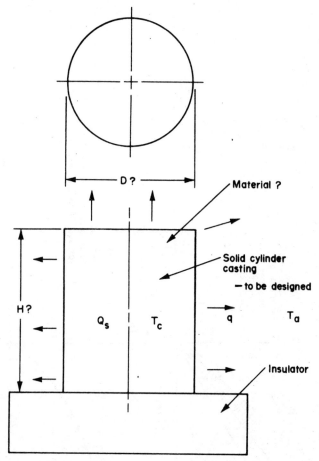

Figure 8.18   Heat-storage cylinder to be designed for maximization of heat flow rate.

assure essentially a uniform temperature distribution within the cylinder, for all practical purposes of design.

Heat flow rate by convection, q_c, is expressed as follows

$$q_c \approx hA_s(T_c - T_a)$$

where q_c is in Btu/hr, A_s is the surface area of heat transfer in ft², h is the film coefficient of heat transfer due to convection in Btu/hr/ft²/°F, and T_c and T_a are cylinder surface and ambient temperatures, respectively, in degrees Rankine.

Heat flow rate by radiation, q_r, is expressed as follows

$$q_r \approx \varepsilon\sigma A_s(T_c^4 - T_a^4)$$

which assumes that the concave area of enclosure for the environment is much larger than the surface area A_s of the cylinder to be designed. In this equation, q_r is the heat flow rate by radiation in Btu/hr, ε is the dimensionless emissivity of the cylinder surface, σ is the Stefan-Boltzmann's natural constant which equals 0.174×10^{-8} Btu/hr/ft^2/°F^4, and T_c and T_a are cylinder surface and ambient absolute temperatures, respectively, in degrees Rankine.

The components of heat flow rate are additive, so our primary design equation becomes (8.41).

$$q \approx A_s[h(T_c - T_a) + \varepsilon\sigma(T_c^4 - T_a^4] \qquad \text{(P.D.E.) (8.41)}$$

All terms in (P.D.E.) (8.41) have now been defined. The objective for optimum design of the solid cylinder is to maximize initial heat flow rate q.

Since stored heat Q_s in the cylinder is limited by our available heating facilities, a subsidiary design equation for this quantity as related to the design parameters must be included in the initial formulation.

$$Q_s = cw \frac{\pi D^2}{4} H(T_c - T_a) \qquad \text{(S.D.E.) (8.42)}$$

where Q_s is the heat stored in the cylinder in Btu, c is the specific heat of the cylinder material in Btu/lb/°F, w is the weight density of the cylinder material in lb/ft.^{3}, D and H are the diameter and height, respectively, of the cylinder in Figure 8.18 in feet, and T_c and T_a are the cylinder and ambient temperatures, respectively, in degrees Rankine. The corresponding regional constraint on permissible values for Q_s as previously discussed is as follows.

$$0 < Q_s \leq (Q_s)_{\text{max}} \qquad \text{(L.E.) (8.43)}$$

where the value for $(Q_s)_{\text{max}}$ is known at this stage of design.

Since height-to-diameter *ratio* of the cylinder is limited to an allowable value R_{max} as previously discussed, in the initial formulation we will include a subsidiary design equation for ratio R as follows.

$$R = H/D \qquad \text{(S.D.E.) (8.44)}$$

where the regional constraint on permissible values for R is

$$0 < R \leq R_{\text{max}} \qquad \text{(L.E.) (8.45)}$$

We assume that an acceptable value for R_{max} has been estimated at this stage of design based on stability considerations previously discussed for avoidance of toppling.

We realize that surface area A_s is mathematically tied together with diameter D and height H of Figure 8.18, so we must include (S.D.E.) (8.46) in our initial formulation.

$$A_s = \pi D H + \frac{\pi D^2}{4} \qquad \text{(S.D.E.)} \ (8.46)$$

where as previously discussed we have the following regional constraint on permissible values for D.

$$0 < D \leq D_{max} \qquad \text{(L.E.)} \ (8.47)$$

Thus, the initial formulation consists of (P.D.E.) (8.41), (S.D.E.)s (8.42), (8.44), and (8.46), and (L.E.)s (8.43), (8.45), and (8.47), which we summarize as follows.

Initial Formulation

$$q \approx A_s[h(T_c - T_a) + \varepsilon\sigma(T_c^4 - T_a^4)] \qquad \text{(P.D.E.)} \ (8.41)$$

$$Q_s = cw \frac{\pi D^2}{4} H(T_c - T_a) \qquad \text{(S.D.E.)} \ (8.42)$$

$$R = H/D \qquad \text{(S.D.E.)} \ (8.44)$$

$$A_s = \pi D H + \frac{\pi D^2}{4} \qquad \text{(S.D.E.)} \ (8.46)$$

$$0 < Q_s \leq (Q_s)_{max} \ ; \ 0 < R \leq R_{max} \ ; \ 0 < D \leq D_{max} \qquad \text{(L.E.)s}$$

Specified or Known: T_c, T_a, σ, $(Q_s)_{max}$, R_{max}, D_{max}, several feasible materials with properties h, ε, c, and w.

Find: Values of D, H, and material for cylinder in Figure 8.18, which maximizes initial heat flow rate q.

From an inspection of the initial formulation, we recognize that the free variables are A_s and H, so $n_f = 2$. Also, there are three (S.D.E.)s, so $N_s = 3$. Thus, in reference to Figure 8.2 we have $n_f = 2 < N_s = 3$, which indicates a case of redundant specifications for this problem. Also, not counting optimization quantity q, the variables are A_s, Q_s, D, H, and R, so $n_v = 5$. Thus, from equation (8.13) we calculate

$$D_{vs} = n_v - N_s + 1 = 5 - 3 + 1 = 3$$

and we should be able to extract the optimum design from a three-dimensional variation study. From equation (8.10) we see that the number of equations in the final formulation will be

$$N_{ff} = N_s - n_f + 1 = 3 - 2 + 1 = 2$$

The constrained variables are Q_s, D, and R, and one of these will have to be chosen as the eliminated parameter with the other two as the related parameters, since by equation (8.18)

$$n_e = N_s - n_f = 3 - 2 = 1$$

and by equation (8.20)

$$n_r = n_c - n_e = 3 - 1 = 2$$

Thus, with the preceding simple calculations we have a better understanding at this time of what we should expect in the final formulation and optimum design variation study which follows.

At this time we must select an approach for derivation of the final formulation. Upon review of the initial formulation we decide to choose Q_s as the eliminated parameter, which we believe should be a straightforward approach. To simplify the manipulation of equations, we define C_1 and C_2 as follows, which are a function of the specified values and material properties. Referring to (P.D.E.) (8.41) and (S.D.E.) (8.42), let

$$C_1 = h(T_c - T_a) + \varepsilon\sigma(T_c^4 - T_a^4) \tag{8.48}$$

$$C_2 = cw\left(\frac{\pi}{4}\right)(T_c - T_a) \tag{8.49}$$

Next we carry through the equation-combination procedure with the (P.D.E.) and (S.D.E.)s of the initial formulation, recognizing that A_s and H are free variables and that Q_s is the eliminated parameter, with R and D as the related parameters. Also, recognition of phantom variables in the initial formulation equations might be of help to some in carrying through the equation-combination procedure. Thus, the initial formulation system of equations could be considered at least mentally as follows, depicting the phantom variables not normally shown.   Ignore the writing of these phantom variables if doing so is more confusing than helpful to you, but at least be aware of their presence during execution of the equation-combination procedure to follow.

$$q \approx A_s C_1 D^0 H^0 R^0 Q_s^0 \qquad \text{(P.D.E.) (8.41a)}$$

$$Q_s = C_2 D^2 H A_s^0 R^0 \qquad \text{(S.D.E.) (8.42a)}$$

$$R = (H/D)Q_s^0 A_s^0 \qquad \text{(S.D.E.) (8.44a)}$$

$$A_s = \left[\pi DH + \frac{\pi D^2}{4}\right]R^0 Q_s^0 \qquad \text{(S.D.E.) (8.46a)}$$

Thus, combining the (S.D.E.)s with the (P.D.E.) by eliminating free variables A_s, H, and eliminated parameter Q_s we obtain the developed (P.D.E.) of the ideal problem (I) as follows.

$$q \approx C_1 \left[\pi D(RD) + \frac{\pi D^2}{4} \right]$$

therefore

$$q \approx C_1 \pi D^2 \left[R + \tfrac{1}{4} \right] \qquad \text{(I)}$$

Similarly, combining (S.D.E.) (8.44) and (8.46) with (8.42) by eliminating free variables A_s and H from the initial formulation equation system, we obtain the relating equation (II) as follows.

$$Q_s = C_2 \, D^2 \, (RD)$$
$$Q_s = C_2 \, D^3 R \qquad \text{(II)}$$

Therefore, for this approach we summarize the final formulation as follows. Note the basic format previously described, suitable for the variation study to follow.

Final Formulation

$$q \approx \pi \, C_1 \, D^2 (R + \tfrac{1}{4}) \qquad \text{(I)}$$
$$Q_s = C_2 \, D^3 R \qquad \text{(II)}$$

where

Q_s = eliminated parameter

D, R = related parameters

(I) = developed (P.D.E.) of ideal problem

(II) = relating equation

From the final formulation now summarized we will make a general variation study for optimum design, with the objective of maximizing q. For the typical three-dimensional variation diagram we first consider optimization quantity q as a function of related parameters D and R from equation (I), imposing typical constraints on D and R in accordance with (L.E.)s (8.47) and (8.45), respectively, Next, on this typical variation diagram we will superimpose the variation of relating equation (II) showing on the D, R plane the typical effects of the constraint on eliminated parameter Q_s in accordance with (L.E.) (8.43). The possibilities for the domain of feasible designs thus defined on the D, R plane will become evident. Hence, based on an investigation of the variation of optimization quantity q within the possible domains of feasible design, the points possible for optimum design are revealed. Such a typical three-dimensional variation diagram is shown in Figure 8.19.

Figure 8.19   Typical three-dimensional variation diagram for maximization of heat flow rate, q, in example 8-5.   Points possible for optimum design are 1, 2, and 3.   (See final formulation for equations I and II.)

In Figure 8.19, recognizing that boundary values R_{max}, D_{max}, and $(Q_s)_{max}$ are independent of each other, we readily see that point 1 is a possibility for optimum design. However, if relating equation (II) with $Q_s = (Q_s)_{max}$ cuts the rectangle bounded by R_{max}, D_{max}, and the R, D axes, we readily recognize from the figure that it is necessary to investigate the variation of q along that boundary curve. Thus, along boundary curve (II) with $Q_s = (Q_s)_{max}$ in Figure 8.19, combining equations (I) and (II) of the final formulation, we have equation (8.50) for the variation of q thereon.

$$q \approx \pi C_1 D^2 (R + \tfrac{1}{4})$$

$$= \pi C_1 D^2 \left[\frac{(Q_s)_{max}}{C_2 D^3} + \frac{1}{4} \right]$$

therefore

$$q = \pi C_1 \left[\frac{(Q_s)_{max}}{C_2 D} + \frac{D^2}{4} \right] \tag{8.50}$$

Thus, the directional derivatives of q tangent to the boundary curve (II) with $Q_s = (Q_s)_{max}$ would be as follows:

$$\frac{dq}{dD} = \pi C_1 \left[-\frac{(Q_s)_{max}}{C_2 D^2} + \frac{D}{2} \right]$$

and

$$\frac{d^2 q}{dD^2} = \pi C_1 \left[\frac{2(Q_s)_{max}}{C_2 D^3} + \frac{1}{2} \right]$$

which is always positive for positive values or $(Q_s)_{max}$, C_2, and D, which would always be the case. Thus, along boundary curve (II) with $Q_s = (Q_s)_{max}$ in Figure 8.19, we conclude that q has a *minimum* value at say point 4, where $dq/dD = 0$ from equation (8.50). Hence from the preceding we find the value of D at point 4 in Figure 8.19 to be

$$\left(\frac{dq}{dD} \right)_4 = \pi C_1 \left[-\frac{(Q_s)_{max}}{C_2 D_4^2} + \frac{D_4}{2} \right] = 0$$

$$D_4 = [2(Q_s)_{max}/C_2]^{1/3} \tag{8.51}$$

which is the point of *minimum* q along that boundary curve. Since our *optimization objective is to maximize* q, point 4 in Figure 8.19 is *not* a point of optimum design. Thus, if the constraints R_{max} and D_{max} should fall relative to point 4 specifically as shown in Figure 8.19, the optimum design would be at either intersection point 2 or 3 depending upon which has the larger value for q. On the other hand, if point 4 should fall above point 3 in Figure 8.19, point 2 would be the optimum design.   Or, if point 4 should fall to the left of point 2 in the figure, point 3 would be the optimum design. From this discussion of the typical variation diagram of Figure 8.19, we conclude that the points possible for optimum design are 1, 2, and 3.

An item of general truth in realistic optimization problems should be mentioned at this time.   For the great majority of cases, the point of optimum design is at the intersection of boundary curves (such as at 1, 2 or 3 in Figure 8.19), where the derivative of the criterion function is *not* zero.   The common belief that the optimum design is always where we set the derivative of optimization quantity to zero is in error.   In fact, in Figure 8.19 at point 4 the directional derivative of q tangent to the boundary curve is zero, but this is the *worst* point of design along that boundary since it is the one of minimum value for q.

In summary, the procedural steps of optimum design as determined in the variation study now made are outlined in the flow chart of Figure 8.20. The logic of its derivation is based on an understanding of the typical variation diagram sketched in Figure 8.19 together with the final formulation previously summarized.   As is typically so in the procedure of optimum design, the last step is one of analyzing the optimum design to determine what has been achieved for the criterion function value, q.   The Figure 8.20 flow chart would be executed for each of the cylinder materials of interest, and the overall optimum design would be the one of greatest q value so calculated from the flow chart.   Hence, the overall optimum design is determined explicitly in relatively few calculations and comparisons.

In reference to the typical variation diagram in Figure 8.19 and to the optimum design flow chart in Figure 8.20, we should mention some more items of interest.   First, from the variation diagram we readily see that there is no possibility for a case of incompatible specifications in this optimization problem, regardless of the particular boundary values.   Hence, there are always an *infinite* number of possible design solutions for each material of interest.   However, the flow chart explicitly leads us directly to the optimum design at point 1, 2, or 3 of Figure 8.19 in few calculations and comparisons.   Time-consuming searching techniques are thus avoided by the method of optimum design.

As an exercise the interested reader should consider the other approaches for derivation of final formulations.   In the presented variation study approach, we chose Q_s as the eliminated parameter with D and R as the related parameters.   However, either D or R could have been selected as the eliminated parameter, giving thereby two other approaches.   Try each, use the simplifications of C_1 and C_2 as defined by equations (8.48) and (8.49), and starting from the initial formulation derive the *final formulations* summarized in Table 8-3.

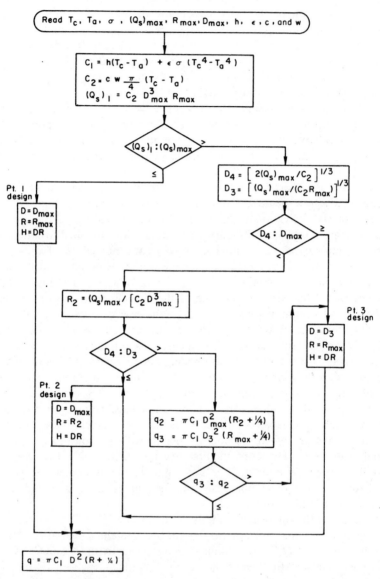

Figure 8.20   Optimum design flow chart for maximization of q in example 8-5.

TABLE 8-3   OTHER FINAL FORMULATIONS FOR EXAMPLE 8-5

For Approach (2)

D = eliminated parameter

$$q \approx \pi C_1 \left[\frac{Q_s}{C_2}\right]^{2/3} \left[R^{1/3} + \frac{1}{4(R)^{2/3}}\right] \qquad \text{(I)}$$

$$D = [Q_s/(C_2 R)]^{1/3} \qquad \text{(II)}$$

where

Q_s, R = related parameters

(I) = developed (P.D.E.) of ideal problem

(II) = relating equation

For Approach (3)

R = eliminated parameter

$$q \approx \pi C_1 \left[\frac{Q_s}{C_2 D} + \frac{D^2}{4}\right] \qquad \text{(I)}$$

$$R = Q_s/(C_2 D^3) \qquad \text{(II)}$$

where

Q_s, D = related parameters

(I) = developed (P.D.E.) of ideal problem

(II) = relating equation

Also, it is suggested that the interested reader sketch a typical three-dimensional variation diagram for each approach showing the points which are possible for optimum design. The results, of course, should be in complete agreement with those of Figures 8.19 and 8.20. Hence, from both approach (2) and (3) summarized above, points corresponding to 1, 2, and 3 of Figure 8.19 should be determined as the only possibilities for optimum design. Finally, for at least one of the approaches derive the optimum design flow chart as a good exercise in that technique.

As an additional exercise in the method of optimum design, assume that cylinder height H is constrained, instead of free, and include (L.E.) (8.52) in the initial formulation.

$$H \leq H_{\max} \qquad \text{(L.E.) (8.52)}$$

Assume that an appropriate value has been estimated for H_{max} at this stage of design because of space constraints, and do not delete any of the other specifications or constraints from the initial formulation previously summarized. Prove that the required variation study will still be three-dimensional. Select an appropriate approach for derivation of a final formulation, sketch a typical variation diagram and derive the optimum design flow chart.

As a final exercise the interested reader should modify the stated problem to one of minimizing the heat flow rate q, with a specified minimum value $(Q_s)_{min}$ for heat stored instead of $(Q_s)_{max}$. Summarize the initial formulation, select an approach for derivation of a final formulation, sketch a typical variation diagram indicating the points possible for optimum design, and derive the optimum design flow chart.

Example 8-6   A curved spring is to be designed for transmitting a force which is to vary with time as shown in Figure 8.21, and force amplitude P of the figure is to be a specified value. Corresponding deflection Δ in Figure 8.21 is to be as large as possible in this particular application, to accommodate the accumulation of tolerances in the total assemblage not shown. Hence, the objective for optimum design of the curved spring is maximization of deflection Δ.

The primary design equation will be derived from strength of materials as typically found in reference 8-9. Thus, strain energy in the curved spring is as follows, referring to Figure 8.21.

$$U = \int_0^\pi \frac{M_b^2(Rd\theta)}{2EI}$$

where bending moment M_b is

$$M_b = PR \sin \theta$$

and rectangular moment of inertia for the spring cross section is

$$I = \frac{1}{12} bt^3$$

Hence, strain energy U in the spring becomes

$$U = \int_0^\pi \frac{(PR \sin \theta)^2 (R\, d\theta)}{2E(bt^3/12)}$$

$$= \frac{6P^2R^3}{Ebt^3} \int_0^\pi \sin^2 \theta\, d\theta$$

Figure 8.21   Curved spring to be designed showing variation of load with time.

and

$$\int_0^\pi \sin^2 \theta \, d\theta = \pi/2$$

Thus,

$$U = \frac{3\pi P^2 R^3}{Ebt^3} \qquad (8.53)$$

By the Theorem of Castigliano from reference 8-9, deflection Δ will be the derivative of strain energy with respect to load.   In this way from equation (8.53) we derive the (P.D.E.) (8.54) which follows.

$$\Delta = \frac{dU}{dP} = \frac{3\pi R^3(2P)}{Ebt^3}$$

$$= \frac{6\pi PR^3}{Ebt^3} \qquad\qquad \text{(P.D.E.) (8.54)}$$

In (P.D.E.) (8.54), referring to Figure 8.21, Δ is deflection of the spring (inches), P is the specified force (lb); R, b, and t are dimensions (inches); and E is the modulus of elasticity of the spring material (psi).

We wish to manufacture the curved spring from standard flat stock material which is readily available. In our particular locality at this time we find that such a material is 0.65–0.80 carbon steel flat stock of hardness 45–48 C Rockwell, which we now select for the design. The safe endurance limit for a unidirectional bending load variation is available from Figure 59 of reference 8-10, and in accordance with the load variation of Figure 8.21 for the curved spring this is the type of loading which we would have. From Figure 59 of reference 8-10 we see that this endurance limit is a function of stock thickness t, and from this available data we have plotted the graph of Figure 8.22 on log-log scales. Hence, by the simple exponential curve fitting technique presented in Chapter 2 of reference 8-1 we derive the following equation

$$S_{bu} = \frac{47,000}{t^{0.25}} \qquad\qquad (8.55)$$

In equation (8.55), S_{bu} is the safe endurance limit of the selected material (psi), for the unidirectional type of bending stress variation depicted in the inset of Figure 8.22, and t is the stock thickness (inches). As seen from Figure 8.22, equation (8.55) is reasonably accurate and conservative over the anticipated range of interest for stock thickness t, which is 0.010 in. $\leq t \leq$ 0.125 in. Incidentally, equation (8.55) is a typical example of how we can account mathematically for size effect on material strength, which is often necessary to do in the derivation of initial formulations for realistic problems of optimum design.

For avoidance of fatigue breakage of our curved spring, amplitude of bending stress σ_b in Figure 8.21 must satisfy the regional constraint

$$\sigma_b \leq S_{bu}/N_e$$

where S_{bu} is the endurance limit material strength from equation (8.55), and N_e is an appropriate factor of safety to account for various uncertainties of design as described in Chapter 6 of reference 8-1 and in reference 8-11.

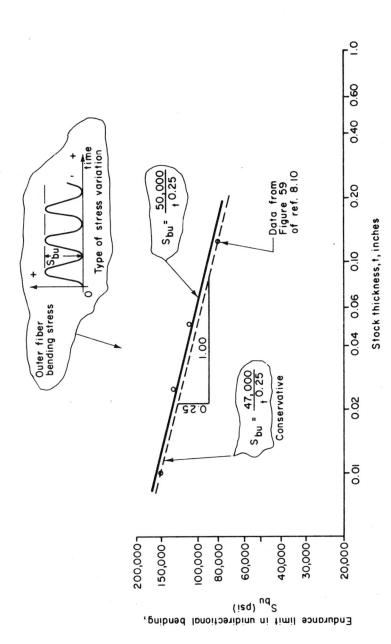

Figure 8.22 Safe endurance limit, S_{bu}, in unidirectional bending of 0.65–0.80 carbon steel flat stock of 45–48 C Rockwell hardness.

We assume that the appropriate value for N_e has been selected by the techniques of the cited references at this stage of design, based on dimensionless estimates of variations anticipated for the various components of load capability and actual load and from a consideration of the significance attached to the occurrence of fatigue breakage in this application. Hence, for avoidance of fatigue breakage, using equation (8.55) we must satisfy the relation

$$\sigma_b \leq \frac{47,000}{t^{0.25}N_e}.$$

Since stock thickness t will be a limited design parameter, we will define a limited stress parameter y as follows:

$$y = \sigma_b t^{0.25} \quad . \tag{8.56}$$

Thus, for avoidance of fatigue breakage we require satisfaction of the constraint

$$y \leq y_{max} = 47,000/N_e \tag{L.E.} \tag{8.57}$$

and we note that y_{max} as so defined is a material property constant at this stage of design.

Maximum bending stress σ_b in the curved spring of Figure 8.21 is expressed in terms of the design parameters by equation (8.58) derived from strength of materials as follows, where we have made the reasonable assumption that any feasible design will satisfy the relation $t << R$.

$$\sigma_b \approx Mc/I$$

$$= \frac{(PR)t/2}{(1/12)bt^3}$$

therefore

$$\sigma_b \approx 6\,PR/(bt^2) \tag{8.58}$$

All terms in equation (8.58) have already been defined.

In manufacturing the curved spring to be designed, we would like to use a standard stock thickness t_s because of practical considerations. Assume that a finite number M of standard stock sizes is available for manufacturing use, such as the standard Birmingham or Stubbs gage sizes listed in Table 24 of reference 8-10 as a specific example. Hence, we have a discrete value constraint for acceptable stock thickness t which must be adhered to in design, expressed as follows:

$$t = t_s \tag{L.E.} \tag{8.59}$$

for $s = 1, \ldots, M$. In addition, we will assume that regional constraints must be adhered to in design for permissible values of dimensions R and b in Figure 8.21.

$$R_{min} \leq R \leq R_{max} \qquad \text{(L.E.) (8.60)}$$

$$b \leq b_{max} \qquad \text{(L.E.) (8.61)}$$

We assume that limit values have been estimated at this stage of design for R_{min}, R_{max}, and b_{max}, from a consideration of space approximately available for the curved spring in the assemblage.

The initial formulation for our problem of optimum design consists of (P.D.E.) (8.54), (S.D.E.)s (8.56) and (8.58), and (L.E.)s (8.57), (8.59), (8.60), and (8.61), which we summarize as follows:

Initial Formulation

$$\Delta = 6\pi PR^3/(Ebt^3) \qquad \text{(P.D.E.) (8.54)}$$

$$y = \sigma_b t^{0.25} \qquad \text{(S.D.E.) (8.56)}$$

$$\sigma_b = 6\,PR/(bt^2) \qquad \text{(S.D.E.) (8.58)}$$

$$\left.\begin{array}{l} y \leq 47{,}000/N_e\,;\ b \leq b_{max} \\ R_{min} \leq R \leq R_{max}\,; \\ t = t_s,\ \text{for } s = 1, \ldots.\ M \end{array}\right\} \qquad \text{(L.E.)s}$$

Specified or Known: P, E, N_e, b_{max}, R_{min}, R_{max}; and standard stock thicknesses t_s, M in number.

Find: Values of b, R, and t which will maximize deflection Δ of Figure 8.21.

From an inspection of the initial formulation, we recognize that the only free variable is σ_b, so $n_f = 1$. Also, there are two (S.D.E.)s, so $N_s = 2$. Thus, in reference to Figure 8.2 we have $n_f = 1 < N_s = 2$, which indicates a case of redundant specifications for this problem. Also, not counting optimization quantity Δ, the variables are R, b, t, y, and σ_b, so $n_v = 5$. Thus, from equation (8.13) we calculate

$$D_{vs} = n_v - N_s + 1 = 5 - 2 + 1 = 4$$

which indicates that a four-dimensional variation study is required for determination of the optimum design. However, we recognize that the discrete value constraint of (L.E.) (8.59) on thickness t is of finite number M. Hence, if for our variation study we hold thickness constant at a standard stock size value, i.e., $t = t_s$, treating t as a constant we now calculate from equation (8.13)

$$D_{vs} = n_v - N_s + 1 = 4 - 2 + 1 = 3$$

In this way we can make the optimum design variation study in three dimensions, and this of course is easy to visualize.   Hence, this approach greatly facilitates our variation study, and the basic technique is a very practical one to follow as will be described later, particularly if a digital computer is available for the making of routine calculations.   Therefore, in the variation study to follow we will assume t constant at a standard size value t_s.   From equation (8.10) we see that the number of equations in the final formulation will be

$$N_{ff} = N_s - n_f + 1 = 2 - 1 + 1 = 2$$

The constrained variables (other than t) are R, b, and y, and for our final formulation one of these will have to be chosen as the eliminated parameter with the other two as the related parameters, which is evident from equation (8.18) as follows.

$$n_e = N_s - n_f = 2 - 1 = 1$$

and by equation (8.20),

$$n_r = n_c - n_e = 3 - 1 = 2$$

Thus, with the simple calculations of this paragraph we have a better understanding at this time of what we should expect in the final formulation and optimum design variation study which follows.

At this time we must select a particular approach for derivation of the final formulation.   Upon review of the initial formulation we decide to choose y as the eliminated parameter, which we believe should be a straightforward approach.   Thus, by eliminating free variable σ_b in the initial formulation system of equations, we derive the final formulation for this approach, which is summarized as follows.   Note the basic format previously described, suitably arranged for the variation study to follow.

Final Formulation

$$\Delta = \frac{6\pi P}{Et_s^3} \frac{R^3}{b} \quad \text{(I)}$$

$$y = \frac{6P}{t_s^{1.75}} \frac{R}{b} \quad \text{(II)}$$

where

    y = eliminated parameter

    R, b = related parameters

    (I) = developed (P.D.E.) of ideal problem

    (II) = relating equation

From the final formulation now summarized we will make a general variation study for optimum design of the curved spring, with the objective of maximizing deflection Δ. For the typical three-dimensional variation diagram we first consider optimization quantity Δ as a function of related parameters R and b from equation (I) imposing typical constraints on R and b in accordance with (L.E.)s (8.60) and (8.61), respectively. Next, on this typical variation diagram we will superimpose the variation of relating equation (II) showing the typical effects of the constraint on eliminated parameter y in accordance with (L.E.) (8.57). The possibilities for the domain of feasible designs thus defined on the R, b plane will become evident. Hence, based on an investigation of the variation of optimization quantity Δ within the possible domains of feasible design, the points possible for optimum design are revealed. Such a typical three-dimensional variation diagram is shown in Figure 8.23.

From the variation diagram of Figure 8.23 we readily see that there are in general an infinite number of feasible designs possible for a standard stock thickness $t = t_s$. Hence, starting from any such typical feasible design point in Figure 8.23, we improve or increase Δ by moving to the boundary line formed by relating equation (II) with $y = 47,000/N_e$. From the variations summarized in the diagram it becomes apparent that we must investigate the variation of Δ along that boundary line. This we can readily do from the final formulation by combining equations (I) and (II) by eliminating b (or R) with $y = y_{max} = 47,000/N_e$, a constant. Thus, along this boundary line

$$\Delta \sim \frac{R^3}{b} \sim \frac{R^3}{R} = R^2$$

and we conclude that Δ increases along this boundary as R increases, and this we can now indicate on Figure 8.23 as shown. Hence, from our understanding of the variation diagram at this time, and recognizing that the boundary values b_{max}, R_{min}, and R_{max} are independent of each other and of boundary line (II) with $y = 47,000/N_e$ in Figure 8.23, we conclude that the only points possible for optimum design are intersection points 1 and 2 in the figure. Specifically, for the standard stock thickness $t = t_s$, point 1 will be optimum if we have the relation $R_{min} \leq R_1 \leq R_{max}$, whereas point 2 will be optimum if $R_1 > R_{max}$. Also, we readily see that there is a case of incompatible specifications for the standard stock size $t = t_s$ if we have the relation $R_1 < R_{min}$.

In summary, the procedural steps of optimum design as determined in the variation study now made are outlined in the flow chart of Figure 8.24. The logic of its derivation is based on an understanding of the typical variation diagram sketched in Figure 8.23 together with the final formulation previously summarized. As indicated in Figure 8.24, the main loop of the flow

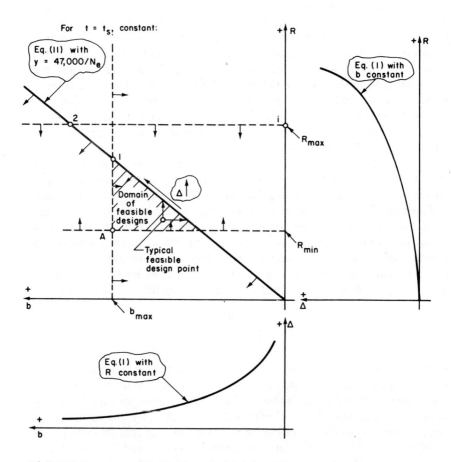

(a) Typical diagram normally sketched

Figure 8.23   Three-dimensional variation diagram for maximization of deflection Δ of the curved spring in example 8-6.   (a) Typical diagram normally sketched.   Points possible for optimum design are 1 and 2.   (For boundary value relationship specifically shown, point 1 would be the optimum design.) (b) Specific possibilities for boundary relations in plan view of the variation diagram in (a).   (Normally only one diagram is sketched with the other possibilities understood.)

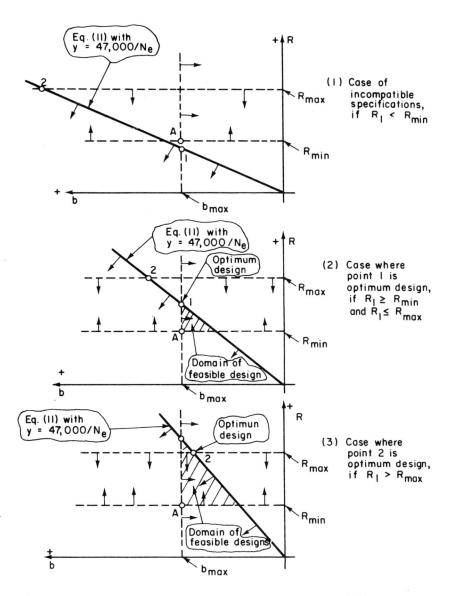

(b) Specific possibilities for boundary relations in plan view of (a)

Figure 8.23 (continued)

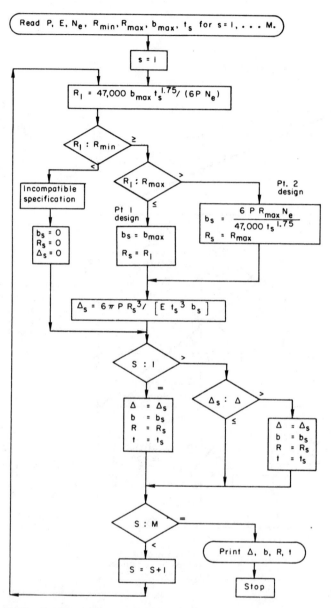

Figure 8.24 Optimum design flow chart for maximizing Δ in the curved spring design of example 8-6.

chart is repeated for each of the standard stock thicknesses t_s of interest, and the overall optimum design is readily recognized as the one of largest Δ_s value.   Execution of the Figure 8.24 flow chart is very rapid using a digital computer, since the optimum design is explicitly determined with relatively few calculations and comparisons.   Time-consuming searching techniques are thus avoided by application of the method of optimum design.

As a numerical example, consider the problem of designing the Figure 8.21 curved spring for maximum deflection Δ, satisfying the following specifications and   constraints: $P = 15.0$ lb; $N_e = 1.5$; $b \leq 1.0$ in.; 1 in. $\leq$ R ≤ 3 in.; $M = 12$ standard stock thicknesses, $t_s = 0.010$ in., 0.014 in., 0.018 in., 0.022 in., 0.025 in., 0.032 in., 0.042 in., 0.058 in., 0.072 in., 0.095 in., 0.109 in., 0.120 in. are readily available of 0.65–0.8 C steel flat stock Rockwell C 45–48 hardness; $E = 30 \times 10^6$ psi.

The computer program written from the Figure 8.24 optimum design flow chart along with the input and output data for the numerical problem is presented in Figure 8.25.   Incidentally, as a minor difference in execution, you will notice that the program of Figure 8.25 is actually written for the use of separate data cards for the input of each stock thickness instead of the equivalent subscripted variable approach implied in the Figure 8.24 flow chart.   Hence, from the output data of Figure 8.25 we see that the overall optimum design gives $\Delta = 0.792$ in., with dimensions $b = 0.861$ in., $R = 3.00$ in., and $t = 0.072$ in.   This maximum deflection value $\Delta = 0.792$ in. is the largest which is possible from any of the infinite number of feasible designs which satisfy all of the specifications and limitations of the numerical problem. Incidentally, this overall optimum design represents a point 2 design referring to Figure 8.23.

As an exercise the interested reader should consider the other approaches for final formulations.   In the presented variation study approach, we held t constant at a standard size value t_s, chose y as the eliminated parameter, and R and b as the related parameters.   However, still holding t constant at a standard size value t_s, there are two other approaches possible.   First, we might choose R as the eliminated parameter and b and y as the related parameters.   Secondly, we might choose b as the eliminated parameter and R and y as the related parameters.   For each approach derive the final formulation and sketch typical three-dimensional variation diagrams, which should be in agreement with Figure 8.23 for the possible points of optimum design. Finally, an exercise in optimum design flow chart derivation should be carried through for each of the approaches, similar to Figure 8.24, and applied to the solution of the given numerical problem using the digital computer. The numerical answers, of course, should completely agree with what we summarized in the text from the presented approach.

```
V-006        $JOB  THEME011,0431RCJ,,059187940
             C CURVED SPRING DESIGN FOR MAXIMIZING SAFE DEFLECTION, EXAMPLE 8.6
      1            READ,P,E,ANE,RMN,RMX,BMX
      2            WRITE (6,1) P,E,ANE,RMN,RMX,BMX
      3          1 FORMAT ('0','SPECS. P,E,ANE,RMN,RMX,BMX'/6E16.7//)
      4            READ, M
      5            WRITE (6,2) M
      6          2 FORMAT ('0','NO. OF STANDARD STOCK SIZES =',3X,I4//)
      7            IS = 1
      8          5 READ, TS
      9            R1 = 47.E3*BMX*TS**1.75/(6.*P*ANE)
     10            IF (R1 - RMN) 10,20,20
     11         10 BS = 0.
     12            RS = 0.
     13            DELTS = 0.
     14            GO TO 30
     15         20 IF (R1 - RMX) 40,40,50
     16         40 BS = BMX
     17            RS = R1
     18            GO TO 60
     19         50 BS = 6.*P*RMX*ANE/(47.E3*TS**1.75)
     20            RS = RMX
     21         60 DELTS = 6.*3.14159*P*RS**3./(E*TS**3.*BS)
     22         30 IF (IS - 1) 80,80,70
     23         80 DELTA = DELTS
     24            B = BS
     25            R = RS
     26            T = TS
     27            GO TO 100
     28         70 IF (DELTS - DELTA) 100,100,80
     29        100 IF (IS - M) 110,120,120
     30        110 IS = IS + 1
     31            GO TO 5
     32        120 WRITE (6,121) B,R,T
     33        121 FORMAT ('0','OPTIMUM DESIGN IS FOR B,R,T =',3X,3E16.7/)
     34            WRITE (6,122) DELTA
     35        122 FORMAT ('0',5X,'WHICH GIVES MAX. POSSIBLE DELTA =',3X,E16.7)
     36            STOP
     37            END

             $ENTRY

      SPECS. P,E,ANE,RMN,RMX,BMX
        0.1500000E 02      0.3000000E 08      0.1500000E 01      0.1000000E 01      0.3000000E 01

        0.1000000E 01

      NO. OF STANDARD STOCK SIZES =         12


      OPTIMUM DESIGN IS FOR B,R,T =        0.8610437E 00      0.3000000E 01      0.7200003E-01

          WHICH GIVES MAX. POSSIBLE DELTA =          0.7919925E 00


   — COMPILATION:     CPU TIME= 1.3192 SECONDS
   — EXECUTION:       CPU TIME= 0.4208 SECONDS
```

Figure 8.25 Computer program for optimum design of the curved spring in numerical example 8-6, written from the flow chart in Figure 8.24.

Concluding Comments on the Method of Optimum Design

The method of optimum design has now been explained and illustrated in examples 8-1 through 8-6, each of which brings out some aspects of general importance for those who are learning the basic technique. The applications for the method of optimum design are virtually unlimited in scope. The only limitation in application can be in execution, if it is not possible because of function characteristics to manipulate the equations of the initial formulation in the equation-combination procedure for the derivation of the final formulation. However, such a limitation if encountered may very well be surmountable with relatively little effort by applying function-conversion techniques for simplification, such as those presented in Chapter 2 of reference 8-1 and specifically illustrated in Chapter 12 therein. The author has yet to encounter a problem in optimum design in his industrial experience for which the equations cannot be manipulated for execution of the method of optimum design if appropriate techniques are applied.

Some additional introductory examples of the method of optimum design are presented in references 8-2 and 8-3, which the interested reader should pursue to gain more experience with the technique. Also, some more examples will be presented in Chapter 9 of this book, which should provide some background for application in more complicated realistic problem situations. Finally, see references 8-25 through 8-28 recently added.

8-3 THE VARIATION OF FUNCTIONS

In the introductory examples presented for the method of optimum design the function variations were simple enough to analyze with no difficulty. In fact, this is generally true for many advanced problems of optimum design, as illustrated in Chapter 9. However, there are some techniques not yet presented which might be of value for an easier understanding of an optimum design variation study in certain cases.

First, the sketching of *contour lines* can be of value for an easier under-standing of some three-dimensional variation diagrams in particular. For example, in the plan view of a diagram such as Figure 8.23 the contour lines for constant value of optimization quantity Δ could be roughly sketched as shown in Figure 8.26, which might make it more clear as to how Δ varies in that view. On the other hand, after some experience with three-dimensional variation diagrams it is usually unnecessary to sketch the contour lines, which in fact might very well clutter the diagram and make it more confusing. For this reason contour lines of constant Δ were not originally shown in the diagram of Figure 8.23.

Secondly, by the theorem of Weierstrass in reference 8-14, we know that a continuous criterion function of optimum design will have maximum and minimum values either in the interior or on the boundary of the closed domain for feasible design solutions. For the great majority of cases which the author has encountered in optimum mechanical design, the maximum or minimum value will occur along the boundary of a feasible design domain at points of function intersection such as illustrated in examples 8-1 through 8-6. How-ever, on rare occasions the maximum or minimum point will occur in the interior of the feasible design domain at what is known as a *stationary point* where the slope of the criterion function is zero in any direction. Hence, from reference 8-15, the *necessary conditions* for a maximum or minimum

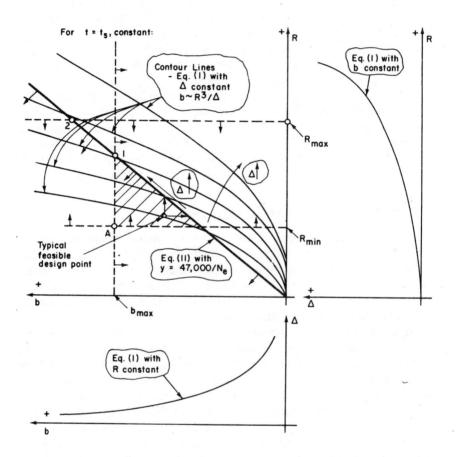

Figure 8.26   Contour lines in the plan view of the variation diagram in Figure 8.23.   (See the final formulation of example 8-6 for equations I and II.)

at a stationary point of a continuous criterion function such as equation (8.1) are satisfaction of equations (8.62).

$$Q = f_1(u, v, w) \qquad (8.1)$$

$$\frac{\partial Q}{\partial u} = 0, \ \frac{\partial Q}{\partial v} = 0, \ \frac{\partial Q}{\partial w} = 0 \qquad (8.62)$$

Satisfaction of equations (8.62) is not sufficient to guarantee that we have a stationary point where the criterion function has at least a local maximum or minimum value, and we will consider this next for the case of three-dimensional variations.

Suppose that we have a three-dimensional criterion function as expressed by equation (8.8) previously discussed.

$$Q = f_4(u, y) \qquad (8.8)$$

If a stationary point exists at the point having coordinates (u_1, y_1), the *necessary conditions* for a local minimum or maximum are satisfaction of equations (8.63)

$$\frac{\partial Q}{\partial u}(u_1, y_1) = 0 \qquad (8.63a)$$

$$\frac{\partial Q}{\partial y}(u_1, y_1) = 0 \qquad (8.63b)$$

From a Taylor series expansion of Q about (u_1, y_1), as presented in reference 8-12, the *sufficient conditions* for a local minimum or maximum which also must be satisfied at (u_1, y_1) are as follows.   For a local minimum

$$\frac{\partial^2 Q}{\partial u^2} > 0 \qquad (8.64a)$$

and

$$\left(\frac{\partial^2 Q}{\partial u^2}\right)\left(\frac{\partial^2 Q}{\partial y^2}\right) - \left(\frac{\partial^2 Q}{\partial u\, \partial y}\right)^2 > 0 \qquad (8.64b)$$

On the other hand, for a local maximum

$$\frac{\partial^2 Q}{\partial u^2} < 0 \qquad (8.65a)$$

and

$$\left(\frac{\partial^2 Q}{\partial u^2}\right)\left(\frac{\partial^2 Q}{\partial y^2}\right) - \left(\frac{\partial^2 Q}{\partial u\, \partial y}\right)^2 \triangleright 0 \qquad (8.65b)$$

If, however, we find that

$$\left(\frac{\partial^2 Q}{\partial u^2}\right)\left(\frac{\partial^2 Q}{\partial y^2}\right) - \left(\frac{\partial^2 Q}{\partial u\, \partial y}\right)^2 < 0$$

the point is neither a maximum nor a minimum but may be what is known as a saddle point. Incidentally, most practical problems involving a stationary point in optimum mechanical design as encountered by the author in over fifteen years of industrial practice can be analyzed conclusively without application of the tests of equations (8.64) and (8.65).

Finally, it should be mentioned that on very rare occasions the maximum or minimum will occur at an interior point which is not a stationary point, where one or more of the first partial derivatives are discontinuous.

Example 8-7 As a specific example consider the following criterion function.

$$U = \frac{21.9 \times 10^7}{E^2 C} + 3.9 \times 10^6 C + 10^3 E \qquad \text{(I)} \qquad (8.66)$$

Equation (8.66) expresses annual operating cost U of an electrical transmission line system being designed, in terms of line voltage E and line conductance C. The total problem is described on pp. 75–79 and 104–107 of reference 8-13, which the interested reader might wish to review, although this is unnecessary for the purpose of the present example. The objective for optimum design is minimization of annual operating cost U, which is to be achieved by proper selection of E and C values.

Equation (8.66) is the developed (P.D.E.) of an ideal problem, (I), and we wish to investigate the variation of U with respect to E and C. An ideal optimum design point i will exist where $\partial U/\partial E = 0$ and $\partial U/\partial C = 0$, as depicted in Figure 8.27. From equation (8.66) it is obvious that this is a minimum point for U, since for $C = C_i$ we see that $U \to \infty$ for $E \to 0$ and $E \to \infty$, and for $E = E_i$ we see that $U \to \infty$ for $C \to 0$ and $C \to \infty$. However, we would also find that the tests of equations (8.64) are satisfied if applied, and this calculation is presented on p. 107 of reference 8.13.

For determination of the coordinate values at point i of Figure 8.27, we apply the necessary conditions for this stationary point to equation (8.66). Thus,

$$\frac{\partial U}{\partial E} = -\frac{43.8 \times 10^7}{E^3 C} + 10^3 = 0$$

$$\frac{\partial U}{\partial C} = -\frac{21.9 \times 10^7}{E^2 C^2} + 3.9 \times 10^6 = 0$$

which when solved simultaneously give E_i and C_i as follows:

$$E_i^3 C_i = 43.8 \times 10^4$$

and

$$E_i^2 C_i^2 = \frac{21.9 \times 10^7}{3.9 \times 10^6} = 56.2$$

Thus,

$$C_i = \frac{\sqrt{56.2}}{E_i} = \frac{7.5}{E_i}$$

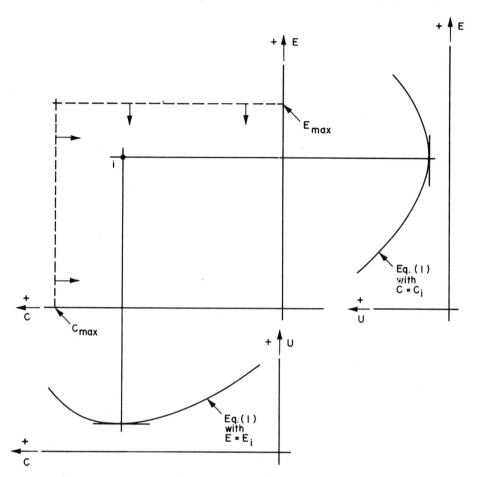

Figure 8.27 Typical three-dimensional variation diagram for the ideal problem in the transmission line design of example 8-7. Ideal point *i* of the optimum design is a stationary point. (Criterion function I is the developed (P.D.E.) of the ideal problem, equation (8.66).)

Therefore

$$E_i^3 C_i = E_i^3 \left(\frac{7.5}{E_i}\right) = 43.8 \times 10^4$$

$$E_i^2 = 5.84 \times 10^4$$

$$E_i = 242 \text{ kilovolts}$$

and

$$C_i = \frac{7.5}{E_i} = \frac{7.5}{242} = 0.0310 \text{ mho}$$

Hence, in Figure 8.27 the values of E and C at point i are

$$E_i = 242 \text{ kV}$$

$$C_i = 0.0310 \text{ mho}$$

We would choose these values for E and C providing that the accessibility of this point i is not excluded because of independent constraints, such as E_{max} and C_{max} in the variation diagram.

For avoidance of some other unacceptable condition in the transmission line system, the effects of appropriate limits would have to be imposed on the (E,C) plane of Figure 8.27.   For example, from p. 104 of reference 8.13, we might wish to avoid excessive corona loss by satisfaction of the regional constraint

$$\phi_1 \geq 0 \qquad \text{(L.E.)} \quad (8.67)$$

where we have the relating equation (II) defining ϕ_1 as a function of E and C.

$$\phi_1 = 63.5 \, C^{1/2} \, (8.52 - \ln C) - E \quad \text{(II)} \qquad (8.68)$$

As an exercise the interested reader might wish to pursue this variation study further by the method of optimum design, from the final formulation system of equations (8.66) and (8.68), adhering to the constraints of (L.E.) (8.67). If done, you will find that point i of Figure 8-27 is excluded because of (L.E.) (8.67).   Also, you will find that the optimum design is on the boundary curve defined by relating equation (8.68) with $\phi_1 = 0$, and the coordinate values at the optimum design point are $E = 167$ kV and $C = 0.0525$ mho.   This we would choose as the optimum design, providing that we did not violate the acceptable values for E_{max} and C_{max}.   If this were the case, we would have to include the constraints $E \leq E_{max}$ and $C \leq C_{max}$ in our optimum design variation study.   Incidentally, solution to the extended problem adhering to $\phi_1 \geq 0$ is presented as example 8-11 using the digital computer.

8-4 METHOD OF LINEAR PROGRAMMING

The method of linear programming is applicable to a certain type of optimization problem characterized by a system of linear equations with certain types of regional constraints. Applications are particularly suited to some problems of optimization with respect to system operations, generally with the objective of either minimizing cost or maximizing profit. On the other hand, applications in mechanical design synthesis are not so prevalent because of nonlinearities, certain types of functional constraints, and the existence of discrete value constraints in the equation system of the initial formulation.

For some nonlinear problems with regional constraints, by transformation an equivalent system of linear equations can be written to which the techniques of linear programming can be applied. As a specific example, a nonlinear equation system solved by the method of optimum design in Chapter 7 of reference 8-1 is transformed and solved by the linear programming technique on pp. 26–19 through 26–21 of reference 8-24. Conversely, many problems which are normally categorized in the linear programming classification often can be solved by the more general method of optimum design. This often provides a more vivid insight to the decisions reached.

Techniques of linear programming are well covered in the literature, such as in reference 8-16. Excellent summaries of significant aspects can also be found in the literature, such as in reference 8-24 and pp. 88–102 of reference 8-13. The latter reference briefly outlines the simplex method of linear programming, which is an algorithm consisting of a routine series of selections, calculations, and tests suitable for solution with the digital computer.

In view of its excellent coverage in other sources, and the fact that there are relatively few applications for its use in the work of mechanical design synthesis, space will not be devoted to further explanation of the linear programming technique in this book. Instead, we will apply the more general technique of the *method of optimum design* to a problem which would normally be solved by linear programming. This will illustrate what was previously discussed, and it will present another introductory example for the method of optimum design which should be of value to those who are learning the technique. Hence, the presentation of example 8-8 is not meant to belittle the linear programming technique in any way, which has many places for appropriate application, particularly outside of the field of mechanical design synthesis.

***Example* 8-8**   An excellent illustrative example for the linear programming technique is the oil refinery problem on pp. 92–101 of reference 8-13. In that problem we wish to decide on the optimum amounts of crude oil to purchase as well as the products to manufacture in order to maximize the profits from operation of our plant. As is typically the case, limitations must be adhered to on available raw materials, manufacturing capabilities, and output production designated by the ϕ_s below. In what follows the Zs represent barrels of various oils associated with the processing plant, and we wish to determine the optimum values of the Zs which will maximize profit U. The Z values which we choose cannot be negative. Solution of the optimization problem by the simplex method of linear programming is presented in detail in reference 8-13. We will now solve the same problem by the method of optimum design.

The system of equations for the *initial formulation* is summarized as follows, taken from p. 93 of reference 8-13.

$$U = -1.8\,Z_1 - 2.0\,Z_2 - 0.2\,Z_3 - 0.3\,Z_4 + 4\,Z_5 + 5\,Z_6 + 6\,Z_7$$
$$\text{(P.D.E.) (8.69)}$$

$$\phi_1 = Z_1 \qquad\qquad \text{(S.D.E.) (8.70)}$$

$$\phi_2 = Z_2 \qquad\qquad \text{(S.D.E.) (8.71)}$$

$$\phi_3 = Z_1 + Z_2 \qquad\qquad \text{(S.D.E.) (8.72)}$$

$$\phi_4 = -0.5\,Z_1 - 0.5\,Z_2 + Z_3 \qquad\qquad \text{(S.D.E.) (8.73)}$$

$$\phi_5 = -0.3\,Z_1 - 0.2\,Z_2 - 0.4\,Z_3 + Z_4 + Z_5 \qquad \text{(S.D.E.) (8.74)}$$

$$\phi_6 = Z_3 + Z_4 \qquad\qquad \text{(S.D.E.) (8.75)}$$

$$\phi_7 = -0.1\,Z_1 - 0.2\,Z_2 - 0.2\,Z_3 - 0.5\,Z_4 + Z_6 \quad \text{(S.D.E.) (8.76)}$$

$$\phi_8 = -0.1\,Z_1 - 0.1\,Z_2 - 0.3\,Z_3 - 0.4\,Z_4 + Z_7 \quad \text{(S.D.E.) (8.77)}$$

$$\left.\begin{array}{l} \phi_1 \le 10{,}000;\ \phi_2 \le 6000; \\[4pt] \phi_3 \le 8000;\ \phi_4 \le 0; \\[4pt] \phi_5 \le 0;\ \phi_6 \le 2500; \\[4pt] \phi_7 \le 0;\ \phi_8 \le 0 \end{array}\right\} \qquad \text{(L.E.)s}$$

Assuming the Zs to be free variables we have in the initial formulation $n_f = 7 < N_s = 8$, which indicates a case of redundant specifications for this problem. In the final formulation, from equation (8.10) we see that there will be $N_{ff} = N_s - n_f + 1 = 8 - 7 + 1 = 2$ equations. The developed (P.D.E.) of the ideal problem will express optimization quantity U in terms of the related parameters, and a single relating equation will express an eliminated parameter in terms of the related parameters. Upon inspection of the initial formulation, we choose the approach of selecting ϕ_3 as the eliminated parameter. We see that the variation study should be fairly simple by this approach, since from (S.D.E.)s (8.70), (8.71), and (8.72) our relating equation will be

$$\phi_3 = \phi_1 + \phi_2 \qquad \text{(II)} \qquad\qquad (8.78)$$

Thus, there will be only two related parameters for the variation study, which will be three-dimensional in degree and easy to handle.

In derivation of the developed (P.D.E.) of the ideal problem, we will combine all of the (S.D.E.)s except (8.72) with the (P.D.E.) (8.69) by eliminating Z_1 through Z_7. This is quite simple if we start with (S.D.E.)s (8.77) and (8.76) eliminating Z_7 and Z_6, and progressing to (S.D.E.)s (8.74), (8.75), (8.73), (8.71), and (8.70) in that order, eliminating Z_7, Z_6, Z_5, Z_4, Z_3, Z_2, and Z_1, respectively, in that order.   Thus, the developed (P.D.E.) of the ideal problem is derived as equation (8.79).   The final formulation summary is as follows.   Again note the basic format previously described, which is suitably arranged for the variation study to follow.

Final Formulation

$$U = 2.3 \, \phi_1 + 2.2 \, \phi_2 + 3.6 \, \phi_4 + 4 \, \phi_5 + 0.6 \, \phi_6 + 5 \, \phi_7 + 6 \, \phi_8 \qquad \text{(I)}$$

(8.79)

$$\phi_3 = \phi_1 + \phi_2 \qquad \text{(II)} \qquad\qquad \text{(8.78)}$$

where

$\phi_4, \phi_5, \phi_6, \phi_7, \phi_8 =$ truly independent parameters

$\phi_3 =$ eliminated parameter

ϕ_1 and $\phi_2 =$ related parameters

(I) = developed (P.D.E.) of ideal problem

(II) = relating equation

With respect to the truly independent parameters in the final formulation, to maximize U in equation (8.79) we wish to choose ϕ_4, ϕ_5, ϕ_6, ϕ_7, and ϕ_8 as large as possible referring to the limits of the initial formulation.   With respect to the optimum values for the eliminated and related parameters ϕ_1, ϕ_2, and ϕ_3, we consider the three-dimensional variation diagram in Figure 8.28 which is drawn directly from the final formulation and limit values of the initial formulation.   Referring to that figure, any point P within the domain of feasible designs can be improved for maximization of U by moving to the boundary line of relating equation (II) with $\phi_3 = 8000$.   However, along this boundary, from the final formulation we note that

$$U \sim 2.3 \, \phi_1 + 2.2(8000 - \phi_1) + C$$
$$= 0.1 \, \phi_1 + C$$

which reveals that U increases as ϕ_1 increases along that boundary line. Therefore, in Figure 8.28 the point of optimum design for maximization of U is designated as A, with $\phi_1 = 8000$ and $\phi_2 = 0$ (corresponding, from the initial formulation, to $Z_1 = 8000$ and $Z_2 = 0$).

As previously described, the values for the truly independent parameters ϕ_4, ϕ_5, ϕ_6, ϕ_7, and ϕ_8 are selected as large as possible, the desirability of which we see from equation (8.79) of the final formulation, keeping in mind that the corresponding Z values must be positive or zero and adhering to the limits of the initial formulation.   Hence, by reversing the equation-combination procedure previously described, we explicitly determine the optimum

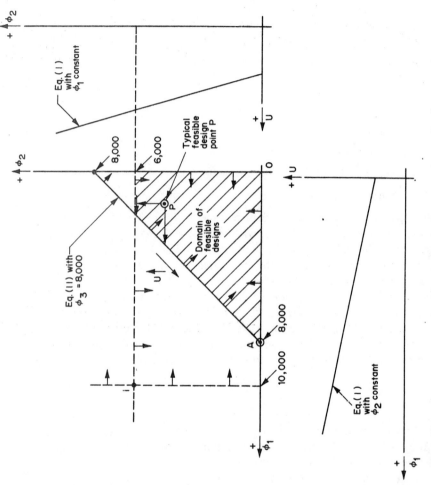

Figure 8.28   Three-dimensional variation diagram for example 8-8.   (See the final formulation for equations I and II.)

values for Z_3 through Z_7, which are: $Z_3 = 2500$, $Z_4 = 0$, $Z_5 = 3400$, $Z_6 = 1300$, and $Z_7 = 1550$. Substituting these values of Z together with the previously determined values $Z_1 = 8000$ and $Z_2 = 0$ in (P.D.E.) (8.69), we obtain the maximum profit achieved as $U = \$14,500$. The results obtained by the method of optimum design, of course, are in complete agreement with those determined by the simplex method of linear programming in reference 8-13. As an exercise the interested reader may wish to carry through the details of equation combination, variation study, and optimum Z value determination as now outlined. If this is done, an interesting decision which you should conclusively reach is that $Z_3 = 2500$ and $Z_4 = 0$ are optimum.

8-5   METHOD OF LAGRANGIAN MULTIPLIERS

The method of Lagrangian multipliers is a technique for finding the coordinates of an ideal optimum design point in some problems. It is applicable to a system of equations consisting of a criterion function

$$Q = f(x_1, x_2, \ldots, x_n) \qquad \text{(P.D.E.) (8.80)}$$

and functional constraints m in number as follows

$$F_1 = f_1(x_1, x_2, \ldots, x_n) = 0$$
$$F_2 = f_2(x_1, x_2, \ldots, x_n) = 0 \qquad \text{(S.D.E.)s (8.81)}$$
$$\vdots \qquad \qquad \vdots$$
$$F_m = f_m(x_1, x_2, \ldots, x_n) = 0$$

It will give the coordinate values of x_1, x_2, $\ldots$, x_n for the point at which differential Q is zero. It assumes that x_1, x_2, $\ldots$, x_n are independent variables which are unconstrained. Hence, it partially solves some realistic optimization problems in that it gives the coordinate values for an ideal optimum design point where $dQ = 0$. However, as illustrated in previous examples, for most realistic problems this is only a small part of the total optimization problem. In fact, for all of the possible points of optimum design in examples 8-1 through 8-6, the method of Lagrangian multipliers would give only point 0 in Figure 8.7 of example 8.2. It would also give point 4 in Figure 8.19 of example 8-5, but this is the poorest design along the boundary curve of the figure and not the best. Therefore, from the discussion now presented we see that the method of Lagrangian multipliers is highly limited in practical problems of optimum design, and for this reason we will not devote much space to the subject. Before presenting some illustrative examples, let us consider the basic theory behind the Lagrangian multiplier technique.

We wish to find the x_k values, where $k = 1, \ldots, n$, for which $dQ = 0$ in the

(P.D.E.) (8.80) while satisfying the functional constraints of the (S.D.E.)s (8.81). Hence, by applying the theory of differentiation to the given system of equations, we must satisfy

$$dQ = \sum_{k=1}^{n} \frac{\partial Q}{\partial x_k} dx_k = 0$$

$$dF_1 = \sum_{k=1}^{n} \frac{\partial F_1}{\partial x_k} dx_k = 0$$

$$dF_2 = \sum_{k=1}^{n} \frac{\partial F_2}{\partial x_k} dx_k = 0$$

$$\vdots \qquad \qquad \vdots$$

$$dF_m = \sum_{k=1}^{n} \frac{\partial F_m}{\partial x_k} dx_k = 0$$

If for *each* of the above equations except the first we multiply through by *an* undetermined parameter called a Lagrangian multiplier, $\lambda_1, \lambda_2, \ldots, \lambda_m$, we obtain the following.

$$\lambda_1 \, dF_1 = \sum_{k=1}^{n} \lambda_1 \frac{\partial F_1}{\partial x_k} dx_k = 0$$

$$\lambda_2 \, dF_2 = \sum_{k=1}^{n} \lambda_2 \frac{\partial F_2}{\partial x_k} dx_k = 0$$

$$\vdots \qquad \qquad \vdots$$

$$\lambda_m \, dF_m = \sum_{k=1}^{n} \lambda_m \frac{\partial F_m}{\partial x_k} dx_k = 0$$

Adding equations now derived we obtain

$$dQ + \lambda_1 \, dF_1 + \lambda_2 \, dF_2 + \cdots + \lambda_m \, dF_m = 0$$

Thus,

$$\sum_{k=1}^{n} \left(\frac{\partial Q}{\partial x_k} + \lambda_1 \frac{\partial F_1}{\partial x_k} + \lambda_2 \frac{\partial F_2}{\partial x_k} + \cdots + \lambda_m \frac{\partial F_m}{\partial x_k} \right) dx_k = 0$$

However, since the $x_k's$ are independent of each other we must in effect satisfy n equations of the form

$$\frac{\partial Q}{\partial x_k} + \lambda_1 \frac{\partial F_1}{\partial x_k} + \lambda_2 \frac{\partial F_2}{\partial x_k} + \cdots + \lambda_m \frac{\partial F_m}{\partial x_k} = 0 \qquad (8.82)$$

for $k = 1, 2, \ldots, n$.

Also, there are m functional constraint equations which are the original (S.D.E.)s (8.81) which must be satisfied. Hence, we have $(m + n)$ equations available and there are $(m + n)$ unknowns, which are the nx_k's $(x_1, x_2, \ldots, x_n)$, and the $m\lambda$'s $(\lambda_1, ., \lambda_m)$.   Therefore, simultaneous solution of the $(m + n)$ equations (8.81) and (8.82) gives values for all of the $(m + n)$ unknowns $x_1, x_2, \ldots, x_n$ and $\lambda_1, \lambda_2, \ldots, \lambda_m$.   Thus, by this method of Lagrangian multi-pliers we have found the x_k values for which $dQ = 0$ while satisfying the given system of equations.   As previously mentioned the optimum design is seldom at such points in realistic problems generally because of regional constraints, as typically illustrated in examples 8-1 through 8-8.   However, let us con-sider some specific examples of the Lagrangian multiplier technique.

Example 8-9   (a): *Ideal Optimum Design for Fenced Yard Problem.*   Refer-ring to the fenced region design problem of example 8-2, we will find point 0 in Figure 8.7 by the method of Lagrangian multipliers.   Our equation system is

$$A = xy \qquad \text{(P.D.E.) (8.22)}$$

$$L = 2(x + y) \qquad \text{(S.D.E.) (8.23)}$$

We wish to find the values of x and y for which A is maximized and for which $dA = 0$, with L a specified constant.   Thus, by the theory of differentiation applied to (P.D.E.) (8.22), we have

$$dA = \frac{\partial A}{\partial x}\, dx + \frac{\partial A}{\partial y}\, dy = 0$$

$$y\, dx + x\, dy = 0 \qquad (8.22a)$$

Also, the functional constraint of (S.D.E.) (8.23) can be expressed as follows, using the notation F of (8.81):

$$F = 2(x + y) - L = 0$$

To this equation we apply the theory of differentiation and multiply through by the undetermined parameter λ to obtain

$$\lambda\, dF = \lambda \frac{\partial F}{\partial x}\, dx + \lambda \frac{\partial F}{\partial y}\, dy = 0$$

Therefore

$$\lambda\, dx + \lambda\, dy = 0 \qquad (8.23a)$$

Hence, we add equations (8.22a) and (8.23a) to obtain

$$(y + \lambda)\, dx + (x + \lambda)\, dy = 0$$

from which we conclude the following, by equating the coefficients of dx and dy to zero.

$$y + \lambda = 0 \qquad (8.83)$$

$$x + \lambda = 0 \qquad (8.84)$$

The simultaneous solution of equations (8.83), (8.84), and (8.23) will give the values of λ, x, and y for which $dA = 0$. Thus, we obtain

$$x + y - (L/2) = 0$$
$$- \lambda - \lambda - (L/2) = 0$$
$$\lambda = - L/4$$
$$x = y = L/4$$

which agrees with what we determined in example 8-2 by equation combination of (8.23) with (8.22) and simple differentiation, giving point 0 in Figure 8.7. In that figure we see that point 1 is also possible for optimum design. Thus, as is generally the case, for the relatively few problems where the method of Lagrangian multipliers can be applied it gives only an ideal point solution which is only part of the variation study for optimum design. Also, such ideal points of optimum design generally can be determined in a simpler fashion by equation combination and setting a derivative equal to zero, as illustrated in example 8-2.

Example 8-9 (b): *Worst Boundary Point for Problem of Maximizing Heat Flow Rate.* In example 8-5 we found that point 4 of Figure 8.19 was the poorest design along the boundary of relating equation (II) with $Q_s = (Q_s)_{max}$. The system of equations are, referring to (8.41a), (8.42a), and (8.46)

$$q = C_1 \left[\pi DII + \frac{\pi D^2}{4} \right] \qquad \text{(P.D.E.) (8.41b)}$$

and, in the form of (8.81) we have,

$$F = C_2 D^2 H - (Q_s)_{max} = 0 \qquad \text{(S.D.E.) (8.42b)}$$

We wish to find the values of D and H for which $dq = 0$ from (P.D.E.) (8.41b), which also satisfies (S.D.E.) (8.42b). C_1, C_2, and $(Q_s)_{max}$ are known constants, C_1 and C_2 having previously been defined by equations (8.48) and (8.49). This will give us minimum q point 4 in Figure 8.19, which is the poorest design along the boundary curve in the figure since we wish actually to maximize q in the optimization problem.

By the Lagrangian multiplier technique, from (P.D.E.) (8.41b) and (S.D.E.) (8.42b), we obtain, as previously described for the method

$$dq = \frac{\partial q}{\partial D} dD + \frac{\partial q}{\partial H} dH$$

$$= C_1 \left[\left(\pi H + \frac{\pi}{2} D \right) dD + (\pi D) \, dH \right]$$

$$= 0$$

$$\lambda \, dF = \lambda \frac{\partial F}{\partial D} dD + \lambda \frac{\partial F}{\partial H} dH$$

$$= \lambda 2 C_2 \, DH \, dD + \lambda C_2 \, D^2 \, dH$$

$$= 0$$

Adding these equations, and separately equating the coefficients of dD and dH to zero we obtain

$$\pi \left(H + \frac{D}{2} \right) + 2C_2 \, DH\lambda = 0 \tag{8.85}$$

$$\pi D + C_2 \, D^2 \lambda = 0 \tag{8.86}$$

Thus, the simultaneous solution of equations (8.85), (8.86), and (8.42b) will give the values of λ, D, and H for which $dq = 0$. Therefore, we obtain

$$C_2 \lambda = \frac{-\pi \left(H + \frac{D}{2} \right)}{2DH} = -\frac{\pi D}{D^2}$$

$$\left(H + \frac{D}{2} \right) = 2H$$

Therefore

$$H = \frac{D}{2}$$

$$C_2 \, D^2 H - (Q_s)_{max} = C_2 \, D^2 \left(\frac{D}{2} \right) - (Q_s)_{max} = 0$$

and

$$D = [2(Q_s)_{max}/C_2]^{1/3}$$

This is in complete agreement with equation (8.51) derived in example 8-5 by the simple equation-combination procedure and setting $dq/dD = 0$.

As shown in the variation diagram in Figure 8.19 of example 8-5, the points possible for optimum design are 1, 2, and 3. The method of Lagrangian multipliers gave us point 4, which is the poorest design along the boundary curve of the figure, for the particular problem of maximizing heat flow rate q in example 8-5. Hence, we should be well aware of the fact that the method of Lagrangian multipliers may give a point which is far from optimum. In general, the technique merely gives us the point where $dQ = 0$ from (P.D.E.) (8.80), which also satisfies the functional constraints of (S.D.E.)s (8.81), and the point may or may not be an ideal optimum design point depending upon what the objective is for optimization.

8-6   METHOD OF DUAL VARIABLES

The method of dual variables presented in reference 8.18 is applicable to the case of a single criterion function of the equation (8.87) form for finding the point of *minimum* Q value of the $dQ = 0$ type, such as is occasionally encountered as an ideal point possible for optimum design in some problems.

$$Q = f_1 + f_2 + \cdots + f_n \qquad (8.87)$$

In equation (8.87), $f_1, f_2, \ldots, f_n$ are positive simple exponential functions of single variable t whose optimum value is sought.

The solution is obtained by defining dual variables $\delta_1, \delta_2, \ldots, \delta_n$ and $x_1, x_2 \ldots x_n$ such that $f_i = \delta_i x_i$ for $i = 1, 2, \ldots, n$. These dual variables are substituted in equation (8.87) as follows. It can be shown mathematically that the following inequality will always be true for this equation (8.88).

$$Q = \delta_1 x_1 + \delta_2 x_2 + \cdots + \delta_n x_n \geq x_1^{\delta_1} x_2^{\delta_2} \cdots x_n^{\delta_n} \qquad (8.88)$$

The minimum value of Q occurs for the case of the equality in the right side of equation (8.88), and this requires that the x_i values be equal *and* that $\delta_1 + \delta_2 + \cdots + \delta_n = 1$. Thus, in effect we must satisfy the following set of equations, using $x_i = f_i/\delta_i$ in accordance with the definition of the dual variables.

$$(f_1/\delta_1) = (f_2/\delta_2)$$
$$(f_1/\delta_1) = (f_3/\delta_3) \qquad (8.89)$$
$$\vdots \qquad \vdots$$
$$(f_1/\delta_1) = (f_n/\delta_n)$$
$$\delta_1 + \delta_2 + \cdots + \delta_n = 1 \qquad (8.90)$$

We note that there are n equations in the (8.89) and (8.90) set with $(n + 1)$ unknowns, which are $\delta_1, \delta_2, \ldots, \delta_n$ *and* t, since all of the f functions are of *t* as previously stated. Therefore, since the number of unknowns exceeds the number of equations by one we must derive one more relationship between these unknowns in order to determine the point sought where $dQ = 0$. Thus, using the equality sign in equation (8.88) and the relations $x_i = f_i/\delta_i$, we obtain equation (8.91) for what we might call a dual function $Q(\delta)$, which conforms to Q in the immediate neighborhood of the point sought. In equation (8.91) we can consider $\delta_1, \delta_2, \ldots, \delta_n$ as undetermined constants of proper value for the solution point sought.

$$Q(\delta) = \left(\frac{f_1}{\delta_1}\right)^{\delta_1} \left(\frac{f_2}{\delta_2}\right)^{\delta_2} \cdots \left(\frac{f_n}{\delta_n}\right)^{\delta_n} \tag{8.91}$$

The additional relationship is derived from (8.91) recognizing that for the point sought $dQ = 0$. Thus, since $f_1, f_2, \ldots, f_n$ are simple exponential functions of *t*, we can collect the exponents of *t* which together must have the sum of zero for $dQ = 0$, thereby giving the additional relationship required which is between $\delta_1, \delta_2, \ldots, \delta_n$. The technique is best illustrated in a specific example, which we will consider next.

The method of dual variables would not be applicable for finding a possible point of optimum design in any of the typical optimization problems so far presented in examples 8-1 through 8-9. The point 0 possibility in Figure 8.7 of example 8-2 is one of the $dA = 0$ type, but since it is one for maximum A the method of dual variables would not be applicable. On the other hand, point 4 in Figure 8.19 of example 8-5 can be determined by this method, although it represents the worst design along the boundary curve of that figure. However, we will present its solution next as an illustrative example of the basic technique.

Example 8-10   In example 8-5 we found that point 4 of Figure 8.19 was the poorest design along the relating equation (II) with $Q_s = (Q_s)_{max}$, which is equation (8.50).

$$q = \pi C_1 \left[\frac{(Q_s)_{max}}{C_2 D} + \frac{D^2}{4}\right] \tag{8.50}$$

We wish to find the value of D for which $dq = 0$, and since this is a point of minimum q the method of dual variables should work. To begin, let us simplify the notation of equation (8.50) as follows:

$$q = \frac{C_3}{D} + C_4 D^2 \tag{8.50a}$$

where

$$C_3 = \pi C_1 (Q_s)_{max}/C_2$$

and

$$C_4 = \pi C_1/4$$

Applying the described method of dual variables to (8.50a), we have specific variable D instead of general variable t, and

$$f_1 = C_3/D$$
$$f_2 = C_4 D^2$$

Thus, applying equation (8.91) to the problem, we obtain for the dual function

$$q(\delta) = \left(\frac{f_1}{\delta_1}\right)^{\delta_1}\left(\frac{f_2}{\delta_2}\right)^{\delta_2} = \left(\frac{C_3}{D\,\delta_1}\right)^{\delta_1}\left(\frac{C_4 D^2}{\delta_2}\right)^{\delta_2}$$

$$= \left(\frac{C_3}{\delta_1}\right)^{\delta_1}\left(\frac{C_4}{\delta_2}\right)^{\delta_2}(D)^{(2\delta_2 - \delta_1)}$$

The sought point of minimum q requires $dq = 0$, and thus it can be shown that the exponent of D above must be zero at that point. Hence, as one equation, we must satisfy the relation (8.92) for the sought point.

$$2\delta_2 - \delta_1 = 0 \qquad\qquad (8.92)$$

Also, applying equations (8.89) and (8.90) to the problem we must satisfy, respectively, (8.93) and (8.94) for determination of D_4, at the sought point 4 of Figure 8.19.

$$(f_1/\delta_1) = (f_2/\delta_2)$$

$$\frac{C_3}{D_4\,\delta_1} = \frac{C_4 D_4{}^2}{\delta_2}$$

$$D_4 = [(C_3/C_4)(\delta_2/\delta_1)]^{1/3} \qquad\qquad (8.93)$$

$$\delta_1 + \delta_2 = 1 \qquad\qquad (8.94)$$

Thus, from (8.92) and (8.94) we determine

$$2\delta_2 - \delta_1 = 2\delta_2 - (1 - \delta_2) = 3\delta_2 - 1 = 0$$

Therefore

$$\delta_2 = 1/3$$

and

$$\delta_1 = 2\delta_2 = 2/3$$

Hence, from equation (8.93) we obtain

$$D_4 = [(C_3/C_4)(\delta_2/\delta_1)]^{1/3}$$

$$= \left[\left(\frac{\pi C_1 (Q_s)_{max}}{C_2 \, \pi C_1/4} \right) \left(\frac{1/3}{2/3} \right) \right]^{1/3}$$

$$D_4 = [2(Q_s)_{max}/C_2]^{1/3} \tag{8.51}$$

and we note that this is equation (8.51) determined previously by simple differentiation in example 8-5, as well as by the method of Lagrangian multipliers in example 8-9(b).

8-7 DIGITAL COMPUTERS IN OPTIMUM DESIGN

Digital computers can be of great value in the work of optimum design, and they should be used appropriately for making routine calculations as part of the solution procedure in optimization problems. Digital computer techniques applicable in such problems can be classified into two categories. The first we will call *exact techniques*, and the second we will designate as *iterative searching techniques*. Both have their place in the work of optimum design, and we will briefly discuss each. However, before doing this we will present an analogy which depicts the difference between the two types of computational techniques.

Suppose that you were interested in making a trip by automobile from locality A to locality B, assumed to be widely separated regions in the United States. Your course of action could follow one of several plans. First, if you had no idea where B was located, you could set up an exhaustive searching pattern covering the entire United States. By systematically covering this pattern you eventually would reach locality B, but the time and effort expended might be tremendous. This strategy would be analogous to an exhaustive searching plan in computational work, which we could classify as the lowest order of the iterative searching technique. Secondly, if you had a knowledge of the location of B relative to A, but no road map, you could with navigational instruments start in the right direction. Then you could stop at appropriate junctions along the way, determine your bearings, and make the necessary corrections in your direction, eventually reaching location B. This would be analogous to a sequential searching plan in computational work, which we could classify as the highest order of iterative searching technique. Thirdly, if you had a good road map available, you could make the trip directly from A to B in a relatively rapid and efficient procedure. This would be analogous to the employment of exact techniques in computational work of optimization. In the computational work of examples 8-6,

9-1, 9-2, and 9-3 this is the approach which we have taken, with the road map or general flow chart being explicitly derived by the method of optimum design. These examples in the text illustrate this new technique, which is relatively unknown at the present time and which has virtually unlimited application in realistic problems of optimum design. On the other hand, the iterative searching techniques are relatively well known and available in the literature, and they would be used only if one could not obtain a good road map. We would not normally force the strategy of an inefficient exhaustive searching plan on a problem of optimization for which a good road map of general value can be readily derived.

Exact Techniques

Exact computational techniques of optimization can be logically programmed into a sequence of calculations and comparisons which explicitly lead to the exact point of optimum design. Digital computer programs for the exact techniques are generally very rapid to execute and highly efficient. An example of an exact computational technique would be a digital computer program written for executing the simplex method of linear programming previously mentioned in section 8-4. The routine calculations, inspections, and comparisons described on pp. 93–101 in reference 8-13 would be programmed on the digital computer which then would rapidly lead to the exact optimum design solution. Another categorical example of an exact computational technique would be a computer program written from an optimum design flow chart as derived by the method of optimum design. Specifically this is illustrated in Figure 8.25, which is derived from the optimum design variation study of example 8-6. The computer programs which are presented in Chapter 9 provide other examples. Hence, we will not present an illustrative example of exact optimization techniques for digital computers in this section. The interested reader can study the cited examples, and if practice is desired he can write the digital computer programs for previous examples from the already derived optimum design flow charts of Figures 8.8, 8.12, 8.16, and 8.20. These are similar to Figure 8.25, which was written from the flow chart of Figure 8.24. Numerical examples, similar to the one in example 8-6, can then be carried through if a digital computer is available. If this is done it is suggested that several feasible materials of known properties be assumed as available in the calculations associated with Figures 8.12 and 8.20, handled in the way described in examples 8-3 and 8-5, respectively. Incidentally, the handling of several feasible materials in optimum design is specifically illustrated in the numerical problem calculations presented for example 9-1 in Chapter 9.

Iterative Searching Techniques

Iterative searching computational techniques of optimization are of the simultaneous or sequential types, and a good reference in this respect is 8-19 of the bibliography.   In the simultaneous approach an exhaustive search is made over the entire domain of feasible designs to determine which point is most favorable.   For example, suppose that we wished to maximize optimization quantity Q in developed (P.D.E.) (8.8), by the proper choice of u and y values.

$$Q = f_4(u, y) \qquad (8.8)$$

Suppose that the acceptable ranges of variation for u and y are estimated as follows.

$$u_{min} \leq u \leq u_{max} \qquad (8.95)$$

$$y_{min} \leq y \leq y_{max} \qquad (8.96)$$

Hence, as illustrated in Figure 8-29, for an exhaustive search approach or simultaneous search plan we would calculate Q values at each net point (u_i, y_j) over the entire domain of feasible designs.   We would then compare

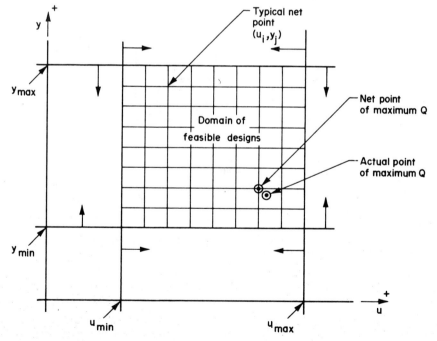

Figure 8.29   Exhaustive search net in the domain of feasible designs.

the calculated Q values and select the greatest one, which would be for the optimum net point. Actually this approach would give us a point in the neighborhood of the optimum design, which we would assume to be close enough for all practical purposes.

In a sequential search plan we use the information from one calculation to control the next calculation, so as to reduce the effort expended in an iterative procedure for converging to the optimum design. For example, we might effectively combine the simultaneous and sequential search plans as follows for the previous problem of equations (8.8), (8.95), and (8.96). First, divide the domain of feasible designs into a coarse net and calculate Q values to determine the approximate location, P_1, of the optimization point as shown in Figure 8.30. Then, divide the neighborhood of P_1 into

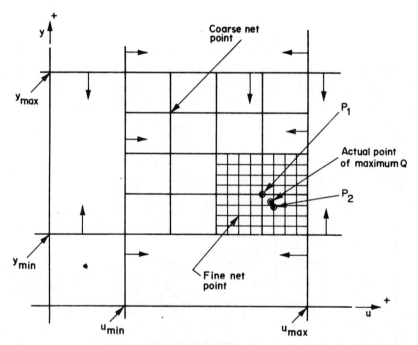

Figure 8.30 Combined simultaneous and sequential search plan with coarse and fine nets in the domain of feasible designs.

a fine net and calculate Q values to determine more accurately P_2 as the location of the optimization point, as shown in Figure 8.30. Hence, by progressing from coarse to fine nets we can accurately and efficiently determine the optimization point in a combined simultaneous and sequential search plan. It is needless to say that the use of a digital computer is a practical necessity in such work, since generally many routine calculations are required before the optimization point can be accurately located.

Various other simultaneous and sequential search plans are explained in references 8-19 and 8-20. They include the techniques of interval halving, Golden section search, multidimensional searches, method of steepest descent, sequential linear search, area elimination, and gradient search, pattern search, and grid search in N-space. These iterative searching techniques can be of value in some problems of optimum design, particularly if analysis of the criterion function is difficult because of its complexity. As a specific example let us consider a sequential searching plan based on the method of iteration as typically described in reference 8-20.

Example **8-11**   As an extension to example 8-7, consider the *final formulation* system of equations (8.66) and (8.68) as follows:

$$U = \frac{21.9 \times 10^7}{E^2 C} + 3.9 \times 10^6 C + 10^3 E \qquad \text{(I)} \qquad (8.66)$$

$$\phi_1 = 63.5\, C^{1/2}(8.52 - \ln C) - E \qquad \text{(II)} \qquad (8.68)$$

The objective for optimum design is minimization of annual operating cost U. We require $\phi_1 \geq 0$, and we wish to determine the optimum values for line voltage E and line conductance C. The ideal point for optimum design of Figure 8.27 was found in example 8-7 to be $E_i = 242$ kV and $C_i = 0.0310$ mho. Substitution of these values in equation (8-68) gives

$$(\phi_1)_i = 63.5(.0310)^{1/2}(8.52 - \ln 0.0310) - 242 = -107$$

which is unacceptable if we require $\phi_1 \geq 0$. Ignoring this constraint on ϕ_1 our typical three-dimensional variation diagram was as sketched in Figure 8.27.

With the knowledge now gained we realize that the domain of feasible designs is actually as sketched in Figure 8.31, by superimposing relating equation (8.68) with $\phi_1 \geq 0$ as required. Also, from equation (8.66) the contour curves of constant U are known in general shape as sketched in Figure 8.31. Obviously the actual optimum design will be along the boundary of (II) with $\phi_1 = 0$ in Figure 8.31, at a point which we designate as A, where $dU = 0$ along that boundary. Along that boundary curve of Figure 8.31 we determine the variation of U by combining equations (8.66) and (8.68) and eliminating E, thereby obtaining (8.97), as follows:

$$U = \frac{21.9 \times 10^7}{[63.5C^{1/2}(8.52 - \ln C)]^2 C} + 3.9 \times 10^6 C + 10^3 [(63.5C^{1/2}(8.52 - \ln C)]$$

Figure 8.31   Three-dimensional variation sketch for example 8-11. (See the final formulation, equations (8.66) and (8.68), for equations I and II, respectively.)

Thus,

$$U = \frac{54{,}300}{C^2(8.52 - \ln C)^2} + 3.9 \times 10^6 C + 63.5 \times 10^3 C^{1/2}(8.52 - \ln C) \quad (8.97)$$

Hence, equation (8.97) expresses the variation of U along the boundary curve (II) with $\phi_1 = 0$ in Figure 8.31, and from it we wish to determine a point A where $dU = 0$ along that boundary. This will be the overall optimum design of minimum U value, which we will determine next by the method of iteration using the digital computer.

From equation (8.97) by the Calculus we derive equation (8.98) for dU/dC.

$$\frac{dU}{dC} = \frac{108,600}{C^3(8.52 - \ln C)^2}\left[\frac{1}{(8.52 - \ln C)} - 1\right]$$

$$+ 3.9 \times 10^6 + \frac{63.5 \times 10^3}{C^{1/2}}\left[\frac{(8.52 - \ln C)}{2} - 1\right] \qquad (8.98)$$

Our solution point A of Figure 8.31 will be for $dU/dC = 0$, and with equation (8.98) set at this value we readily see that it is too complex for an explicit solution of the optimum C value. However, we recognize that the solution should be determinate by iteration, solving for C^3 in equation (8.98) with $dU/dC = 0$. Thus, we derive equation (8.99).

$$C_{k+1} = \left\{\frac{108,600\left[\frac{1}{(8.52 - \ln C_k)} - 1\right]}{-\left\{3.9 \times 10^6 + \frac{63.5 \times 10^3}{C_k^{1/2}}\left[\frac{(8.52 - \ln C_k)}{2} - 1\right]\right\}(8.52 - \ln C_k)^2}\right\}^{1/3}$$

$$(8.99)$$

As a starting value for C_k we will use the ideal optimum design coordinate value already calculated, i.e., $C = 0.0310$ mho, which is therefore at point P in Figure 8.31. With this initial value for C_k we can calculate C_{k+1} from equation (8.99). The procedure is repeated until C_{k+1} and C_k are sufficiently close, and based on our judgement we have selected the acceptable range for stopping the iterative calculations as

$$0.999 \le \left(\frac{C_{k+1}}{C_k}\right) \le 1.001 \qquad (8.100)$$

Therefore, having now determined the value of C at point A in Figure 8.31, we now calculate the corresponding value of E from equation (8.68) with $\phi_1 = 0$. Finally, we analyze the optimum design by equation (8.66) to determine what value of U has been achieved, and this will be the smallest possible value from any of the infinite number of feasible designs represented by points in the domain of acceptability in Figure 8.31.

The digital computer program and results obtained are given in Figure 8.32. Hence, we summarize the optimum design at point A in Figure 8.31

Figure 8.32 Iterative determination by digital computer of optimum design point A in Figure 8.31 of example 8-11. (a) Flow chart for iterative solution of equation (8.99). (b) Computer printout from program of part (a).

as follows:

$$C = 0.0525 \text{ mho}$$

$$E = 166.8 \text{ kV}$$

$$\phi_1 = 0$$

$$U = \$521{,}449 \text{ annual operating cost}$$

Incidentally, the annual operating cost for ideal point i of Figure 8.31 would be $U = \$484{,}000$, which is appreciably less than the above value for point A, but as previously determined at point i we would have $\phi_1 = -107$, which is unacceptable. Hence, point A summarized above is the optimum design

```
V=005      $JOB  THEMEOA1,0531RCJ,,059167940  OPTIMUM DESIGN PT. A IN FIG. 8.31
    1            CK = .0310
    2            I = 1
    3        10 A = 8.52 - ALOG(CK)
    4            B = 63.5E3/CK**.5
    5            D = 1.086E5*((1./A)-1.)
    6            CK1 = (D/(-(3.9E6 + B*(A/2 - 1.))*A**2.))**(1./3.)
    7            R = CK1/CK
    8            IF (R-.999) 20,40,40
    9        20 IF (I - 100) 30,50,50
   10        30 CK = CK1
   11            I = I + 1
   12            GO TO 10
   13        40 IF (R - 1.001) 50,50,20
   14        50 C = CK1
   15            PHI = 0.
   16            E = 63.5*C**.5*(8.52 - ALOG(C))
   17            U = 21.9E7/(E**2.*C) + 3.9E6*C + 1.E3*E
   18            WRITE (6,60) I,R,PHI,C,E
   19        60 FORMAT ('0','SOL. IS FOR I,R,PHI,C,E =',2X,I4,4E16.7/)
   20            WRITE (6,61) U
   21        61 FORMAT ('0',5X,'GIVING FOR PT. A MIN U =',2X,E16.7)
   22            STOP
   23            END

          $ENTRY

SOL. IS FOR I,R,PHI,C,E =      4   0.1000592E 01   0.0000000E 00   0.5248908E-01   0.1668260E 03

GIVING FOR PT. A MIN U =      0.5214493E 06

          COMPILATION:   CPU TIME= 0.7129 SECONDS, REAL TIME= 5 SECONDS,
          EXECUTION:     CPU TIME= 0.1742 SECONDS, REAL TIME= 1 SECONDS,
          CODE REQUIRED: OBJECT CODE= 1,272 BYTES, ARRAYS= 0 BYTES, UNUSED= 58,728 BYTES.
          I/O REQUESTS:  CARDS READ= 25, CARDS PUNCHED= 0, LINES PRINTED= 31.
```

Figure 8.32   (continued)

within the confines of all the listed constraints, and we assume that the determined values of E and C are acceptable for design.   If not, the constraints E_{max} and C_{max} would be imposed on the variation diagram of Figure 8.31, and based on an understanding of the variation study now made the decisions of optimum design would be straightforward.

Before leaving our brief discussion on digital computers in optimum design we should mention that there is much literature available as background material on the subject.   In addition to references 8-19 and 8-20 already cited, we suggest typically references 8-21, 8-22, and 8-23 as worthwhile for additional study.   Also, see Appendix C recently added.

Also, it should be mentioned that computational subprograms of both the exact and iterative searching techniques can be stored in the memory of a digital computer as subroutines subject to call when needed in the calculations of optimum design.   The procedure of writing a subroutine is recommended for computational techniques when general application is anticipated, and some illustrative examples of iterative searching techniques so used in optimum design are presented in the Iowa Cadet subroutines in Chapter 5 and Appendixes 2 and 3 of reference 8-19.   Since the subject is so well covered in the listed references, the material will not be repeated here for lack of space.   Instead, the examples of application for digital computers presented in this book will be primarily for exact computational techniques with programs written from flow charts explicitly derived by the method of optimum design.   Being a very new technique of optimization, computer programs so derived are not readily found elsewhere in the literature, and the basic technique will be illustrated specifically in examples 8-6, 9-1, 9-2, and 9-3. As previously mentioned, computer programs so written are generally very rapid to execute and highly efficient, being void of time-consuming iterative-type calculations of searching procedures. Also, being derived from general variation studies in the method of optimum design, computer programs so written are based on satisfaction of a total equation system of an initial formulation, which for a realistic problem includes a consideration of the criterion function, functional constraints, regional constraints, and discrete value constraints.

8-8 SUMMARY COMPARISON OF TECHNIQUES FOR OPTIMUM MECHANICAL DESIGN

Various techniques for optimum mechanical design have been briefly discussed in the preceding sections. They include the method of optimum design, the simplex method of linear programming, the method of Lagrangian multipliers, the method of dual variables, and the use of digital computers. All have their particular applications and limitations in the work of optimum design, and some of the characteristics of importance have been brought out in the illustrative examples. A brief summary of some of these characteristics is presented in Table 8-4 for purposes of comparing the applications and limitations of the techniques described in this chapter.

8-9 IMPROVEMENT BY MODIFICATION OF BOUNDARY VALUES

There are two types of situations encountered in optimum design work where it is necessary to modify the boundary values of the problem. For one, upon analyzing the optimum design initially determined we might find that the optimization quantity Q is not of acceptable value. In other words, although we have found the most favorable design possible within the constraints of the problem, the design achieved is not good enough. Hence, to improve the design further, we must relieve some of the boundary values or specifications. Thus, we face the problem of determining which boundary values should be relieved, and which of these would be most effective in improving the value of optimization quantity Q? The second type of situation occurs if we encounter a case of incompatible specifications. The problem which we then face is to determine which boundary values must be relieved and how much change should be made for each in order to give us a feasible design solution, which then will be also the optimum design. The techniques for handling the two types of situations now described are best explained by an illustrative example which follows.

TABLE 8-4 COMPARISON OF SOME TECHNIQUES FOR OPTIMUM MECHANICAL DESIGN

Technique	Refer to Section	Typical Equation System, Restrictions if any, and type of optimization problem	Comments
Method of optimum design	8-2 (also, see examples 8-1–8-8)	(1) Primary Design eq. (8.1) (2) Subsidiary design eqs. (8.2), (8.3), etc. (3) Regional and discrete value constraints, eqs. (8.4), (8.5), (8.6), etc. No restrictions in general; gives point of optimum design in real problems of many constraints; maximum or minimum Q	Possibilities are unlimited as long as functions can be combined and the variations determined as required in execution of the general method; other techniques listed may be used to solve part of problem, such as for an ideal point where $dQ=0$
Simplex method of linear programming	8-4 (also, see pp. 88–102 of ref. 8-13)	Linear equations only, for (1) Primary design equation (2) Subsidiary design equations with (3) Regional constraints only Minimization or maximization of optimization quantity	Many applications for the determination of the optimum operation of a system such as in a production plant, for maximizing profit or minimizing cost; relatively few applications in mechanical design synthesis because of linear equations required and restriction to regional type constraints
Method of Lagrangian multipliers	8-5 (also, see example 8-9)	(1) Primary design equation (8.80) (2) Functional constraints, eq. (8.81) No regional or discrete value constraints can be included; gives point of $dQ=0$, for maximum or minimum Q	Practical applications are relatively few; generally such problems can be solved by simple equation combination and differentiation; technique may give worst design as in example 8-9(b), or else only an ideal point for optimum design as in example 8-9(a)
Method of 1 variables	8-6 (also, see example 8-10)	(1) Single criterion function eq. (8.87) of single variable t No other constraints or limits included, unless combined with eq. (8.87); gives only minimum Q point of $dQ=0$ type	Practical applications are very few; generally such problems can be solved by simple differentiation; technique may give worst design as in example 8-10, or else only an ideal point for optimum design
Digital computer techniques	8-7 (also, see example 8-11) and Appendix C	(1) Optimum design programmed from *systems of equations* by exact computational techniques, such as for sections (8-2) and (8-4) above (2) Optimum design from *criterion function* by iterative searching computational techniques	Possibilities are unlimited as a tool of computation for realistic problems of optimum design; exact techniques once programmed are very rapid and efficient to execute; iterative techniques are necessary in variation studies of complicated functions

Example 8-12 In example 8-6 we wished to design the curved spring of Figure 8.21 for maximization of deflection Δ. The initial formulation is summarized in the example, and for the approach selected the *final formulation* is repeated below.

$$\Delta = \frac{6\pi PR^3}{Et_s^3 b} \quad (I)$$

$$y = \frac{6PR}{t_s^{1.75} b} \quad (II)$$

where

y = eliminated parameter

R,b = related parameters

The typical three-dimensional variation diagram which follows is sketched in Figure 8.23 with points 1 and 2 determined as possibilities in general for optimum design. The procedural steps of optimum design thereby derived are presented in the flow chart of Figure 8.24, and the boundary values of the problem are summarized in the READ block at the top of that figure. For the numerical problem presented in example 8-6 we found that the optimum design required that $t = 0.072$ in., $b = 0.861$ in., and $R = 3.00$ in. The resulting value of the optimization quantity, Δ, is 0.792 in. We noted that the optimum design was at intersection point 2 of Figure 8.23, with $b < b_{max}$ and $R = R_{max}$. Let us now determine what modifications in boundary values would result in an improved design of greater Δ value, and which boundary values would be most effective for change.

At point 2 of Figure 8.23, referring to the final formulation and to the variation diagram, we derive the following equation (8.101) for optimization quantity Δ.

$$\Delta_2 = \frac{6\pi PR_{max}^3}{Et_s^3 b_2} = \frac{6\pi PR_{max}^3}{Et_s^3 \left[\dfrac{6PR_{max}}{t_s^{1.75} 47,000/N_e}\right]}$$

$$\Delta_2 = \frac{47,000\pi R_{max}^2}{Et_s^{1.25} N_e} \quad (8.101)$$

From equation (8.101) we see that an increase in Δ_2 can be achieved by boundary value modifications, e.g., either increasing R_{max} and/or decreasing N_e. We assume that steel of standard stock thickness must be used, so we cannot change E, and t_s can be changed only in discrete step values. Also note from equation (8.101) that changing the boundary values of P, R_{min}, or b_{max} will not improve the value of Δ_2 that can be achieved. Hence, we would be wasting our time in changing these, if we wish to increase Δ for a point 2 design.

Assume that we have some freedom in design in being able to change the boundary values R_{max} and N_e. Hence, in the neighborhood of point 2 the change $d\Delta_2$ resulting from changes dR_{max} and dN_e is derived by Calculus from equation (8.101) as follows:

$$d\Delta_2 = \frac{\partial \Delta_2}{\partial R_{max}} dR_{max} + \frac{\partial \Delta_2}{\partial N_e} dN_e$$

$$= \frac{94{,}000\pi R_{max} \, dR_{max}}{E t_s^{1.25} N_e} - \frac{47{,}000\pi R_{max}^2 \, dN_e}{E t_s^{1.25} N_e^2}$$

Dividing this equation by (8.101) we obtain

$$\left(\frac{d\Delta_2}{\Delta_2}\right) = 2\left(\frac{dR_{max}}{R_{max}}\right) - \left(\frac{dN_e}{N_e}\right) \tag{8.102}$$

Hence, from equation (8.102) we see that Δ_2 is more sensitive to changes in R_{max} than to changes in N_e. In other words, increasing R_{max} by 5% giving approximately a 10% increase in Δ_2, would be nearly twice as effective in increasing Δ_2 as would be decreasing N_e by 5%. We might consider this study of sensitivity to change as a guide for deciding how much to change the boundary values of R_{max} and/or N_e. Once the modifications have been made for R_{max} and/or N_e, the flow chart of Figure 8.24 would be executed again to determine the new optimum design values for b, R, t, and Δ.

A similar approach could be followed for modifying boundary values if the optimum design were found in another numerical problem to exist at point 1 of Figure 8.23. Hence, at point 1 of the figure, referring to the final formulation and to the variation diagram, we derive the following equation (8.103) for optimization quantity Δ.

$$\Delta_1 = \frac{6\pi P R_1^3}{E t_s^3 b_{max}} = \frac{6\pi P}{E t_s^3 b_{max}} \left[\frac{47{,}000 \, t_s^{1.75} b_{max}}{N_e} \frac{}{6P}\right]^3$$

$$\Delta_1 = \frac{\pi}{36}(47{,}000)^3 \frac{t_s^{2.25} b_{max}^2}{E N_e^3 P^2} \tag{8.103}$$

From equation (8.103) we see that an increase in Δ_1 can be achieved by boundary value modifications, e.g., either increasing b_{max} and/or by decreasing N_e and/or P. Again we assume that steel of standard stock thickness must be used, so we cannot change E, and t_s can be changed only in discrete step values. Also, note from equation (8.103) that changing the boundary values of R_{min} or R_{max} will not improve the value of Δ_1 which can be achieved.

If we assume that some freedom in design exists for changing boundary values of P, b_{max}, and N_e. from equation (8.103) by the application of Calculus we derive equation (8.104),

$$d\Delta_1 = \frac{\partial \Delta_1}{\partial b_{max}} \, db_{max} + \frac{\partial \Delta_1}{\partial N_e} \, dN_e + \frac{\partial \Delta_1}{\partial P} \, dP$$

$$\left(\frac{d\Delta_1}{\Delta_1}\right) = 2\left(\frac{db_{max}}{b_{max}}\right) - 3\left(\frac{dN_e}{N_e}\right) - 2\left(\frac{dP}{P}\right) \qquad (8.104)$$

Hence, from equation (8.104) we see that Δ_1 is more sensitive to changes in N_e than to changes in either b_{max} or P. Thus, decreasing N_e by 5% would increase Δ_1 by approximately 15%, whereas either increasing b_{max} by 5% or decreasing P by 5% would increase Δ_1 by only approximately 10%. Again we might consider this study of sensitivity to change as a guide for deciding how much to modify the boundary values of b_{max}, N_e, and/or P. Once the modifications have been made for b_{max}, N_e, and/or P, the flow chart of Figure 8.24 would be executed again to determine the new optimum design values for b, R, t, and Δ.

An important observation of general significance should be brought out at this time. By properly combining equations (I) and (II) of the final formulation, referring to the variation diagram of Figure 8.23(a), we derived equation (8.103) from which the conclusion was drawn that Δ_1 could be increased by decreasing the specified value of P. However, if we had looked at equation (I) by itself, we might have concluded that the specified value of P should be increased rather than decreased, and of course this would have been a valid conclusion in analyzing a fixed design. For improvement in optimum design, the error of such an analysis is really twofold. For one, at this stage the design is not fixed but variable, the design possibilities of which correspond to the infinite number of points in the neighborhood of point 1 in Figure 8.23(a). Secondly, instead of considering only part of the problem such as equation (I) by itself, it is necessary to consider the entire system of equations (I) and (II) in the final formulation together with the pertinent boundary conditions of the problem; this is what we have done in deriving equation (8.103), from which the valid conclusion was reached. Hence, this is an example of how erroneous conclusions of optimum design can be reached when only part of a problem, rather than the total system of equations and boundary conditions, is considered. The method of optimum design presented in this chapter provides a systematic way of properly handling such systems of equations and boundary conditions, in a practical manner which is easy to visualize and to understand once the basic technique is mastered.

Let us now briefly consider the second type of situation requiring a change in boundary values. Suppose a numerical problem were such that we had the case of incompatible specifications shown in the flow chart of Figure 8.24, with $R_1 < R_{min}$. To alleviate this situation we must at least obtain $R_1 = R_{min}$ for points 1 and A coincident in Figure 8.23. This we can do in one or both of two ways, considering that R_1 is initially less than R_{min} in Figure 8.23. First, we can decrease the value of R_{min}. Secondly, we can increase the value of R_1, which is expressed in terms of the boundary values by equation (8.105); this equation is derived from the final formulation referring to the variation diagram of Figure 8.23.

$$R_1 = 47,000 b_{max} t_s^{1.75}/(6PN_e) \tag{8.105}$$

Hence, from equation (8.105) we can increase R_1 either by increasing b_{max} and/or by decreasing P and/or N_e. As before, we apply Calculus to derive the sensitivity to change equation (8.106), which might be of help in making the decisions of modification.

$$dR_1 = \frac{\partial R_1}{\partial b_{max}} db_{max} + \frac{\partial R_1}{\partial P} dP + \frac{\partial R_1}{\partial N_e} dN_e$$

$$\left(\frac{dR_1}{R_1}\right) = \left(\frac{db_{max}}{b_{max}}\right) - \left(\frac{dP}{P}\right) - \left(\frac{dN_e}{N_e}\right) \tag{8.106}$$

Thus, from equation (8.106) we see that R_1 is equally as sensitive to changes in b_{max}, P, and N_e. In other words, a 5% increase in R_1 would be achieved either by increasing b_{max} by 5% or by decreasing either P or N_e by 5%. The required changes in boundary values would have to satisfy at least the relationship $R_1 = R_{min}$, which by equation (8.105) is as follows:

$$R_{min} = 47,000 b_{max} t_s^{1.75}/(6PN_e) \tag{8.107}$$

For a case of incompatible specifications, the *least change* in boundary values for obtainment of *a* design solution results from selecting the largest standard stock thickness available, since then by equation (8.105) R_1 would be greatest and closest to R_{min}. Thus, required percent changes in R_{min}, b_{max}, P, and/or N_e would be the least possible for obtaining satisfaction of equation (8.107).

8-10 PLANNING THE ATTACK FOR DESIGN OPTIMIZATION

We will conclude this Chapter with some general philosophy which has been helpful to the beginner in application of design optimization. First, be sure that the problem is critical in nature, where the time spent on optimization techniques is justified. Remember, minimization of engineering time and cost is generally achieved if the design arrived at is acceptable from all standpoints by the cut-and-try approach and analysis.

Secondly, if the application of optimization techniques appears to be justified, use your good engineering judgement as well as theory when deriving the *initial formulation*. To start, include only the specifications and constraints which logically apply for the specific problem at hand. There should be a logical explanation for all of the required boundary conditions listed. Inclusion of specifications or constraints which are not pertinent will unnecessarily complicate the effort of design optimization, and also this may result in a solution which is not as good as it might be. Any specifications or constraints of significance which are erroneously left out will be revealed in the end analysis. Then, you will be in a good position to include the effects of such an additional requirement in a second approach for design optimization.

Third, carefully review the initial formulation equation system to decide which optimization technique is most appropriate. If engineering time is critical, and a thorough insight into the solution is not necessary, perhaps an available computer code as explained in Appendix C would be expedient. In fact, to minimize the possibility of mistakes, you might take such an approach in parallel with the *method of optimum design* to check your final results.

Finally, if the *method of optimum design* appears appropriate, plan your attack carefully. Going from the *initial formulation* to the *final formulation* can be difficult to the beginner. Carefully decide what the specified items are and consider which of the determinate approaches would be simplest to take. Use a circle diagram as illustrated in Table 8-2 of Example 8-4 to reveal the possible approaches to take. Also, summarize the final formulation in a general format equation system before it is specifically derived. This can be an invaluable guide indicating what to do and where you are heading prior to detailed derivation of the final formulation. Thus, as a specific illustration, the following general format might have been helpful to write after the initial formulation of Example 8-4.

$$T_{80} = f_1(R,C) \qquad \text{(I)}$$
$$E = f_2(R,C) \qquad \text{(II)}$$
$$W_c = f_3(R,C) \qquad \text{(III)}$$

Of course, we realize that in an equation of the final formulation a related parameter might be a phantom variable. This was the case for C in equation (II) above, where we had $E = I(R + R_o)$.

BIBLIOGRAPHY

8-1 R. C. Johnson, *Optimum Design of Mechanical Elements*, John Wiley & Sons, Inc., New York, 1961, 535 pp.

8-2 R. C. Johnson, "Three-Dimensional Variation Diagrams for Control of Calculations in Optimum Design," *Transactions of the ASME, J. Eng. Ind.*, Aug. 1967, pp. 391–398.

8-3 J. P. Vidosic, *Elements of Design Engineering*, The Ronald Press Company, New York, 1969, Chap. 12.

8-4 Ernest Rabinowicz, *Friction and Wear of Materials*, John Wiley & Sons, Inc., New York, 1965, 244 pp.

8-5 R. C. Binder, *Fluid Mechanics*, Prentice-Hall, Inc., Englewood Cliffs, N.J., 1962, p. 167.

8-6 Ernest Rabinowicz, "Surface Energy Approach to Friction and Wear," *Product Engineering*, McGraw-Hill Book Co., Inc., New York, March 15, 1965, pp. 95–99.

8-7 S. I. Pearson and G. J. Maler, *Introductory Circuit Analysis*, John Wiley & Sons, Inc., New York, 1965, 546 pp.

8-8 Max Jakob and G. A. Hawkins, *Elements of Heat Transfer*, 3rd ed., John Wiley & Sons, Inc., New York, 1957, 317 pp.

8-9 S. Timoshenko, *Strength of Materials*, Part I, D. Van Nostrand Company, Inc., Princeton, N.J., 1955, 442 pp.

8-10 *Handbook of Mechanical Spring Design*, Associated Spring Corporation, Bristol, Connecticut, 1964, 83 pp.

8-11 R. C. Johnson, "Predicting Part Failures," Parts 1 and 2, *Machine Design*, Penton Publishing Company, Cleveland, Ohio; Jan. 7, 1965, pp. 137–142 and Jan. 21, 1965, pp. 157–162.

8-12 G. Leitmann, *Optimization Techniques With Applications to Aerospace Systems*, Academic Press, New York, 1962, 453 pp.

8-13 M. Asimow, *Introduction to Design*, Prentice-Hall, Inc., Englewood Cliffs, N.J., 1962, 135 pp.

8-14 R. Courant and D. Hilbert, *Methods of Mathematical Physics*, Interscience Publishers, Inc., New York, 1953, p. 164.

8-15 I. S. Sokolnikoff and R. M. Redheffer, *Mathematics of Physics and Modern Engineering*, McGraw-Hill Book Co., Inc., New York, 1958, p. 247.

8-16 G. B. Dantzig, *Linear Programming and Extensions*, Princeton University Press, Princeton, N.J., 1963, 625 pp.

8-17 J. R. Dixon, *Design Engineering: Inventiveness, Analysis, and Decision Making*, McGraw-Hill Book Co., Inc., New York, 1966, 345 pp.

8-18 C. Zener, "A Mathematical Aid in Optimizing Engineering Design," *Proc. Nat Acad. Sci.* **47**, 537–539 (1961).

8-19 C. R. Mischke, *An Introduction to Computer-Aided Design*, Prentice-Hall, Inc., Englewood Cliffs, N.J., 1968, 211 pp.

8-20 K. S. Kunz, *Numerical Analysis*, McGraw-Hill Book Co., Inc., New York, 1957, 381 pp.

8-21 S. S. Kuo, *Numerical Methods and Computers*, Addison-Wesley Co., Reading, Mass., 1965, 341 pp.

8-22 R. Beckett and J. Hurt, *Numerical Calculations and Algorithms*, McGraw-Hill Book Co., Inc., New York, 1967, 298 pp.

8-23 D. D. McCracken, *Fortran With Engineering Applications*, John Wiley & Sons, Inc., New York, 1967, 237 pp.

8-24 R. E. Machol, *System Engineering Handbook;* Chapter 25 by G. B. Dantzig, "The Simplex Method," and Chapter 26 by A. Charnes and W. W. Cooper, "Elements of a Strategy for Making Models in Linear Programming"; McGraw-Hill Book Co. Inc., New York, 1965.

8-25 R. C. Johnson, "Design Synthesis—", *Machine Design*, Penton Publishing Co., Cleveland, Ohio; Part 1, Oct. 18, 1973, pp. 149-155; Part 3, Dec. 13, 1973, pp. 164-169; Part 4, Dec. 27, 1973, pp. 52-57.

8-26 J. Ellis, "Johnson's Method of Optimum Design Applied to a Problem with Simple Functional Relationships", *Computer Aided Design*, Vol. 8, No. 1, Science and Technology Press, Ltd., I.P.C. House, Surrey, England, Jan. 1976, pp. 9-12.

8-27 J. Ellis, "An Introductory Problem in Mechanical Design Optimization", *International Journal of Mechanical Engineering Education*, Vol. 5, No. 2, I. Mech. E. & UMIST 1977, pp. 121-126.

8-28 M. A. Murray Lasso and E. Chicurel Uziel, *Aplicaciones de Computacion a la Ingenieria*, Editorial Limusa, S. A., Mexico City, 1975, pp. 167-174.

8-29 J. N. Siddall, *Analytical Decision-Making in Engineering Design*, Prentice-Hall, Inc., Englewood Cliffs, N.J., 1972, 431 pp.

8-30 W. F. Stoecker, *Design of Thermal Systems*, McGraw-Hill Book Company, New York, 1971, Chapters 6-11.

8-31 G. R. Walsh, *Methods of Optimization*, John Wiley & Sons, Inc., New York, 1975, 200 pp.

Additional references are listed at the end of Appendix C on "Computers in Automated Optimal Design".

PROBLEMS

Many suggested exercises are distributed throughout the text in this Chapter, which should provide ample practice for the student, particularly in learning the *method of optimum design.*

9

Advanced Design of Elements and Systems

9-1 INTRODUCTION TO SYSTEM DESIGN

In general, the basic purpose of design is to arrive at a physical system which will satisfactorily perform various functions for the solution of a specific problem that is complex in nature. Often a particular piece of equipment in such a system is subjected to various modes of operation. Hence, any particular element must be designed to satisfy the total system requirements in all of the modes of operation anticipated.

In the examples which follow in this chapter we will try to illustrate the total systems aspect of design for some complex mechanical elements and mechanisms. Of course, the broad aspects of system requirements are only an important part of the problem of design. Broad considerations external to the system being designed must be kept in mind, such as limitations on available design time, available design manpower and talent, and budgetary allowances. Also, more confined considerations internal to the system being designed must be carefully made, being sure that all elements are designed to perform their specific functions as well as possible, and making design decisions within the boundaries of limitations on space available, on feasible materials, and on manufacturing facilities as typical examples.

The particular process of decision making that should be followed in the design of a system is not straightforward at the present time.   Essentially, the Figure 1.3 morphology would be followed with initial emphasis on the synthesis of system configurations, based on satisfaction of the basic function requirements.   Often the function requirements are tied in with the steps of processing for a product, such as is illustrated in Figure 9.34 of section 9-5 for the processing of raw cotton into yarn.   After the synthesis of the basic function requirements, effort must be spent on the synthesis of basic machines and components.   Often this also involves the synthesis of arrangements of machines for obtainment of optimum system performance, as is illustrated in Figure 9.40 of section 9-6 for the processing of long staple synthetic materials into coarse yarns.   Next, effort must be spent on the synthesis of the machines themselves, followed by the synthesis of components and mechanisms.   This latter stage would be tied in more specifically with the Figure 1.3 morphology.

A simplified summary of the process of system synthesis is presented in Figure 9.1.   Obviously, the specific decision-making process followed in

Figure 9.1   Typical system synthesis flow chart.

execution of the flow chart relies heavily on experience, ingenuity, and the exercising of good engineering judgement together with the making of order-of-magnitude feasibility calculations at appropriate places.   Unfortunately, at this time, explicit techniques are not available for the solution of system synthesis problems in the early stages of the Figure 9.1 process.   In essence, the logical building-block·approach described in Chapter 4 must be followed. However, in the latter stages of the synthesis process of Figure 9.1, we can often apply the more explicit techniques of synthesis as described in Chapters 2, 3, 4, and 5.   In any of the stages of system synthesis in Figure 9.1, the central theme of optimization is the influence of control for the making of design decisions.

In the remaining part of the chapter we will present some advanced design examples of complex mechanical elements and mechanisms.   Basic techniques of execution will be illustrated for the Figure 1.3 morphology in realistic design situations.   Also, as much as feasible, we will try to illustrate the necessity of considering the total system requirements in the design of specific elements and components.   Hopefully, the examples which follow will help to tie together some of the techniques presented in earlier chapters.

SOME COMPLEX ELEMENTS IN SYSTEM OPERATIONS

9-2   TORSION BAR DESIGN FOR A VEHICLE SUSPENSION

In the design of the automobile of Figure 9.2 we are presently concerned with the front-wheel suspension design.   To avoid as much as possible problems of excessive tire wear and skidding tendency, we have decided to incorporate the independent suspension configuration of Figure 9.3(b). Also, from a review of various spring types possible, such as those described in reference 9-1, we have selected the torsion-bar configuration of Figure 9.4. This decision was reached based on considerations of compatibility with our suspension system as illustrated in Figure 9.5, space available, and previous success with this type of spring.

For deflections Δ of the wheel relative to the frame in Figure 9.5, the wheel will remain in its approximately vertical orientation, as illustrated in Figure 9.3(b), because of the parallelogram suspension linkage.   For mounting stability of our wheel, we desire to keep suspension points A and B in approximately the same positions relative to the tire, as shown in Figure 9.5.   Hence, angle θ is virtually a specified value of $\theta = \tan^{-1}(H/V)$, referring to Figure 9.5 for θ, H, and V.   However, we do have some freedom in design for location of pivot points C and D in the figure, enabling us to establish a regional constraint on r as described next.

(a) Real model

(b) Analytical model for
front-wheel suspension

Figure 9.2   Automobile for torsion-bar design problem.

(a) Rigid axle front-wheel suspension

(b) Independent front-wheel suspension

Figure 9.3   Some front-wheel suspension possibilities.

Referring to Figure 9.6, we recognize that our torsion bar must fit within space constraints.   Hence, with the notation of Figure 9.4 we must satisfy the regional constraints given by limit equations (9.1) and (9.2).

$$L_{min} \leq L \leq L_{max} \tag{9.1}$$

$$r_{min} \leq r \leq r_{max} \tag{9.2}$$

The boundary values for these ranges are determined from a consideration of the surroundings of the torsion bar in Figure 9.6, estimating perhaps with some graphical layout work the feasible extremes for L and r in design of the automobile.   Similarly, in considerations of practicality we estimate an acceptable range for bar diameter d as follows.

$$d_{min} \leq d \leq d_{max} \tag{9.3}$$

Figure 9.4   Torsion-bar configuration selected.

For satisfactory performance of our automobile, several modes of operation must be considered. First, in loading the vehicle we must be sure that static deflection is not excessive, as caused by passengers, cargo, etc. Hence, on a per wheel basis we estimate the *maximum* anticipated live load increment, W_i, and the corresponding *allowable* wheel deflection, Δ_i, relative to the frame. In this way we determine the smallest acceptable value for spring constant or force gradient, expressed as k_{min} in equation (9.4).

$$k_{min} = W_i/\Delta_i \qquad (9.4)$$

The units for k_{min} are lb/in., load increment W_i being expressed in lb and allowable deflection Δ_i in in.

The second mode of operation is encountered in driving our vehicle over normal road conditions. We want to be sure that at least a reasonably smooth ride is provided for our passengers, and that they do not feel every normal bump in the road. As illustrated in Figure 9.7, we would like to design our suspension system for operation in the soft-spring region if possible. To some extent at least, we can control this in design of the torsion bar by being sure that our spring stiffness is not excessive. Also, for a

Figure 9.5   Torsion-bar spring in independent wheel suspension shown in position of normal load.

typical transient road disturbance, as shown in Figure 9.8, our passengers will be subjected to the smallest peak acceleration by having spring stiffness k of low value. From considerations such as these, supplemented with knowledge from our practical experience, we could reach a decision on maximum allowable stiffness, k_{max}, for adequate isolation characteristics of our automobile.

Based on the considerations now discussed, we have established a regional constraint (9.5) on permissible values for force gradient k of the suspension system we are designing.

$$k_{min} \leq k \leq k_{max} \tag{9.5}$$

The boundary values in limit equation (9.5) have been determined from the various normal modes of system operation anticipated to this point for our vehicle.

The third mode of operation which we anticipate for our automobile is in

Figure 9.6   Torsion-bar front-wheel suspension in setting of automobile being designed (steering mechanism, shock absorbers, etc., not shown).

unusual shock loading as encountered from adverse road conditions or careless driving tactics.   Of course, we will eventually incorporate shock absorbers in our suspension design, but we want to be certain as much as possible that our torsion-bar springs do not break under the worst unpredictable road conditions or operational modes encountered.   Hence, we select as an appropriate objective for optimum design of our torsion bar the *maximization of energy-absorption capability*, without exceeding the allowable shearing stress of the material, S_s/N.

The torsion-bar configuration of Figure 9.4 will be subjected to virtually a pure twisting load if we decide from mounting accuracy requirements of the suspension system to place the support bearing close to the suspension link as shown.   For a torsion bar, twisting moment, M_t, will vary linearly with angle of twist, θ, as shown in Figure 9.9, where from strength of materials such as given in reference 9-3, angle of twist is expressed as follows:

$$\theta = M_t L/(JG) \tag{9.6}$$

In equation (9.6), θ is the angle of twist of the torsion bar in radians, M_t is the twisting moment in in.-lb, G is the modulus of rigidity of the material in

Figure 9.7   Desired relationship of road excitation frequencies to response diagram of spring-suspended mass for a smooth ride in automobile (typically, see references 9-2 and 9-5).

For $c = 0$:

$$y = S \left[1 - \cos \omega_n t \right]$$

$$\therefore \quad \ddot{y} = S \omega_n^2 \cos \omega_n t$$

where $\omega_n = \sqrt{k/M}$

$$\therefore \quad (\ddot{y})_{max} = S\, k/M, \text{ peak acceleration of vehicle}$$

Figure 9.8   Effect of a typical transient disturbance in the road on peak acceleration of the vehicle (typically, see reference 9-5).

psi, J is the polar moment of inertia of the circular cross-section in in.^4, expressed by equation (9.7), and L is length (see Figure 9.4) in inches.

$$J = \pi d^4/32 \qquad (9.7)$$

In equation (9.7), d is the diameter of the torsion bar shaft shown in Figure 9.4, in inches.

Strain energy U in the torsion bar is expressed as follows, referring to Figure 9.9.

$$U = \int_0^\theta M_t \, d\theta = M_t \theta / 2$$

By combination with equation (9.6) and elimination of θ, we obtain

$$U = M_t^2 L/(2JG) \qquad (9.8)$$

where strain energy U is expressed in in.-lb and all other terms are as previously defined.   However, referring again to Figure 9.5, we express twisting moment M_t in terms of wheel force F_w as follows.

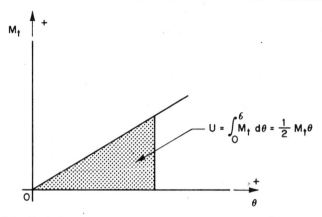

Figure 9.9   Twisting momer t, M_t, versus angle of twist, θ, for a torsion bar.

$$M_t = (F_{3/4})(r \cos \theta)$$

$$= (F_{3/4} \cos \theta)r = F_w r$$

$$M_t = F_w r \tag{9.9}$$

In equation (9.9), F_w is wheel force shown in Figure 9.5 in lb and r is the suspension link length of the figure in in.   Equation (9.9) is based on the reasonable assumption that wheel inertia force is small compared with wheel road force F_w.

Combining equations (9.7) and (9.9) with (9.8) by eliminating J and M_t we obtain (9.8a) as the primary design equation or criterion function.

$$U = (F_w r)^2 L/(2G(\pi d^4/32))$$

$$U = 16 F_w^2 r^2 L/(\pi G d^4) \tag{9.8a}$$

The theorem of Castigliano as presented in reference 9-3 can be applied directly to equation (9.8a) for derivation of force gradient equation (9.10) as follows, referring to Figure 9.5.

$$\Delta = \frac{dU}{dF_w} = 32 F_w r^2 L/(\pi G d^4)$$

therefore

$$k = F_w/\Delta = \pi G d^4/(32r^2 L) \tag{9.10}$$

Finally, critical shear stress, τ_{max}, will be in the stress concentration splined regions of Figure 9.4, expressed as follows from strength of materials, using equations (9.7) and (9.9).

$$\tau_{max} = K_t M_t c/J$$

$$= K_t(F_w r)(d/2)/(\pi d^4/32)$$

$$\tau_{max} = 16 K_t F_w r/(\pi d^3) \qquad (9.11)$$

In equation (9.11), maximum shear stress, τ_{max}, is in psi, K_t is the dimensionless theoretical stress concentration factor at the splined regions of Figure 9.4, and all other terms are as previously described. The value of K_t depends primarily on the geometric proportions of the splined regions, such as determined from reference 9-6 or 9-7. With good design, as in the proportions of Figure 9.4, we would have $K_t \approx 1.2$.

For a safe design we impose a regional constraint on permissible values of τ_{max} as follows.

$$\tau_{max} \le S_s/N \qquad (9.12)$$

In limit equation (9.12), S_s is the shear strength of the material in psi, estimated typically from information in a source such as reference 9-8. Also, we will assume that appropriate factor of safety N is a determined value at this stage of design, having been estimated as typically explained in references 9-9 (Chapter 6) and 9-10.

A summary of the initial formulation for this problem is now presented as follows:

Initial Formulation

$$U = 16 F_w^2 r^2 L/(\pi G d^4) \qquad \text{(P.D.E.) (9.8a)}$$

$$k = \pi G d^4/(32 r^2 L) \qquad \text{(S.D.E.) (9.10)}$$

$$\tau_{max} = 16 K_t F_w r/(\pi d^3) \qquad \text{(S.D.E.) (9.11)}$$

$$\left.\begin{aligned}
r_{min} &\le r \le r_{max} \\
k_{min} &\le k \le k_{max} \\
L_{min} &\le L \le L_{max} \\
d_{min} &\le d \le d_{max} \\
\tau_{max} &\le S_s/N
\end{aligned}\right\} \qquad \text{(L.E.)s}$$

Known or Specified: r_{min}, r_{max}, k_{min}, k_{max}, L_{min}, L_{max}, d_{min}, d_{max}, K_t, N, and properties G and S_s for the available feasible materials of interest.

Find: Values of r, L, and d and the material for the design which will maximize strain energy capacity, U_c, of the bar.

Inspection of the initial formulation reveals a case of redundant specifications, since with only one free variable, F_w, we have $n_f = 1 < N_s = 2$,
referring to Figure 8.2.   Also, not counting optimization quantity U, the
design variables other than material properties are F_w, r, L, d, k, and τ_{max},
so $n_v = 6$.   Thus, since $N_s = 2$, from equation (8.13) we calculate

$$D_{vs} = n_v - N_s + 1$$
$$= 6 - 2 + 1 = 5$$

as the number of dimensions for the variation study, which may be reduced
if we have truly independent parameters in the final formulation.

Proceeding directly to the final formulation, we combine the (S.D.E.) (9.11)
with (P.D.E.) (9.8a) by eliminating the free variable F_w.

$$U = \frac{16[\pi d^3 \tau_{max}/16K_t r]^2 r^2 L}{\pi G d^4}$$

$$U = \pi \tau_{max}^2 d^2 L r^0/(16K_t^2 G) \qquad (9.8b)$$

In developed (P.D.E.) of the ideal problem (9.8b) we have purposely shown
the phantom variable r^0, which has *not* been eliminated from the equation
system by the combination procedure.   It just happened to cancel out in the
derivation of equation (9.8b) because of the particular functions involved.

From an inspection of the initial formulation, we see that a simple approach
for a final formulation will be to choose k as the eliminated parameter.   Thus,
our final formulation system of equations would consist of (9.8b) and (9.10),
as (I) and (II), respectively, arranged in the basic format as follows.

Final Formulation

$$U = \frac{\pi \tau_{max}^2 d^2 L r^0}{16K_t^2 G} \qquad \text{(I)} \qquad \qquad (9.8b)$$

$$k = \pi G \, d^4/(32r^2 L) \quad \text{(II)} \qquad \qquad (9.10)$$

where

k = eliminated parameter

τ_{max} = truly independent parameter

d, L, r = related parameters

(I) = developed (P.D.E.) of ideal problem

(II) = relating equation

Hence, at this time we are justified in making a design decision on τ_{max},
which from equation (I) should be as large as possible for maximization of
strain energy U.   Thus, referring to regional constraint (9.12), we now make

the decision to place $\tau_{max} = S_s/N$, and equation (I) of the final formulation becomes (9.8c) which follows for strain energy *capacity*.

$$U_c = \frac{\pi S_s^2 d^2 L r^0}{16 N^2 K_t^2 G} \quad \text{(I)} \tag{9.8c}$$

In reference to our final formulation system of equations, the remaining problem is to select the values of d, L, r, and k which will maximize strain energy capacity, U_c. With three related parameters, d, L, and r, we readily recognize that a four-dimensional variation study will be required, which is less by one than the value $D_{vs} = 5$ previously calculated, because of truly independent parameter τ_{max}. The variation study can be reduced to three dimensions if we hold either d, L, or r constant for that step. We choose to hold diameter d constant, since standard sizes are generally desired for such a dimension, such as in $\frac{1}{32}$-in. increments. Thus, we introduce the discrete value constraint $d = d_s$ for $s = 1, \ldots, m$ standard sizes within our range of interest. Hence, at this time we decide to replace regional constraint (9.3) with the discrete value constraint (9.13) expressed as follows.

$$d_{min} \leq d = d_s \leq d_{max} \tag{9.13}$$

$$\text{for } s = 1, \ldots, m$$

In this way we can hold d constant at a value d_s for a three-dimensional variation study, thereby deriving an explicit flow chart for optimum design. We can have our digital computer execute the calculations of this flow chart program for each of the standard size values d_s. Thus, by comparison of criterion function values, the technique of which was illustrated previously in example 8-6, we can explicitly determine the overall optimum design in relatively short order.

A *typical* three-dimensional variation diagram for a specific material of interest with d constant at d_s is sketched in Figure 9.10 from the final formulation. The points possible for optimum design are any accessible point on the line segment 1–2 and point 3, which we readily see from Figure 9.10. Also, there are two possibilities for incompatible specifications, i.e., if $r_B > r_{max}$ or $L_3 < L_{min}$ in the figure.

If the constraints fell as specifically shown in Figure 9.10, there would be an infinite number of points possible for optimum design. Any point on the line segment $A-2$ would have equally as large a value for strain energy capacity, U_c. In a situation such as this, where there are an infinite number of points possible for optimum design, we often can profitably introduce a secondary objective of optimization. For this particular automobile suspension problem, we might wish to obtain the smoothest possible ride as the secondary objective for optimum design. In reference to Figures 9.7 and

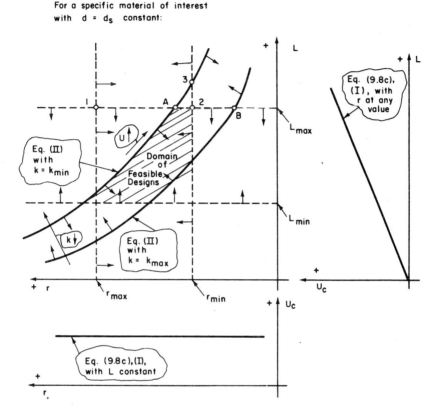

For a specific material of interest
with d = d$_s$ constant:

Figure 9.10   Typical three-dimensional variation diagram for the torsion
bar design problem.   The objective is maximization of strain energy capacity,
U_c.   Points possible for optimum design are any accessible point on line
segment 1–2 and point 3.   (See final formulation for equations I and II.)

9.8, the smoothest ride would be obtained by having force gradient k as small
as possible.   Thus, for the constraints as specifically shown in Figure 9.10,
we would choose point A for the optimum design with $k = k_{min}$.   This particular design would have the greatest possible strain energy capacity as well
as the smoothest ride characteristics.   On the other hand, if the boundary
values were such that $r_A > r_{max}$, with $r_B < r_{max}$, we choose point 1 in
Figure 9.10 for maximization of strain energy capacity and smoothest ride

characteristics.   Hence, with the primary objective of maximizing strain energy capacity and the secondary objective of minimizing spring constant k for smoothest ride characteristics, we would have only three points possible for optimum design in Figure 9.10.   They are points 1, A, and 3.

Incidentally, we should mention that our final design would incorporate damping in the form of a shock absorber at pivot D in Figure 9.5.   We would want to introduce sufficient damping but not an excessive amount, i.e., damping ratio $(c/c_c) \approx 0.4$ in Figure 9.7.   This would require a shock absorber damping value of $c \approx 0.4$ $c_c = 0.4(2\sqrt{kM}) = 0.8\sqrt{kM}$.   Hence, our secondary objective of minimizing k would also minimize the required amount of damping c.   Thus, for a realistic transient disturbance in the road, referring to Figure 9.8 as an extreme ideal example, having minimum damping value for c would also minimize the peak force transmitted to M and therefore the corresponding peak acceleration which the passengers would feel.   Therefore, we see that the secondary objective of having the smoothest possible ride is satisfied by choosing spring constant k as small as feasible, even for transient disturbances in the final design where adequate shock absorbers are introduced.

From an understanding of the variation study now made, referring to the diagram of Figure 9.10 and to the final formulation, we can program the procedure of optimum design as presented in the Flow Chart of Figure 9.11. This optimum design flow chart would be executed for each of the materials of interest, with the overall optimum design being readily recognized in the end from a simple comparison of U_c values.   A numerical problem is presented next together with the solution obtained by a digital computer.

Numerical Example 9-1   A torsion bar spring is to be designed for the front-wheel suspension of an automobile in accordance with the discussion now presented, for the primary objective of maximizing strain energy capacity with the secondary objective of providing the smoothest ride possible.   For the unloaded vehicle a 500-lb weight is assigned to the front wheel, giving $M = 500/g$.   For the fully loaded vehicle a maximum weight increment of $W_i = 200$ lb is estimated, referred to the front wheel, and for this loading the allowable deflection should not exceed $\Delta_i = 2$ in.   Hence, from equation (9.4) we estimate

$$k_{min} = \frac{W_i}{\Delta_i} = \frac{200}{2} = 100 \text{ ib/in.}$$

Next, from a consideration of typical road imperfections and transient disturbances, supplemented with our engineering judgement, we must estimate a value for greatest allowable spring stiffness k_{max}.   For instance, for an ideal 2-in. step in the road, suppose that we do not wish M to be subjected to

329

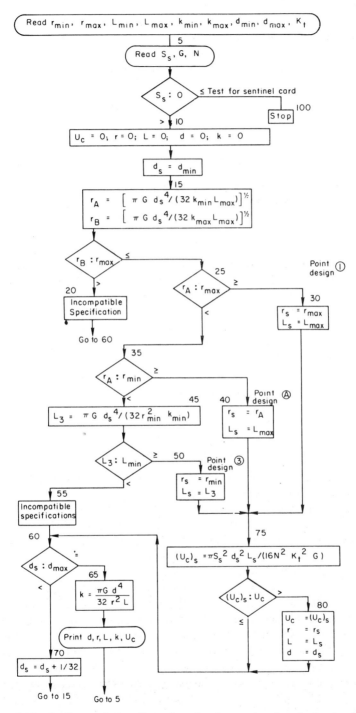

Figure 9.11 Optimum design flow chart for the torsion-bar design problem.

an acceleration greater than approximately 0.8 of the gravitational constant g. Of course, passengers in the vehicle would be further isolated from the road because of the seat-cushion and tire spring-effect. Thus, referring to Figure 9.8 we estimate

$$(\ddot{y})_{max} \approx 0.8g \gtrsim Sk/M = Skg/W$$

therefore

$$k_{max} \approx 0.8 \ W/S = (0.8)(500/2)$$
$$= 200 \ \text{lb/in.}$$

We decide that this would be an acceptable maximum value for spring constant k of our suspension spring. Hence, at this time we have established the following regional constraint on permissible values for k.

$$100 \ \text{lb/in.} \le k \le 200 \ \text{lb/in.}$$

In reference to Figure 9.4, we believe that manufacturing quantities will be sufficient to warrant the forging process. Hence, the splined region proportions of Figure 9.4 can be specified, with $K_t \approx 1.2$ resulting for the governing stress concentration factor referred to diameter d of the figure. Also, in consideration of the space available for the torsion bar in the automobile being designed, as depicted in Figure 9.6, we estimate the following regional constraints on permissible values for r, L, and d. The lower limit of $d_{min} = 0.000$ in. was selected since for this particular application we are willing to accept any value of $d \le 1.500$ in.

$$9 \ \text{in.} \le r \le 12 \ \text{in.}$$
$$24 \ \text{in.} \le L \le 36 \ \text{in.}$$
$$0.000 < d \le 1.500 \ \text{in.}$$

Also, we assume that any acceptable value for d will be a fractional size in $\frac{1}{32}$-in. increments within the above range.

Several materials are available for manufacturing the bar. Of the steels we need merely calculate for the one of greatest shear strength value, S_s, since G would be the same for all of the steels, and therefore the boundaries in the (r, L) plane of Figure 9.10 would be unaffected by changing steels, as seen from the diagram with the final formulation. However, the criterion function surface is of greatest level for the steel of greatest S_s value, as seen from ideal developed (P.D.E.) (9.8c). Because of this, we need make calculations only for the steel of greatest S_s value, which we will assume to be the AISI 2340 one from a list of feasible steels in our locality at this time. Suppose also that a titanium alloy is of interest for the torsion bar, and that we are able to estimate its allowable shear strength with an application of the maximum shear stress theory of fatigue failure, as presented in reference

9-9. Hence, for the two materials of interest at this time we summarize the significant properties as follows, together with our selected factor of safety, N.

	$S_s\,(psi)$	$G(psi)$	N
AISI 2340 steel	78,000	12×10^6	1.25
Titanium alloy	52,000	6×10^6	1.25

With boundary values now estimated for the torsion-bar design problem, we are ready to make calculations in accordance with the optimum design flow chart of Figure 9.11. The digital computer program so written is presented in Figure 9.12, together with the input and output data for the numerical example. The overall optimum design is determined from a comparison of U_c values for the best designs of each material; i.e., comparing $U_c = 1400$ in.-lb for the best design of the AISI 2340 steel with $U_c = 1792$ in.-lb, for the best design of the titanium alloy, referring to the output data of Figure 9.12. Incidentally, both of these designs happen to be point 1 designs in the Figure 9.10 variation diagram, which is readily recognized from the Figure 9.12 output data since $r = r_{max} = 12$ in. and $L = L_{max} = 36$ in. for both cases. Thus, from the numerical results obtained, we recommend the following as the overall optimum design for the torsion bar:

Optimum Design for Numerical Example

Material:	titanium alloy
Dimensions:	$d = 1.125$ in.
	$r = 12$ in.
	$L = 36$ in.
	Splined region proportions of Figure 9.4
Characteristics:	$k = 182$ lb/in.
	$U_c = 1792$ in.-lb strain energy capacity

On the other hand, if in the final analysis we decide that the titanium alloy is an unacceptable material because of cost or manufacturing considerations, the *next choice* referring to Figure 9.12 would be the best steel design summarized as follows:

Material:	AISI 2340 steel
Dimensions:	$d = 0.9375$ in.
	$r = 12$ in.
	$L = 36$ in.
	Splined region proportions of Figure 9.4

```
 1        READ,RMN,RMX,ALMN,ALMX,AKMN,AKMX,DMN,DMX,AKT
 2        WRITE (6,1) RMN,RMX,ALMN,ALMX,AKMN,AKMX
 3      1 FORMAT ('0','SPECS. RMN,RMX,LMN,LMX,KMN,KMX ='/6E14.5)
 4        WRITE (6,2) DMN,DMX,AKT
 5      2 FORMAT ('0',5X,'DMN,DMX,KT =',3E14.5)
 6      5 READ,SS,G,AN
 7        WRITE (6,6) SS,G,AN
 8      6 FORMAT ('0','MATERIAL PROPERTIES SS, G AND N =',1X,3E16.7)
        C TEST FOR SENTINEL CARD
 9    100 IF (SS .LT. 0.0) STOP
10     10 UC = 0.
11        R = 0.
12        AL = 0.
13        D = 0.
14        AK = 0.
15        DS = DMN
16     15 RA = (3.14159*G*DS**4/(32.*AKMN*ALMX))**.5
17        RB = (3.14159*G*DS**4/(32.*AKMX*ALMX))**.5
18     20 IF (RB .GT. RMX) GO TO 60
19     25 IF (RA .LT. RMX) GO TO 35
20     30 RS = RMX
21        ALS = ALMX
22        GO TO 75
23     35 IF (RA .LT. RMN) GO TO 45
24     40 RS = RA
25        ALS = ALMX
26        GO TO 75
27     45 AL3 = 3.14159*G*DS**4/(32.*RMN**2*AKMN)
28     55 IF (AL3 .LT. ALMN) GO TO 60
29     50 RS = RMN
30        ALS = AL3
31     75 UCS = 3.14159*SS**2*DS**2*ALS/(16.*AN**2*AKT**2*G)
32        IF (UCS .LE. UC) GO TO 60
33     80 UC = UCS
34        R = RS
35        AL = ALS
36        D = DS
37     60 IF (DS .LT. DMX) GO TO 70
38     65 AK = 3.14159*G*D**4/(32.*R**2*AL)
39        WRITE (6,85) D,R,AL,AK,UC
40     85 FORMAT ('0','BEST DESIGN IS FOR D,R,L,K,UC ='/5E16.7)
41        GO TO 5
42     70 DS = DS + 1./32.
43        GO TO 15
44        END
```

```
      $ENTRY

 SPECS. RMN,RMX,LMN,LMX,KMN,KMX =
   0.90000E 01    0.12000E 02    0.24000E 02    0.36000E 02    0.10000E 03    0.20000E 03

   DMN,DMX,KT =    0.00000E 00    0.15000E 01    0.12000E 01

 MATERIAL PROPERTIES SS, G AND N =    0.7800000E 05    0.1200000E 08    0.1250000E 01

 BEST DESIGN IS FOR D,R,L,K,UC =
   0.9375000E 00    0.1200000E 02    0.3600000E 02    0.1755500E 03   | 0.1399911E 04 |

 MATERIAL PROPERTIES SS, G AND N =    0.5200000E 05    0.6000000E 07    0.1250000E 01

 BEST DESIGN IS FOR D,R,L,K,UC =
   0.1125000E 01    0.1200000E 02    0.3600000E 02    0.1820101E 03   | 0.1791886E 04 |
```

Figure 9.12   Digital computer program from the flow chart in Figure 9.11, giving optimum design of the torsion bar of example 9-1 (compilation time, 8.00 sec; execution time, 2.86 sec).

Characteristics: $k = 176$ lb/in.

$U_c = 1400$ in.-lb strain energy capacity

In conclusion, our solution of optimum design for the torsion-bar spring in the numerical example has satisfied all of the modes of system operation anticipated for the automobile being designed.   The objective of maximizing strain energy capacity is in effect one of maximizing reliability for the otherwise elusive and critical operational mode of unusual dynamic loads.   Minimizing the chance of failure in this way is often an appropriate technique for the design of critical elements in a system where there is uncertainty in loading because of factors beyond your control. For this particular example we cannot be certain of road conditions or operational tactics in the handling of the automobile once it is placed in the field. The torsion-bar spring now designed should have the minimum chance of failure for this situation of uncertainty, and yet it adheres to all of the constraints of significance as previously summarized.

9-3 DESIGN OF A HELICAL SPRING FOR A CAM-DRIVEN SYSTEM

A helical compression load spring is to be designed for the high-speed cam mechanism shown in Figure 9.13. At this stage of design we have estimated the value for follower mass M, and we have decided on the motion character-istics depicted in Figure 9.14 for the most critical region of the cam cycle. The helical spring must be designed to satisfy the system requirements for all conditions of operation.

In reference to Figure 9.13, spring force F must be sufficient to maintain contact between cam and follower without excessive loading. Knowing the motion characteristics of Figure 9.14, mass times acceleration may be plotted as the Ma curve of Figure 9.15. Applying Newton's second law from elementary dynamics to Figure 9.13, we express cam contact force F_C as the sum of spring force F and Ma, assuming that cam pressure angle will be reasonably small for good design.

$$F_C \approx F + Ma \qquad (9.14)$$

We know that spring force F will be linearly related to follower displacement s of Figure 9.14. Thus, based on satisfaction of equation (9.14), we now know the basic shape of cam contact force F_C versus cam angle, as shown in Figure 9.15. Hence, at this stage of design we could estimate quite accurately an appropriate value for maximum spring force F_{mx} which without being excessive would yet assure adequate minimum cam contact force $(F_C)_{min}$, both forces of which are depicted in Figure 9.15. Also, at this time we could estimate an appropriate value for minimum spring force F_{mn}. In fact, we could express an acceptable range for F_{mn} which is $F_S \leq F_{mn} \leq F_G$, where the boundary values are typically shown in Figure 9.15. Smallest acceptable value F_S for minimum spring force F_{mn} would be determined to assure adequate contact force F_C in the initial region at the far left of Figure 9.15. On the other hand, greatest acceptable value F_G for minimum spring force F_{mn} would be based on avoidance of an excessive value for the peak contact force $(F_C)_{max}$ shown in Figure 9.15. Thus, associated with the acceptable

Figure 9.13 Cam mechanism in uppermost position for design of the helical compression spring.

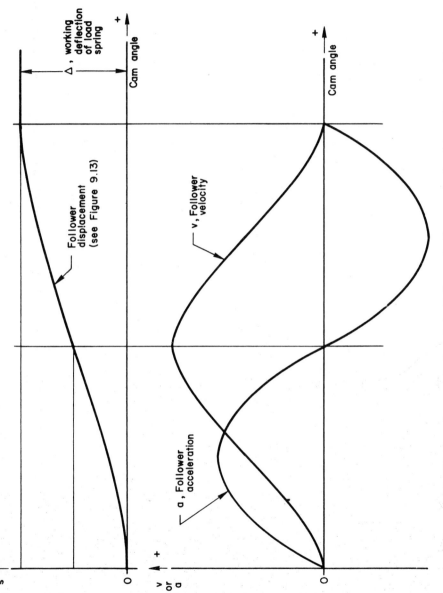

Figure 9.14   Motion characteristics of the follower in the cam mechanism of Figure 9.13.

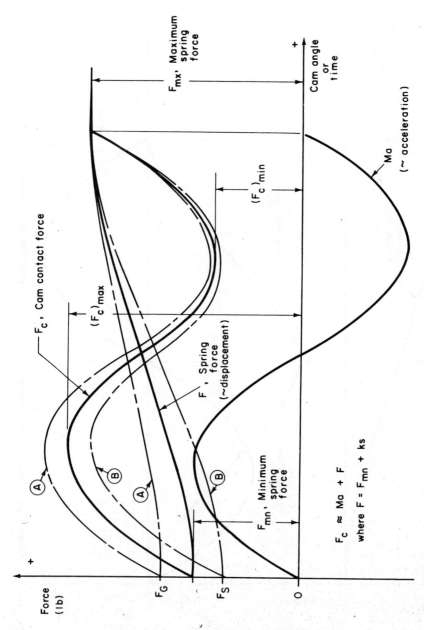

Figure 9.15   Dynamic force diagram for the cam mechanism of Figure 9.13 with the motion characteristics of Figure 9.14.

range $F_S \le F_{mn} \le F_G$, by equation (9.14) we have established bands bounded by curves A and B for spring force F and cam contact force F_C in Figure 9.15. From a force-deflection standpoint, the acceptable range now determined is shown in Figure 9.16.   From a time-variation standpoint, we know that

Figure 9.16   Acceptable force-deflection characteristics for the load spring of the cam mechanism.

spring force F will vary periodically as typically shown in Figure 9.17 for more than one cycle of the cam.

In addition to satisfying its functional requirement of delivering proper spring force characteristics to the cam mechanism to assure cam contact without excessive loading, the helical compression spring being designed must satisfy many other system requirements.   For one, it must fit within space constraints on length and diameter.   Also, it must use a standard wire size $d = d_s$ of which there are M in number, and the ratio of mean coil diameter to wire diameter must be within a certain range for acceptable practical proportions, expressed as $u_{min} \le u = D/d \le u_{max}$.   Finally, as much as possible we must avoid all of several modes of failure, since reliability of operation is of utmost importance in this particular application.

The most serious failure phenomenon in this particular application we

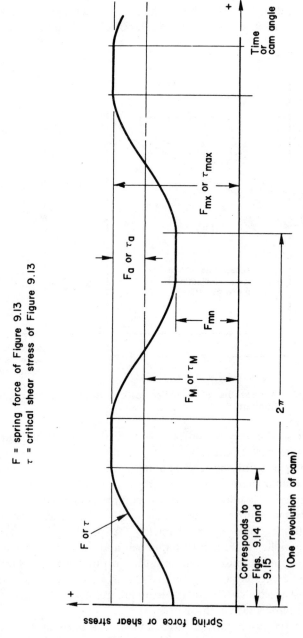

F = spring force of Figure 9.13
τ = critical shear stress of Figure 9.13

Figure 9.17   Spring force or critical stress variation with respect to time for the cam mechanism.

decide would be the occurrence of fatigue breakage, the probability of which we wish to minimize by optimum design. This we can relate to factor of safety as described in Chapter 6 of reference 9-9. Hence, our objective for optimum design will be to maximize reliability or to maximize factor of safety N_e against fatigue breakage. The failure criterion will be for the special case of pure torsion fatigue as typically presented on pp. 180 and 296 of reference 9-11, which is substantiated by the experimental data in Figure 7.8 of reference 9-12. Thus, based on this appropriate fatigue-failure criterion we can write primary design equation (9.15) for factor of safety N_e.

$$N_e = S_{se}/\tau_a \qquad \text{(P.D.E.) (9.15)}$$

In (P.D.E.) (9.15), S_{se} is the reversed shearing stress endurance limit of the material, and τ_a is the critical shear stress amplitude depicted in Figure 9.17. In reference 9-14, Zimmerli reported a value of $S_{se} \approx 45,000$ psi for any of the commonly used unpeened spring steels in sizes under $\frac{3}{8}$ in., such as for music wire, carbon valve-spring wire, chromium vanadium valve-spring wire, and chromium-silicon valve-spring wire. For peened steel wire the value becomes $S_{se} \approx 67,500$ psi.

Regarding the critical shear stress amplitude τ_a of Figure 9.17 in psi, we introduce (S.D.E.) (9.16), which is derived in many references on machine design, such as 9.8, 9.11, and 9.13.

$$\tau_a = K_{ts}\, 8\, F_a\, D/(\pi d^3) \qquad \text{(S.D.E.) (9.16)}$$

In this equation K_{ts} is the dimensionless Wahl correction factor which can be approximated in simple exponential form by (S.D.E.) (9.17) (from p. 480 of ref. 9-9), which is valid with good accuracy within the practical range of interest $4 \le (D/d) \le 20$.

$$K_{ts} \approx 1.60/(D/d)^{0.140} \qquad \text{(S.D.E.) (9.17)}$$

In this equation as well as in (S.D.E.) (9.16), D and d are mean coil diameter and wire diameter, respectively, in in., referring to Figure 9.13. Finally, in (S.D.E.) (9.16), F_a is force amplitude in lb, referring to Figure 9.17. Hence,

$$F_a = (F_{mx} - F_{mn})/2 \qquad \text{(S.D.E.) (9.18)}$$

where F_{mx} and F_{mn} are maximum and minimum spring forces, respectively, in lb, referring to Figure 9.17.

Another failure phenomenon which we wish to avoid is the occurrence of permanent set or yielding. Hence, referring to Figure 9.17, we require satisfaction of the regional constraint on maximum shear stress $\tau_{max} \le S_{sy}/(N_y)_{min}$, where S_{sy} is the shearing yield strength of the material in psi and $(N_y)_{min}$ is the smallest acceptable value for factor of safety against yielding, which has been estimated at this stage. For steel wire, shear strength S_{sy} is a function of diameter d in inches expressed in general by equation (9.19), where C_1 and C_2 are material properties readily determined.

$$S_{sy} \approx C_1/d^{C_2} \tag{9.19}$$

As a specific example, for A.S.5 music wire we can derive the relation

$$S_{sy} \approx 80,800/d^{0.145} \tag{9.19a}$$

For derivation of equation (9.19a) we use the distortion energy theory for the relation $S_{sy} \approx 0.577\, S_{ty}$, as suggested on p. 292 of reference 9-11; $S_{ty} \approx 0.7\, S_t$ and $S_t \approx 200,000/d^{0.145}$ are presented as equations (13.7) and (13.10), respectively, in reference 9-9 for A.S.5 music wire. Thus, for avoidance of permanent set or yielding in the spring we define a limited positive stress parameter y as follows, together with the acceptable regional constraint having a constant $C_1/(N_y)_{min}$ for the limit value:

$$\tau_{max} \leq S_{sy}/(N_y)_{min} \approx \frac{C_1}{(N_y)_{min}\, d^{C_2}}$$

therefore

$$y = \tau_{max}\, d^{C_2} \tag{S.D.E.)(9.20}$$

where for an acceptable design we require $y \leq C_1/(N_y)_{min}$. Maximum shear stress τ_{max} of Figure 9.17 is expressed by (S.D.E.) (9.21) in psi; this is similar to (S.D.E.) (9.16) except for the use of maximum spring force F_{mx} instead of force amplitude F_a.

$$\tau_{max} = K_{ts}\, 8\, F_{mx}\, D/(\pi d^3) \tag{S.D.E.)(9.21}$$

In this equation all terms have been defined, and F_{mx} is in lb.

Another failure phenomenon which we wish to avoid in this high-speed mechanism is excessive internal longitudinal vibrations, often called spring surging. This requires lowest natural frequency f_{n1} to be high enough to avoid resonance with any of the significant excitation frequencies of the given cam motion. Thus, we require $f_{n1} \geq (f_{n1})_{min}$; in the design of automotive valve springs, as a typical example, the value of $(f_{n1})_{min}$ is often taken as ten or twelve times the greatest cam shaft speed. Lowest natural frequency f_{n1} is derived on pp. 286–290 of reference 9-8, giving the following equation for steel compression springs excited at one end with the other end fixed, which is the case in Figure 9.13.

$$f_{n1} = 14,100\, d/(D^2 N) \tag{S.D.E.)(9.22}$$

In this equation f_{n1} is the lowest natural frequency for internal longitudinal vibrations in cps, d and D are wire and mean coil diameters, respectively, in in., and N is the number of active coils as depicted in Figure 9.13.

The final failure phenomenon which we wish to avoid in this particular application is the occurrence of buckling for our compression spring. An excellent approach to the problem is derived on pp. 76–78 of reference 9-13. Thus, conservatively assuming the case of hinged ends, and applying the terminology of Figure 9.13, for avoidance of buckling we require

$$L \leq \frac{11.15(D/2)^2}{(F_{mx}/k)}$$

However, referring to Figure 9.13, minimum active working length L in inches is related to allowable clearance between coils, which must be expressed by geometrical (S.D.E.) (9.23).

$$L = Nd\,(1 + c) \qquad \text{(S.D.E.) (9.23)}$$

In this equation N is the number of active coils, d is the wire diameter in inches, and c is the dimensionless percentage of d which is allowable for minimum practical clearance between coils, as depicted in Figure 9.13. Hence, at this time we can define a limited buckling parameter x as follows, which must satisfy the regional constraint $x \leq 1$ to guarantee avoidance of the buckling phenomenon.

$$L = Nd(1 + c) \leq 2.79D^2k/F_{mx}$$

$$x = \frac{Nd(1 + c)F_{mx}}{2.79D^2k} \qquad \text{(S.D.E.) (9.24)}$$

where for an acceptable design we require $x \leq 1$.

In (S.D.E.) (9.24), k is the well-known spring constant or force gradient, derived in many machine design books such as references 9-8, 9-11, and 9-13, giving the following equation.

$$k = Gd^4/(8D^3N)$$

For steel wire with the modulus of rigidity $G = (11.5)10^6$ psi, the equation becomes

$$k = (1.44)10^6 \, d^4/(D^3N) \qquad \text{(S.D.E.) (9.25)}$$

In (S.D.E.) (9.25), k is the force gradient in lb/in., an unspecified quantity in this particular application, and all other terms have already been defined.

Referring to Figure 9.16, we recognize that force gradient k is mathematically tied to constrained parameters F_{mx}, F_{mn}, and Δ as follows.

$$F_{mn} = F_{mx} - k\Delta \qquad \text{(S.D.E.) (9.26)}$$

All terms in this equation have already been defined except for cam throw, Δ, in inches, shown in Figures 9.13, 9.14, and 9.16.

As previously mentioned, practical proportions for (D/d) and space constraints on diameter and spring pocket length must be satisfied, and typically they are $u_{min} \leq u = (D/d) \leq u_{max}$, $D_0 \leq (D_0)_{max}$, $D_i \geq (D_i)_{min}$, $L_{pmx} \leq (L_{pmx})_{max}$, respectively.   Hence, referring to Figure 9.13 we must include the following (S.D.E.)s in the Initial Formulation system of equations.

$$u = D/d \qquad \text{(S.D.E.) (9.27)}$$

$$D_0 = D + d \qquad \text{(S.D.E.) (9.28)}$$

$$D_i = D - d \qquad \text{(S.D.E.) (9.29)}$$

$$L_{pmx} = L + 2d + \Delta \qquad \text{(S.D.E.) (9.30)}$$

In these geometric (S.D.E.)s all terms are obviously defined in Figure 9.13, except perhaps L_{pmx}, which is the spring pocket length with follower in the lowest position of the figure. L_{pmx} can be related to spring pocket length space available in the mechanism setting, as expressed by the regional constraint on L_{pmx} previously given.   Finally, we should mention that there is an acceptable·minimum value, N_{min}, for active number of coils N, such as 3, which must be satisfied for obtainment of a practical spring.   Hence, we must also adhere to the regional constraint $N \geq N_{min}$ on permissible values for N, in order to obtain a manufacturable spring whose characteristics can be adequately controlled.

At this point we have briefly described the system of equations which our helical spring design must satisfy, with the objective of maximizing factor of safety N_e against fatigue breakage in a critical design situation where reliability of operation is of utmost importance.   This Initial Formulation system of equations is summarized in Table 9-1.   It assumes the use of steel wire, and (9.19) is not included since it merely defines material constants C_1 and C_2.

From the Initial Formulation equation system given in Table 9-1 we must decide on an approach for the making of an optimum design variation study, in accordance with the general plan outlined in Figure 8.2 of Chapter 8. To facilitate our initial efforts in this respect, we will make some very simple calculations from equations of Chapter 8, which will indicate at this point what lies ahead.   First, from inspection of the initial formulation summary we see that there are fourteen (S.D.E.)s, so $N_s = 14$.   Next, from that summary we also note that there are seven variables without constraints directly imposed, which are the free variables τ_a, K_{ts}, F_a, D, L, k, and τ_{max}.   Hence, for the number of free variables we have $n_f = 7$.   Thus, in reference to Figure 8.2, we have the case of redundant specifications since $n_f = 7 < N_s = 14$.

In reference to the initial formulation equation system from Table 9-1 we see that there are ten variables with constraints directly imposed, which are the constrained variables d, F_{mn}, y, f_{n1}, N, x, u, D_o, D_i, and L_{pmx}.   Hence, the number of constrained variables is $n_c = 10$, and from equation (8.7) the

TABLE 9-1 INITIAL FORMULATION SUMMARY

$$N_e = S_{se}/\tau_a \qquad \text{(P.D.E.) (9.15)}$$

$$\tau_a = K_{ts} 8 F_a D/(\pi d^3) \qquad \text{(S.D.E.) (9.16)}$$

$$K_{ts} \approx 1.60/(D/d)^{0.140} \qquad \text{(S.D.E.) (9.17)}$$

$$F_a = (F_{mx} - F_{mn})/2 \qquad \text{(S.D.E.) (9.18)}$$

$$y = \tau_{max}\, d^{C2} \qquad \text{(S.D.E.) (9.20)}$$

$$\tau_{max} = K_{ts} 8 F_{mx}\, D/(\pi d^3) \qquad \text{(S.D.E.) (9.21)}$$

$$f_{n1} = 14{,}100\, d/(D^2 N) \qquad \text{(S.D.E.) (9.22)}$$

$$L = Nd\,(1 + c) \qquad \text{(S.D.E.) (9.23)}$$

$$x = Nd\,(1 + c)\, F_{mx}/(2.79\, D^2 k) \qquad \text{(S.D.E.) (9.24)}$$

$$k = (1.44)\, 10^6 d^4/(D^3 N) \qquad \text{(S.D.E.) (9.25)}$$

$$F_{mn} = F_{mx} - k\Delta \qquad \text{(S.D.E.) (9.26)}$$

$$u = D/d \qquad \text{(S.D.E.) (9.27)}$$

$$D_o = D + d \qquad \text{(S.D.E.) (9.28)}$$

$$D_i = D - d \qquad \text{(S.D.E.) (9.29)}$$

$$L_{pmx} = L + 2d + \Delta \qquad \text{(S.D.E.) (9.30)}$$

$$\left.\begin{array}{l} d_{min} \le d = d_s \le d_{max} \text{ for } s = 1, \ldots, M; \\[4pt] F_S \le F_{mn} \le F_G;\ y \le C_1/(N_y)_{min}; \\[4pt] f_{n1} \ge (f_{n1})_{min};\ N \ge N_{min}; \\[4pt] x \le 1;\ u_{min} \le u \le u_{max}; \\[4pt] D_o \le (D_o)_{max};\ D_i \ge (D_i)_{min}; \\[4pt] L_{pmx} \le (L_{pmx})_{max} \end{array}\right\} \quad \text{(L.E.)s}$$

Specified or Known: S_{se}, F_{mx}, d_s standard wire sizes M in number, F_S, F_G, C_1, C_2, $(N_y)_{min}$, $(f_{n1})_{min}$, N_{min}, c, Δ, u_{min}, u_{max}, $(D_o)_{max}$, $(D_i)_{min}$, $(L_{pmx})_{max}$

Find: Design of helical compression spring which maximizes N_e.

---

total number of variables not counting the optimization quantity is $n_v = n_f + n_c = 7 + 10 = 17$. Therefore, from equation (8.13) we calculate for the number of dimensions D_{vs} required for the variation study

$$D_{vs} = n_v - N_s + 1 = 17 - 14 + 1 = 4$$

which may possibly be reduced if there are truly independent parameters in the final formulation. From equation (8.10) we see that the number of equations in the final formulation would be

$$N_{ff} = N_s - n_f + 1 = 14 - 7 + 1 = 8$$

Also, from equation (8.18) we see that the number of eliminated parameters would be

$$n_e = N_s - n_f = 14 - 7 = 7$$

The theoretical number of approaches which we might take for derivation of a final formulation is calculated from equation (8.19) as follows

$$A_t = \frac{n_c!}{n_e!(n_c - n_e)!} = \frac{10!}{7!(10 - 7)!} = \frac{(10)(9)(8)}{(3)(2)} = 120$$

To save on space and time, we will not investigate the 120 theoretical approaches to find how many of these are indeterminate. Instead, we will be more practical and attempt to find a workable approach which is determinate.

The number of dimensions required for the variation study would be reduced to three if we would hold one of the variables constant, such as wire diameter $d = d_s$ at a standard size value. Since this discrete value constraint on d is of finite number, we could employ the digital computer for finding the overall optimum design, as previously illustrated in examples 8-6 and 9-1. Therefore, at this time we make the decision to hold wire diameter constant at $d = d_s$ for our optimum design variation study. Thus, on review of the initial formulation we now realize that a further simplification is thereby gained, as will now be explained. Hence, with $d = d_s$ constant, from (S.D.E.)s (9.20), (9.21), (9.17), (9.27), (9.28), and (9.29) we see that the constraints $y \le C_1/(N_y)_{min}$, $u_{min} \le u \le u_{max}$, $D_o \le (D_o)_{max}$, and $D_i \ge (D_i)_{min}$ can all be expressed as direct constraints on the single variable of mean coil diameter D_s, with subscript s merely designating the standard wire size under consideration. More specifically, from (S.D.E.)s (9.20), (9.21), and (9.17) we have

$$y = \tau_{max} d_s^{C_2} = \frac{K_{ts} 8 F_{mx} D_s}{\pi d_s^{(3-C_2)}}$$

$$= \frac{1.60}{(D_s/d_s)^{0.140}} \frac{8 F_{mx} D_s}{\pi d_s^{(3-C_2)}}$$

$$= \frac{4.07 F_{mx} D_s^{0.86}}{d_s^{(2.86-C_2)}} \le \frac{C_1}{(N_y)_{min}}$$

thus

$$D_s \leq \left[\frac{C_1 d_s^{(2.86 - C_2)}}{4.07 F_{mx}(N_y)_{min}} \right]^{1.163}$$

Hence, mean coil diameter D_s must satisfy the constraint

$$D_s \leq D_{ymx} \qquad \text{(L.E.) (9.31)}$$

for avoidance of permanent set or yielding, where D_{ymx} is a constant for d_s, defined by equation (9.32) as follows from the preceding derivation.

$$D_{ymx} = \left[\frac{C_1 d_s^{(2.86 - C_2)}}{4.07 F_{mx}(N_y)_{min}} \right]^{1.163} \qquad (9.32)$$

Similarly, from (S.D.E.) (9.27) we obtain the following acceptable range for D_s as governed by requirements of practical proportions.

$$u_{min} \leq u = D_s/d_s \leq u_{max}$$

therefore

$$D_{umn} \leq D_s \leq D_{umx} \qquad \text{(L.E.) (9.33)}$$

where

$$D_{umn} = u_{min} d_s \qquad (9.34)$$

$$D_{umx} = u_{max} d_s \qquad (9.35)$$

Likewise, from (S.D.E.)s (9.28) and (9.29) we obtain the following acceptable range for D_s as governed by space constraints on D_o and D_i.

$$D_o = D_s + d_s \leq (D_o)_{max}$$

$$D_i = D_s - d_s \geq (D_i)_{min}$$

therefore

$$D_{min} \leq D_s \leq D_{max} \qquad \text{(L.E.) (9.36)}$$

where

$$D_{min} = (D_i)_{min} + d_s \qquad (9.37)$$

$$D_{max} = (D_o)_{max} - d_s \qquad (9.38)$$

Of course we realize that *all* three of the regional constraints (9.31), (9.33), and (9.36) on D_s must be satisfied for an acceptable spring. The right sides of these (L.E.)s will determine a *governing* upper limit, D_{mx}, on permissible values for D_s. Likewise, the left sides of (L.E.)s (9.33) and (9.36) will determine a *governing* lower limit D_{mn} on permissible values of D_s. This acceptable range for D_s is expressed as follows.

$$D_{mn} \leq D_s \leq D_{mx} \qquad \text{(L.E.) (9.39)}$$

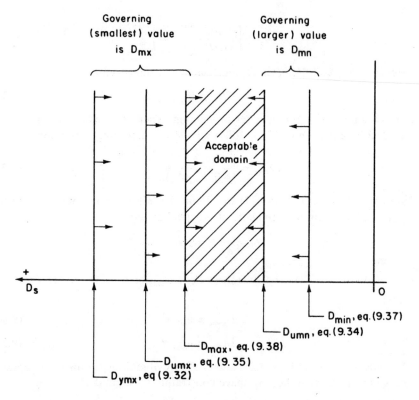

Figure 9.18 Typical relationship of constraints on D_s for determination of D_{mn} and D_{mx} or incompatible specifications.

The acceptable range for D_s now described is also explained graphically in Figure 9.18. Hence, we readily recognize the possibility of incompatible specifications, if $D_{mn} > D_{mx}$. Also, the flow chart for determining D_{mn} and D_{mx} and this possibility of incompatible specifications is thereby readily derived as presented in Figure 9.19, with the station numbers at the various blocks of that figure corresponding to the computer program statement numbers in Figure 9.28. Thus, we have simplified the Initial Formulation by replacing the constraints $C_1/(N_y)_{min}$, u_{min}, u_{max}, $(D_i)_{min}$, and $(D_o)_{max}$ with the constraints D_{mn} and D_{mx} on the single variable D_s, which we see also enables us to disregard from further consideration the (S.D.E.)s (9.20), (9.21), (9.27), (9.28), and (9.29). These equations are not mathematically tied in any other way to the rest of the initial formulation.

Similarly, upon further review of the initial formulation we note that the constraint on L_{pmx} may be replaced by one on L, since for our variation study we now have constant $d = d_s$. This replacement we readily see from (S.D.E.)

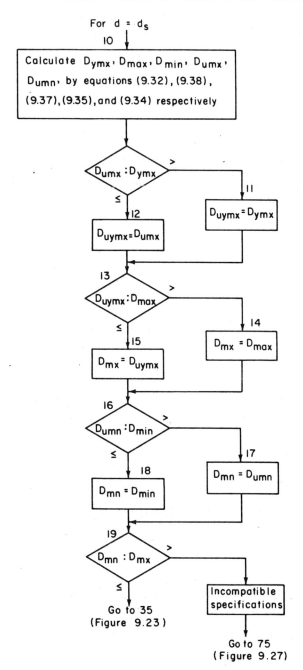

For d = d_s

10

Calculate D_{ymx}, D_{max}, D_{min}, D_{umx}, D_{umn}, by equations (9.32), (9.38), (9.37), (9.35), and (9.34) respectively

$D_{umx} : D_{ymx}$

11    $D_{uymx} = D_{ymx}$

12    $D_{uymx} = D_{umx}$

13    $D_{uymx} : D_{max}$

14    $D_{mx} = D_{max}$

15    $D_{mx} = D_{uymx}$

16    $D_{umn} : D_{min}$

17    $D_{mn} = D_{umn}$

18    $D_{mn} = D_{min}$

19    $D_{mn} : D_{mx}$

Go to 35 (Figure 9.23)

Incompatible specifications

Go to 75 (Figure 9.27)

Figure 9.19   Flow chart for determining D_{mn} and D_{mx} or incompatible specifications.

(9.30). Hence, again using subscript s to designate the wire size under consideration, we can consider L_s constrained as follows, (S.D.E.)(9.30) having no further use in the initial formulation.

$$L_s \leq (L_s)_{max} \qquad \text{(L.E.) (9.40)}$$

where from (S.D.E.) (9.30) we have

$$(L_s)_{max} = (L_{pmx})_{max} - 2d_s - \Delta \qquad (9.41)$$

With the simplifications now proposed, our initial formulation for a standard wire size of interest with $d = d_s$ constant has been reduced from Table 9-1 to the following:

Initial Formulation for $d = d_s$

$$N_e = S_{se}/\tau_a \qquad \text{(P.D.E.) (9.15)}$$

$$\tau_a = K_{ts}\, 8\, F_a\, D_s/(\pi d_s^3) \qquad \text{(S.D.E.) (9.16a)}$$

$$K_{ts} \approx 1.60/(D_s/d_s)^{0.140} \qquad \text{(S.D.E.) (9.17a)}$$

$$F_a = (F_{mx} - F_{mn})/2 \qquad \text{(S.D.E.) (9.18)}$$

$$f_{n1} = 14{,}100\, d_s/(D_s^2 N) \qquad \text{(S.D.E.) (9.22a)}$$

$$L_s = N d_s(1 + c) \qquad \text{(S.D.E.) (9.23a)}$$

$$x = N d_s(1 + c)\, F_{mx}/(2.79\, D_s^2 k) \qquad \text{(S.D.E.) (9.24a)}$$

$$k = (1.44)\, 10^6 d_s^4/(D_s^3 N) \qquad \text{(S.D.E.) (9.25a)}$$

$$F_{mn} = F_{mx} - k\Delta \qquad \text{(S.D.E.) (9.26)}$$

$$\left.\begin{array}{l} F_S \leq F_{mn} \leq F_G; f_{n1} \geq (f_{n1})_{min}; \\ N \geq N_{min}; x \leq 1; L_s \leq (L_s)_{max}; \\ D_{mn} \leq D_s \leq D_{mx} \end{array}\right\} \qquad \text{(L.E.)s}$$

Specified or Known: $S_{se}, F_{mx}, d_s, F_S, F_G, (f_{n1})_{min}, N_{min}, c, \Delta, D_{mn}, D_{mx}, (L_s)_{max}.$

Find: Design of helical compression spring which maximizes N_e (i.e., best design for the value $d = d_s$ under consideration).

Similar to previous calculations, we will apply equations (8.7), (8.10), (8.13), (8.18), and (8.19) to the initial formulation for $d = d_s$ constant, the results of which are summarized next.

$$n_f = 4 \qquad (\tau_a, K_{ts}, F_a, k)$$
$$n_c = 6 \qquad (F_{mn}, f_{n1}, N, x, D_s, L_s)$$
$$n_v = n_c + n_f = 6 + 4 = 10$$

Also,
$$N_s = 8$$
therefore
$$D_{vs} = n_v - N_s + 1$$
$$= 10 - 8 + 1 = 3$$
$$N_{ff} = N_s - n_f + 1$$
$$= 8 - 4 + 1 = 5$$
$$n_e = N_s - n_f = 8 - 4 = 4$$
$$A_t = \frac{n_c!}{n_e!\,(n_c - n_e)!} = \frac{6!}{4!\,(6-4)!}$$
$$= (6)(5)/2 = 15$$

From the above calculations we see that the optimum design variation study problem has been appreciably simplified by holding $d = d_s$ constant, which later can be varied in discrete steps by use of the digital computer.

For derivation of the final formulation we will combine the initial formulation for the $d = d_s$ system of equations by eliminating free variables τ_a, K_{ts}, F_a, and k. Of the fifteen theoretical number of approaches possible, a fairly simple determinate one is obtained by choosing of the constrained variables F_{mn}, f_{n1}, N, and x as the four eliminated parameters, with D_s and L_s becoming the related parameters in the final formulation format which we will now derive, to be set up as follows in general notation:

$$N_e = f_1(D_s, L_s) \tag{I}$$
$$f_{n1} = f_2(D_s, L_s) \tag{II}$$
$$x = f_3(D_s, L_s) \tag{III}$$
$$F_{mn} = f_4(D_s, L_s) \tag{IV}$$
$$N = f_5(D_s, L_s) \tag{V}$$

The specific functions f_1, f_2, f_3, f_4, and f_5 will be derived next by the equation-combination procedure, which was explained and illustrated many times in Chapter 8 for the method of optimum design.

For derivation of developed (P.D.E.) of the ideal problem (I), we refer to the initial formulation for $d = d_s$ and combine (S.D.E.)s (9.18), (9.26), (9.25a), (9.23a), (9.16a), and (9.17a) with (P.D.E.) (9.15), eliminating F_{mn}, k, N, τ_a, F_a, and K_{ts}, respectively. For derivation of relating equation (II) we merely combine (S.D.E.)s (9.23a) with (9.22a), eliminating N. For derivation of relating equation (III) we combine (S.D.E.)s (9.25a) and (9.23a) with (9.24a), eliminating k and N, respectively.   For derivation of relating equation (IV) we combine (S.D.E.)s (9.25a) and (9.23a) with (9.26), eliminating k and N, respectively.   For derivation of relating equation (V) we merely solve (S.D.E.) (9.23a) explicitly for N.   In this way we derive our final formulation for $d = d_s$, which is summarized next in the proper format outlined previously for a three-dimensional variation study.

Final Formulation for $d = d_s$

$$N_e \approx \frac{S_{se} D_s^{2.14} L_s}{(2.935)10^6 \Delta (1 + c) d_s^{2.14}} \qquad \text{(I)}$$

$$f_{n1} = 14{,}100\,(1 + c) d_s^2 / (D_s^2 L_s) \qquad \text{(II)}$$

$$x = F_{mx} L_s^2 D_s / [(4.02)10^6 (1 + c) d_s^5] \qquad \text{(III)}$$

$$F_{mn} = F_{mx} - \frac{(1.44)10^6 (1 + c)\Delta d_s^5}{D_s^3 L_s} \qquad \text{(IV)}$$

$$N = L_s / [d_s (1 + c)] \qquad \text{(V)}$$

where

f_{n1}, x, F_{mn}, N = eliminated parameters

D_s, L_s = related parameters

(I) = developed (P.D.E.) of ideal problem

(II), (III), (IV), (V) = relating equations

For the final formulation system of equations for $d = d_s$, the arrangement is in the proper format for the making of an optimum design variation study. For the ideal problem, from developed (P.D.E.) (I) we sketch the three-dimensional variation diagram of Figure 9.20, where point 1 ideally would be best for maximization of N_e.   The typical effects of the regional constraints on the related parameters D_s and L_s, from (L.E.)s (9.39) and (9.40), are shown thereon.   They are $D_{mn} \le D_s \le D_{mx}$ and $L_s \le (L_s)_{max}$, respectively.   However, in the total picture we must also include the effects of constraints on the eliminated parameters f_{n1}, x, F_{mn}, and N.   Hence, following the method of optimum design described in Chapter 8, we will superimpose the relating equation function variations of (II), (III), (IV), and (V) on the diagram, placing the eliminated parameters along the same axis as the optimization quantity N_e.   The constraints $f_{n1} \ge (f_{n1})_{min}$, $x \le 1$, $F_S \le F_{mn} \le F_G$, and $N \ge N_{min}$, when imposed on these three-dimensional function variations, establish boundary curves of significance as shown typically in the plan view of the total diagram typically sketched in Figure 9.21.   If the constraints fell as specifically shown in that figure, point 12 would be the best design for maximization of N_e with that particular wire size d_s.   However, in general there

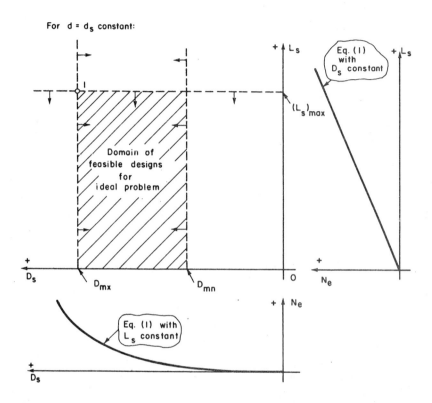

Figure 9-20 Typical three-dimensional variation diagram for the ideal problem for optimum design of the helical compression spring. The ideal point of optimum design is 1, for maximization of N_e.

are nine points *possible* for optimum design, which are indicated as 1, 2, 3, 4, 5, 6, 7, 11, and 12 in Figure 9.21. This may not be so obvious without some further explanation.

The fact that point 1 is the best design for maximization of N_e is obvious from Figure 9.20, and this point we would choose if it were accessible in the total picture. Let us next impose relating equations (II) and (III) on the ideal problem diagram, with the regional constraints $f_{n1} \geq (f_{n1})_{min}$ and $x \leq 1$ resulting in two boundary curves of significance, as shown in the typical plan view diagram of Figure 9.22(a). These regional constraints, together with the corresponding relating equations (II) and (III), clearly indicate which side of each boundary curve is acceptable, as shown by the arrows from each curve

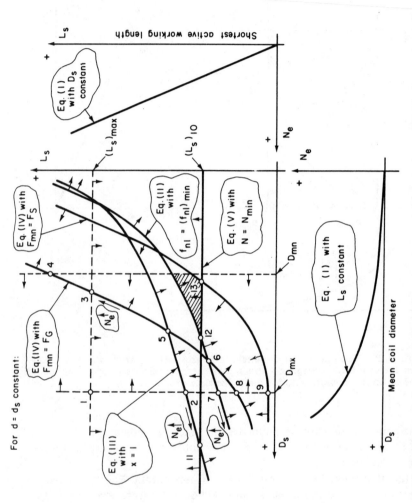

Figure 9.21 Typical three-dimensional variation diagram for optimum design of helical compression spring with objective of maximizing N_e. Points possible for optimum design are 1, 2, 3, 4, 5, 6, 7, 11, or 12. (See final formulation summary for equations I, II, III, IV, and V. See flow chart in Figure 9.19 for determination of D_{mn} and D_{mx}.)

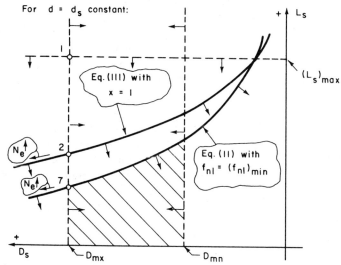

(a) Typical plan view of Figure 9.20 variation diagram with relating equations (II) and (III) and constraints on x and f_{nl} imposed.

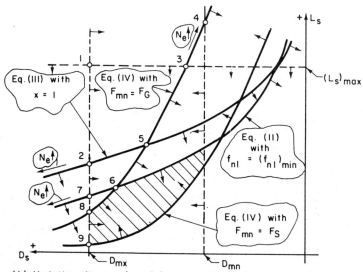

(b) Variation diagram from (a) above with relating equation (IV) and constraints on F_{mn} imposed.

Figure 9.22 Typical plan view diagrams used for explanation of Figure 9.21. (a) Typical plan view of the variation diagram in Figure 9.20 with relating equations II and III and constraints imposed on x and f_{nl}. Points possible for optimum design are 1, 2, and 7, for maximizing N_e. **(b)** Variation from (a) with relating equation IV and constraints imposed on F_{mn}. Points possible for optimum design are 1, 2, 3, 4, 5, 6, and 7.

in Figure 9.22(a).   Along each boundary curve of that diagram, N_e would be maximized by going in the direction of increasing D_s as indicated thereon. This fact is readily concluded from the final formulation first by combining (II) having $f_{n1} = (f_{n1})_{min}$ with (I) and eliminating L_s, showing that $N_e \sim D_s^{2.14}L_s \sim D_s^{2.14}(1/D_s^2) = D_s^{0.14}$.   From this proportionality, we see that N_e increases as D_s increases along that boundary curve.   Next, by combining (III) having $x = 1$ with (I) of the final formulation by eliminating L_s gives $N_e \sim D_s^{2.14}L_s \sim D_s^{2.14}(1/D_s^{1/2}) = D_s^{1.64}$: so we see that N_e increases as D_s increases along that boundary curve.   Thus, in the plan view diagram of Figure 9.22(a) we now recognize that the points possible for optimum design are 1, 2, and 7, with the one of *smallest* L_s value being the governing choice.

Let us now carry the explanation further by imposing relating equation (IV) from the final formulation on the diagram in Figure 9.22(a); the regional constraint $F_S \leq F_{mn} \leq F_G$ results in the two additional boundary curves of significance as typically shown in Figure 9.22(b).   For these simple exponential curves thus sketched on that diagram merely consider relating equation (IV) in the following form

$$(1.44)10^6(1 + c)\Delta d_s^5/(D_s^3 L_s) = (F_{mx} - F_{mn})$$

and realize from Figure 9.16 that we have the relationship $F_{mx} > F_G > F_S$, so $(F_{mx} - F_S) > (F_{mx} - F_G)$.   Hence, the upper of these two curves in Figure 9.22(b) is for relating equation (IV), where $F_{mn} = F_G$.   Along that boundary we have, referring first to (P.D.E.) (I),

$$N_e \sim D_s^{2.14}L_s \sim D_s^{2.14}(1/D_s^3) = 1/D_s^{0.86}$$

which shows that N_e increases as D_s decreases along the curve, as indicated on the diagram.   Therefore, along that boundary curve the greatest acceptable N_e value would be for either point 3, 4, 5, or 6, the governing value being for the one of those points having the smallest L_s value.   Also, it should be mentioned that the boundary curves for relating equations (II), (III), and (IV) intersect with relative slopes as shown in Figure 9.22(b).   This is obvious from the final formulation relating equations (II), (III), and (IV) where for the boundary curves of Figure 9.22(b) we have $L_s \sim 1/D_s^2$, $L_s \sim 1/D_s^{1/2}$ and $L_s \sim 1/D_s^3$, respectively.   The exponents of D_s reveal these relative slopes.

From an understanding of the variation study carried through to Figure 9.22(b), which is Figure 9.21 without the effects of relating equation (V) imposed, we can program the procedure of optimum design as presented in the flow chart of Figure 9.23.   The various calculations and comparisons required are obvious from the variation diagram of Figure 9.21, using the particular boundary values and relating equations indicated thereon.   More specifically, the various calculations so determined for particular points of interest are described in Table 9-2.   Hence, at the conclusion of the flow

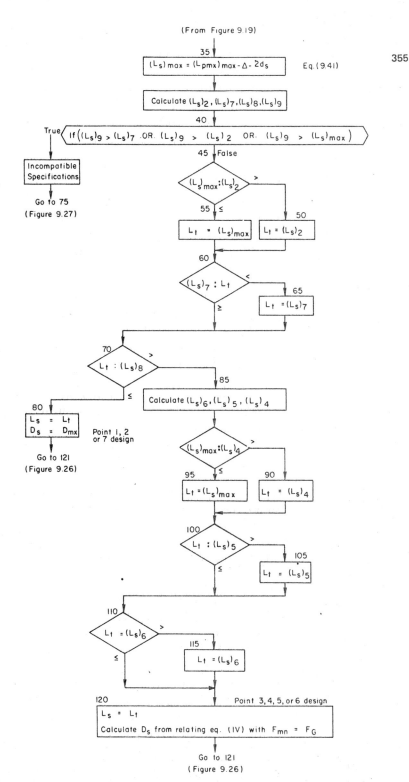

Figure 9.23 Flow chart for determination of optimum design in Figure 9.22(b). (See Table 9-2 for calculations.)

TABLE 9-2   CALCULATION OF PERTINENT VALUES IN FIGURE 9.21

Point	Value	Calculated by: (see Figure 9.21 and Final Formulation)
—	D_{mx}	Figure 9.19 flow chart
—	D_{mn}	Figure 9.19 flow chart
2	$(L_s)_2$	Relating equation (III) with $x = 1$ and $D_s = D_{mx}$
7	$(L_s)_7$	Relating equation (II) with $f_{n1} = (f_{n1})_{min}$ and $D_s = D_{mx}$
8	$(L_s)_8$	Relating equation (IV) with $F_{mn} = F_G$ and $D_s = D_{mx}$
9	$(L_s)_9$	Relating equation (IV) with $F_{mn} = F_S$ and $D_s = D_{mx}$
6	$(L_s)_6$	Simultaneous solution of relating equation (II) with $f_{n1} = (f_{n1})_{min}$, and relating equation (IV) with $F_{mn} = F_G$
5	$(L_s)_5$	Simultaneous solution of relating equation (III) with $x = 1$, and relating equation (IV) with $F_{mn} = F_G$
4	$(L_s)_4$	Relating equation (IV) with $F_{mn} = F_G$ and $D_s = D_{mn}$
11	$(D_s)_{11}$	Simultaneous solution of relating equation (III) with $x = 1$, and relating equation (V) with $N = N_{min}$
12	$(D_s)_{12}$	Simultaneous solution of relating equation (II) with $f_{n1} = (f_{n1})_{min}$, and relating equation (V) with $N = N_{min}$
13	$(D_s)_{13}$	Simultaneous solution of relating equation (IV) with $F_{mn} = F_S$, and relating equation (V) with $N = N_{min}$

chart in Figure 9.23 we have determined which of the points 1, 2, 3, 4, 5, 6, or 7 is optimum or if incompatible specifications exists for the particular d_s value under consideration, ignoring temporarily the effects of relating equation (V), where $N \geq N_{min}$. Also, it should be mentioned that the station numbers at the blocks of the flow chart in Figure 9.23 correspond to the statement numbers in the computer program in Figure 9.28. Incidentally, in Figure 9.23, terminology associated with digital computer work, such as with the Fortran IV language, is again employed as in previous flow charts. Also, it should be mentioned that L_t in Figure 9.23 is used merely for designating a temporary value for L_s.

Let us now carry the explanation of Figure 9.21 to conclusion by imposing relating equation (V) (from the final formulation) on the diagram in Figure 9.22(b); the regional constraint $N \geq N_{min}$ results in a boundary line of significance as typically shown in the plan view of Figure 9.21. From this equation and regional constraint we readily see that the domain of acceptability is above that boundary line, as indicated by arrows in Figure 9.21. Obviously some interesting possibilities result because of this final step in the variation study, as we will now explain.

Relating equation (V) with $N = N_{min}$ results in the particular boundary line of constant L_s value which we will designate $(L_s)_{10}$, calculated thereby from equation (9.42).

$$(L_s)_{10} = N_{min} d_s (1 + c) \tag{9.42}$$

If we impose this boundary line on the ideal problem variation diagram of Figure 9.20, we obtain the typical diagram shown in Figure 9.24.   Obviously, with maximization of N_e in mind, the only effect of this constraint in the ideal problem would be possibly to exclude the obtainment of ideal optimum design point 1 in Figure 9.24, with incompatible specifications existing if $(L_s)_{10} > (L_s)_{max}$.   Let us next impose the constraint of this boundary line, $L_s = (L_s)_{10}$, on the variation diagram of Figure 9.22(a), which we typically show in the variation diagram of Figure 9.25.   Obviously, this constraint might possibly prevent us from reaching either optimum design point 7 or 2 in the diagram,

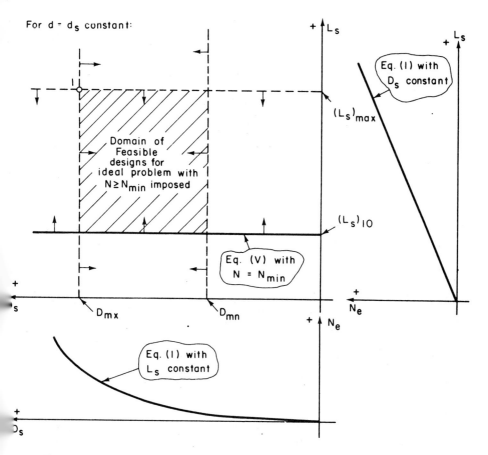

Figure 9.24   Typical effect of relating equation (V), with $N \sim N_{min}$ imposed on the ideal problem variation diagram of Figure 9.20. The only point possible for optimum design is 1. (See final formulation summary for equations (I) and (V).   See the flow chart in Figure 9.19 for determination of D_{mn} and D_{mx}.)

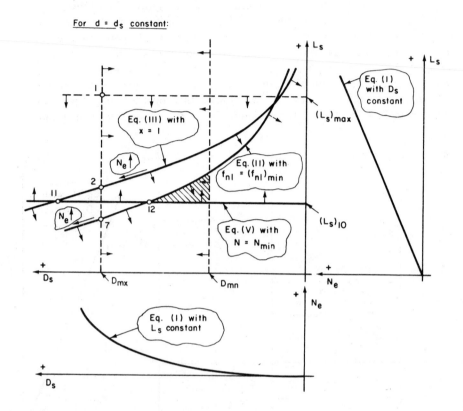

Figure 9.25 Typical effect of relating equation (V) with $N \geq N_{min}$ imposed on the variation diagram of Figure 9.22(a). Points possible for optimum design are 1, 2, 7, 11, and 12.

requiring us to settle for either optimum design point 12 or 11, respectively. If the constraints fell as specifically shown in Figure 9.25 (or in Figure 9.21), point 12 would be the best possible design for maximization of N_e with that particular value of d_s under consideration. Also, from the typical diagram of Figure 9.25 we recognize that possibilities of incompatible specifications exist if either $(D_s)_{11} < D_{mn}$ or $(D_s)_{12} < D_{mn}$. Let us finally impose the constraint of boundary line $L_s = (L_s)_{10}$ on the variation diagram of Figure 9.22(b), which results in the total diagram as typically shown in Figure 9.21. Obviously the intersection of this line with either one of the two boundary curves of relating equation (IV) that are shown in the figure does not result in a point of optimum design, since N_e increases as L_s increases along those curves. However, the boundary line of $L_s = (L_s)_{10}$ might possibly prevent us from reaching either point 6, 5, 3, or 4, which are possible ones for optimum design as previously discussed in the explanation of Figure 9.22(b). The first two of these possibilities for exclusion could result in either point 12 or 11, respectively, being the optimum design, unless incompatible specifications existed where either $(D_s)_{13} > (D_s)_{12}$ or $(D_s)_{13} > (D_s)_{11}$. On the other hand, the last two of these possibilities for exclusion would result in incompatible specifications where either $(L_s)_{10} > (L_s)_{max} = (L_s)_3$ or $(L_s)_{10} > (L_s)_4$.

The possibilities of optimum design or incompatible specifications introduced by relating equation (V) with $N > N_{min}$ required, as now discussed, are summarized in the flow chart of Figure 9.26 presented as a continuation of the flow chart of Figure 9.23. Hence, we start the flow chart with the particular governing L_s value determined from Figure 9.23, from either optimum design point 1, 2, 7, 3, 4, 5, or 6 of Figure 9.22(b). As before, the station numbers at the blocks of the flow chart in Figure 9.26 correspond to the statement numbers in the digital computer program of Figure 9.28. Also, similar to the previous flow charts, the subscript t in Figure 9.26 merely designates a temporary value, in this case for D_s. Also, we should mention that stations 75 and 125 will be described in the overall flow chart of Figure 9.27. As before, the calculations for $(D_s)_{11}$, $(D_s)_{12}$, and $(D_s)_{13}$ in Figure 9.26 are made referring to Figure 9.21 and the final formulation, and these are more specifically described in Table 9-2.

Before considering a numerical example, let us tie together the results of our variation study as summarized in the flow charts of Figures 9.19, 9.23, and 9.26. Hence, we present the overall optimum design flow chart in Figure 9.27. In brief description, the input boundary values must first be read. For each of the standard wire sizes of interest, d_s, the components of Figure 9.27 (Figures 9.19, 9.23, and 9.26) must be executed. Pertinent values are retained from the particular best design of largest N_e value which has been encountered thus far in the calculations. Finally, after all standard wire sizes of interest have been considered for determination of their best designs, a sentinel card of negative d_s value is encountered which concludes this stage of calculations. Then, characteristics of interest are calculated for the overall optimum design starting at station 205 of Figure 9.27. The pertinent results are printed in the end for the design of greatest possible reliability or N_e value. Of course, if a case of incompatible specifications exists for all of the standard wire sizes considered, there is no design solution which satisfies all of the constraints of the particular numerical problem, and the discovery of such a situation would then be recorded.

The digital computer program of Figure 9.28 was written in Fortran

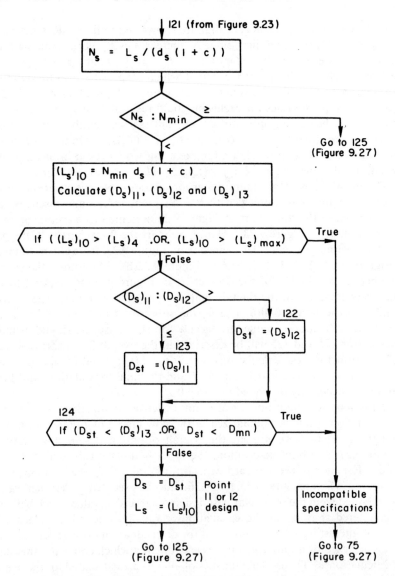

Figure 9.26   Flow chart for inclusion of effects from relating equation (V) with $N \geq N_{min}$ when imposed on Figure 9.21. (See Table 9-2 for calculations.)

language from the flow chart in Figure 9.27.   It is valid for use in the optimum design of steel wire helical compression springs, giving very rapidly the specifications for the spring of greatest possible reliability or factor of safety, N_e, against fatigue breakage.   The program can be applied to the design of

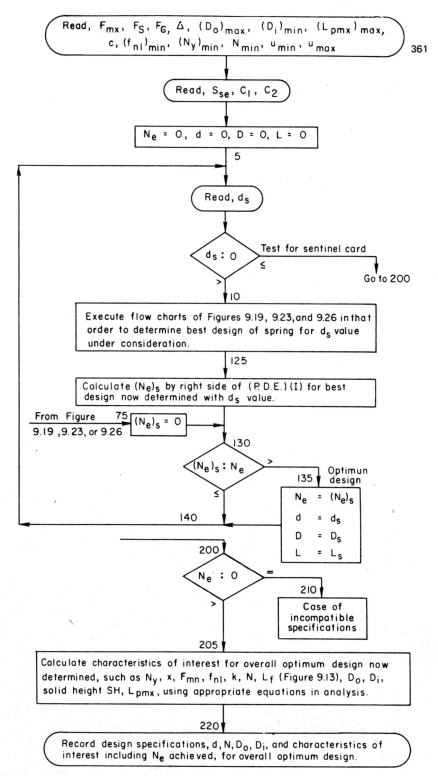

Figure 9.27 Overall optimum design flow chart for helical compression spring.

```
 1          READ,FMX,FS,FG,DEL,DUMX,DIMN,ALPMXX,C,FN1MN,ANYMN,ANMN,UMN,UMX
 2          WRITE (6,1)FMX,FS,FG,DEL,DUMX,DIMN
 3        1 FORMAT ('0','SPECS. FMX,FS,FG,DEL,DUMX,DIMN'/6E16.7/)
 4          WRITE (6,2) ALPMXX,C,FN1MN,ANYMN,ANMN,UMN,UMX
 5        2 FORMAT ('0','SPECS. LPMXX,C,FN1MN,NYMN,NM',UMN,UMX'/7E14.5/)
 6        3 READ,SSE,C1,C2
 7          WRITE (6,4) SSE,C1,C2
 8        4 FORMAT ('0','MATERIAL PROPERTIES SSE,C1,C2='/3E16.7/)
 9          ANE = 0.
10          DW = 0.
11          D = 0.
12          AL = 0.
13        5 READ, DWS
   C TEST FOR SENTINEL CARD
14          IF (DWS .LT. 0.0) GO TO 200
15       10 DYMX= (C1*DWS**(2.86-C2)/(4.07*ANYMN*FMX))**1.163
16          DMAX = DUMX - DWS
17          DMIN= DIMN + DWS
18          DUMX = UMX*DWS
19          DUMN = UMN*DWS
20          IF (DUMX - DYMX) 12,12,11
21       11 DUYMX = DYMX
22          GO TO 13
23       12 DUYMX = DUMX
24       13 IF (DUYMX - DMAX) 15,15,14
25       14 DMX = DMAX
26          GO TO 16
27       15 DMX = DUYMX
28       16 IF (DUMN - DMIN) 18,18,17
29       17 DMN = DUMN
30          GO TO 19
31       18 DMN = DMIN
32       19 IF (DMN - DMX) 35,35,75
33       35 ALSMX = ALPMXX - DEL - 2.*DWS
34          ALS2 = (4.02E6*(1.+C)*DWS**5/(FMX*DMX))**.5
35          ALS7 = 1.41E4*(1.+C)*DWS**2/(FN1MN*DMX**2)
36          ALS8 = 1.44E6*(1.+C)*DEL*DWS**5/(DMX**3*(FMX-FG))
37          ALS9 = 1.44E6*(1.+C)*DEL*DWS**5/(DMX**3*(FMX-FS))
38       40 IF(ALS9.GT.ALS7.OR.ALS9.GT.ALS2.OR.ALS9.GT.ALSMX) GO TO 75
39       45 IF (ALSMX - ALS2) 55,55,50
40       55 ALT = ALSMX
41          GO TO 60
42       50 ALT = ALS2
43       60 IF (ALS7 - ALT) 65,70,70
44       65 ALT = ALS7
45       70 IF (ALT - ALS8) 80,80,85
46       80 ALS = ALT
47          DS = DMX
48          GO TO 121
49       85 ALS6 = 1.35*(1.+C)*(FMX-FG)**2/(DEL**2*DWS**4*FN1MN**3)
50          ALS5 = (3.56E4/FMX)**.6*(1.+C)**.4*DWS**2*((FMX-FG)/DEL)**.2
51          ALS4 = 1.44E6*(1.+C)*DEL*DWS**5/(DMX**3*(FMX-FG))
52          IF (ALSMX - ALS4) 95,95,90
53       90 ALT = ALS4
54          GO TO 100
55       95 ALT = ALSMX
56      100 IF (ALT - ALS5) 110,110,105
57    *   105 ALT = ALS5
58      110 IF (ALT - ALS6) 120,120,115
59      115 ALT = ALS6
60      120 ALS = ALT
61          DS = (1.44E6*(1.+C)*DEL*DWS**5/(ALS*(FMX-FG)))**(1./3.)
62      121 ANS = ALS/(DWS*(1.+C))
63          IF (ANS .GE. ANMN) GO TO 125
64          ALS10 = ANMN*DWS*(1.+C)
65          DS11 = 4.02E6*(1.+C)*DWS**5/(FMX*ALS10**2)
66          DS12 = (1.41E4*(1.+C)*DWS**2/(ALS10*FN1MN))**.5
67          DS13 = (1.44E6*(1.+C)*DEL*DWS**5/(ALS10*(FMX-FS)))**(1./3.)
68          ALS4 = 1.44E6*(1.+C)*DEL*DWS**5/(DMN**3*(FMX-FG))
69          ALSMX = ALPMXX - DEL - 2.*DWS
70          IF (ALS10 .GT. ALS4 .OR. ALS10 .GT. ALSMX) GO TO 75
71          IF (DS11 - DS12) 123,123,122
72      122 DST = DS12
73          GO TO 124
74      123 DST = DS11
75      124 IF (DST .LT. DS13 .OR. DST .LT. DMN) GO TO 75
76          DS = DST
77          ALS = ALS10
78      125 ANES = SSE*DS**2.14*ALS/(2.935E6*DEL*(1.+C)*DWS**2.14)
79          GO TO 130
80       75 ANES = 0.
81      130 IF (ANES - ANE) 140,140,135
82      135 ANE = ANES
83          DW = DWS
84          D = DS
85          AL = ALS
86      140 GO TO 5
```

Figure 9.28 (a) Digital computer program written from the flow chart in Figure 9.27 for maximization of N_e. (b) Input and output data for numerical example 9-2.

```
 87    200 IF (ANE-0.) 210,210,205
 88    210 WRITE (6,211)
 89    211 FORMAT ('0','CASE OF INCOMPATIBLE SPECIFICATIONS')
 90        GO TO 225
 91    205 ANY = C1*DW**(2.86-C2)/(4.07*FMX*D**.86)
 92        X = FMX*AL**2*D/(4.02E6*(1.+C)*DW**5)
 93        FMN = FMX - 1.44E6*(1.+C)*DEL*DW**5/(D**3*AL)
 94        FN1 = 1.41E4*(1.+C)*DW**2/(D**2*AL)
 95    215 AK = (FMX - FMN)/DEL
 96        AN = 1.44E6*DW**4/(D**3*AK)
 97        ALF = AL + DEL + (FMN/AK) + 2.*DW
 98        DO = D + DW
 99        DI = D - DW
100        SH = (AN + 2.)*DW
101        ALPMX = AL + 2.*DW + DEL
102    220 WRITE (6,221)DW,AN,DO,DI,ALPMX,ALF,SH
103    221 FORMAT ('0','OPT. DES. IS FOR DW,N,DO,DI,LPMX,LF,SH ='/7E14.5/)
104        WRITE (6,222) AK,ANY,X,FMN,FN1,ANE
105    222 FORMAT ('0',2X,'WITH CHARACTERISTICS K,NY,X,FMN,FN1,NE ='/6E16.7)
106    225 STOP
107        END
```

```
SPECS, FMX,FS,FG,DEL,DOMX,DIMN
  0.3000000E 02   0.1000000E 02   0.2000000E 02   0.2500000E 00   0.1500000E 01   0.7500000E 00


SPECS, LPMXX,C,FN1MN,NYMN,NMN,UMN,UMX
  0.15000E 01   0.40000E 00   0.50000E 03   0.12000E 01   0.30000E 01   0.40000E 01   0.20000E 02


MATERIAL PROPERTIES SSE,C1,C2=   0.4500000E 05   0.8080000E 05   0.1450000E 00


OPT. DES. IS FOR DW,N,DO,DI,LPMX,LF,SH =
  0.10500E 00   0.30000E 01   0.10985E 01   0.88848E 00   0.90100E 00   0.11552E 01   0.52500E 00


WITH CHARACTERISTICS K,NY,X,FMN,FN1,NE =
  0.5950084E 02   0.1464432E 01   0.8069670E-01   0.1512479E 02   0.5000000E 03   0.2368915E 01

COMPILE TIME=   18.41 SEC,EXECUTION TIME=   3.18 SEC,OBJECT CODE=   6320 BYTES
```

Figure 9.28 (continued)

helical springs in any problem situation where the basic specifications and constraints summarized in the initial formulation must be satisfied.   In fact, it may be applied to problems of design where some of the listed constraints are insignificant.   For example, in a slow-speed application where surging is of no concern we may arbitrarily take a very low value for $(f_{n1})_{min}$, such as one cycle per second.   Or as another example, if constraint on D_i is of no concern, we may arbitrarily take $(D_i)_{min} = 0$ as part of the input data.

Next, we will illustrate application of results from our variation study in a numerical example.

***Numerical Example* 9-2**   Referring to Figures 9.13 through 9.18 for notation, the problem is to design a helical compression spring for a high-speed application with the objective of maximizing reliability or factor of safety, N_e, against fatigue breakage.   Specifications and constraints to be satisfied are as follows, which have been determined at this stage of design as previously described from a consideration of the total system requirements.

Specifications and Constraints:

$F_{mx} = 30.0$ lb; 10.0 lb $\leq F_{mn} \leq 20.0$ lb;

$\Delta = 0.250$ in.; $D_o \leq 1.500$ in.; $D_i \geq 0.750$ in.;

$L_{pmx} \leq 1.500$ in.; $c = 0.4$ (i.e., minimum clearance between coils $= 0.4\,d$);

$f_{n1} \geq 500$ cps; $N_y \geq 1.20$;

$N \geq 3.0$; $4 \leq (D/d) \leq 20$.

Use unpeened A.S. 5 Music Wire. Therefore, as previously stated in the text, reversed shear fatigue strength is $S_{se} = 45,000$ psi, and from equations (9.19) and (9.19a) the yield strength material constants are $C_1 = 80,800$ and $C_2 = 0.145$. Standard size wire diameters to be considered are as follows (in in.): $d_s = 0.005, 0.006, 0.007, 0.008, 0.009, 0.010, 0.012, 0.014, 0.016, 0.018,$ 0.020, 0.024, 0.026, 0.028, 0.032, 0.036, 0.042, 0.048, 0.055, 0.063, 0.072, 0.080, 0.092, 0.105, 0.120, 0.135, 0.156, 0.188.

Find: The helical compression spring design of greatest reliability or factor of safety N_e against fatigue breakage, which satisfies all of the specifications and constraints as now stipulated.

Solution: Application of the digital computer program in Figure 9.28 gives the following results in a matter of seconds for the overall optimum design. All values are found in the output data at the end of Figure 9.28.

$d = 0.105$ in., wire size

$N = 3.00$, active coils

$D_o = 1.0985$ in., outside diameter

$D_i = 0.8885$ in., inside diameter

$L_{pmx} = 0.901$ in., pocket length of Figure 9.13

$L_f = 1.1552$ in., free length

$SH = 0.525$ in., solid height

(understood: one dead coil each end, squared and ground flat)

Some pertinent characteristics of this overall optimum design spring are (also from Figure 9.28 output data):

$k = 59.5$ lb/in., force gradient

$N_y = 1.464$, factor of safety against yielding

$x = 0.0807$, for the buckling parameter (which is a very safe value, since $x << 1$).

$F_{mn} = 15.125$ lb, minimum spring force for Figures 9.15–9.17

$f_{n1} = 500$ cps, lowest natural frequency for internal longitudinal vibrations

$N_e = 2.369$, factor of safety against fatigue breakage (which is the greatest possible value for the infinite number of feasible designs)

For the overall optimum design now determined we see that it happens to be the point 12 design of Figure 9.21, since for it we have found $N = 3 = N_{min}$ and $f_{n1} = 500$ cps $= (f_{n1})_{min}$, which uniquely determine that point in the figure. However, it should be emphasized that for other numerical values of specifications and constraints the overall optimum design might very well be for a different wire size at any one of the nine possible optimum design points designated as 1, 2, 3, 4, 5, 6, 7, 11, or 12 in Figure 9.21.

In conclusion, by application of the method of optimum design we have very rapidly found the overall optimum design in the numerical example, for maximization of reliability with the value for factor of safety of $N_e = 2.369$ achieved. We must now decide whether or not this value is acceptable for our particular application. As a guide we might reach this decision by anticipating the various uncertainties of design and applying the techniques of Chapter 6 in reference 9-9 and reference 9-10, thereby calculating a minimum acceptable value for N_e which can be compared with the highest possible design value of $N_e = 2.369$ now achieved. If the best possible design now determined is not acceptable we must alleviate some of the boundary values, such as by using $S_{se} = 67,500$ and specifying shot peened springs. Also, in such a situation where the best design is not good enough, modification of appropriate boundary values in Figure 9.21 could be made by following the general technique described and illustrated in section 8-9 of Chapter 8.

SOME MECHANISMS FOR HIGH-SPEED MACHINE SYSTEMS

9-4 OPTIMUM DESIGN OF A HIGH-SPEED INDEXING MECHANISM BY DIGITAL COMPUTER

In this section we will very briefly consider the optimum design of a Geneva mechanism for use in a high-speed machine. The basic purpose for this example is to illustrate the application of the digital computer with the method of optimum design for the solution of a high-speed mechanism design problem. Manipulation of the equation system for the making of the variation study and derivation of the optimum design flow chart with the resulting computer program will be presented in sufficient detail for understanding by the reader. However, derivation of the initial formulation equations and background will not be presented here since the interested reader can find this material published elsewhere, specifically in references 9-15, 9-16, and pp. 499–503 of reference 9-9.

The four-station Geneva mechanism of Figure 9.29 is to be designed for the intermittent motion drive of the given inertial load, J_L. For the high-speed machine being designed, the objective of optimum design for this critical Geneva mechanism is maximization of safe speed for long life operation. Based on experience and feasibility calculations, we believe that the critical region of the well-lubricated Geneva mechanism will be at the

Figure 9.29 Geneva mechanism to be designed for the intermittent motion of a given inertial load.

contact of drive pin and wheel slot for the position of maximum force, F_{max}, shown in Figure 9.29. For this position of interest, subsurface fatigue will be of primary concern, similar to the type of wear failure discussed in references 9-19 and 9-20. Hence, for a given design the long-life maximum operational speed, ω_{max}, will be limited by the reaching of an allowable shear stress endurance limit, S_{se}, on maximum shear stress, τ_{max}, slightly beneath the surface of contact. Thus, our primary design equation will be based in part on the equation for τ_{max} derived as follows from p. 382 of reference 9-21 and from Case 4 in Table XIV of reference 9-22, referring to Figure 9.29 of the present book for notation.

$$\tau_{max} \approx 0.304\sigma_c$$

$$= (0.304)(0.798) \sqrt{\frac{F_{max}}{Td\left[\frac{1 - v_1^2}{E_1} + \frac{1 - v_2^2}{E_2}\right]}}$$

therefore

$$\tau_{max} = k_1 \sqrt{F_{max}/(Td)} \qquad (9.43)$$

where

$$k_1 \approx 0.2426 \Bigg/ \sqrt{\frac{(1 - v_1^2)}{E_1} + \frac{(1 - v_2^2)}{E_2}} \qquad (9.44)$$

In equation (9.43), τ_{max} is the maximum shear stress in psi slightly beneath the surface of the pin-to-slot contact for the position of greatest force, as shown in Figure 9.29, F_{max} is this greatest force in lb, T and d are dimensions in inches shown in the figure, and k_1 is a material constant consisting of the dimensionless Poisson's ratio v terms and the modulus of elasticity E terms in psi for the contacting materials.

Maximum dynamic contact force, F_{max}, depends on the Geneva wheel design and the speed of operation ω, which we derive from elementary dynamics of rotating rigid bodies as follows, referring to Figure 9.29.

$$\sum T_0 = J\ddot{\beta}$$

therefore

$$F_{max} kD \approx (J_G + J_L)\omega^2 \left(\frac{\partial^2 \beta}{\partial \alpha^2}\right)_{max}$$

thus

$$F_{max} \approx \frac{(J_G + J_L)\omega^2}{kD} \left(\frac{\partial^2 \beta}{\partial \alpha^2}\right)_{max} \qquad (9.45)$$

In equation (9.45), F_{max} is the maximum pin-to-slot contact force in lb; J_G and J_L are mass moments of inertia for Geneva wheel and load inertia, respectively, in lb-sec^2-in.; ω is speed of operation for drive shaft in radians per second; $(\partial^2\beta/\partial\alpha^2)_{max}$ is the dimensionless second derivative as stated, for the position of F_{max} in Figure 9.29, and obtained approximately from Table 2 in Part 1 of reference 9-15; D is Geneva wheel diameter of Figure 9.29 in inches; and k is a dimensionless constant of the basic mechanism taken directly from Table 1 of reference 9.16.

Combining equations (9.43) and (9.45) by eliminating unspecified force F_{max} we obtain (P.D.E.) (9.46) as follows, and our objective in design of the Geneva mechanism will be to minimize the ratio $(\tau_{max}/\omega)^2$. By choosing this objective we really are solving two problems of optimization. First, for a given level of acceptable stress in the material it maximizes operational speed ω, which we initially stated as the basic goal. Secondly, for a given operational speed ω, this objective will also minimize critical stress τ_{max}, and in that way it will minimize the chance of failure due to subsurface fatigue wear.

$$\tau_{max} = k_1\sqrt{F_{max}/(Td)}$$

$$\approx k_1\sqrt{(J_G + J_L)\omega^2\left(\frac{\partial^2\beta}{\partial\alpha^2}\right)_{max}\Big/(kDTd)}$$

$$(\tau_{max}/\omega)^2 \approx k_1^2(J_G + J_L)\left(\frac{\partial^2\beta}{\partial\alpha^2}\right)_{max}\Big/(kDTd) \qquad \text{(P.D.E.) (9.46)}$$

All terms in the preceding equation have been previously defined.

The Geneva wheel will be designed for the optimum proportions shown in Figure 9.29, in accordance with recommendations of reference 9-16. Values of the dimensionless proportionality constants k_2 and k_4 will be selected using our practical judgment and Table 1 of reference 9-16 as a guide. Hence, we have

$$d = k_2 D \qquad \text{(S.D.E.) (9.47)}$$

as a geometric subsidiary design equation to satisfy. Also, with these optimum proportions in the plan view of Figure 9.29, we can express the mass moment of inertia, J_G, of the Geneva wheel as follows.

$$J_G = k_6 TD^4 \qquad \text{(S.D.E.) (9.48)}$$

For rapid evaluation of constant k_6 in this equation we can use the values of k_3 and k_5 from Table 1 of reference 9-16 and calculate k_6 from equation (9.49), which is derived from that background.

$$k_6 = w\,k_5/(k_3 g) \qquad (9.49)$$

In equation (9.49), w is the weight density of the Geneva wheel in lb./in.^{3}, g is the gravitational constant whose value is 386 in./sec^2, and k_3 and k_5 are taken directly from Table 1 of reference 9-16. Hence, if T and D of Figure 9.29 are in inches, J_G from (S.D.E.) (9.48) will be in lb-sec^2 in.

For a satisfactory design of the Geneva mechanism in Figure 9.29, generally we must adhere to space constraints on D and T.   Hence, we have the limit equations (9.50) and (9.51) to adhere to.

$$D \leq D_{max} \qquad \text{(L.E.) (9.50)}$$

$$T \leq T_{max} \qquad \text{(L.E.) (9.51)}$$

Incidentally, the constraint D_{max} might be imposed by height as well as lateral constraints on the entire mechanism of Figure 9.29.   In addition, our Geneva mechanism must be within a range of acceptability for practical proportions.   For a given wheel diameter D, it cannot be too thin.   Hence, we must satisfy the regional constraint $(T/D) \geq k_7$, where we must exercise our practical judgement in deciding on the acceptable lower limit value of k_7, considering material and manufacturing characteristics in the process.   Also, the length to diameter ratio (T/d) of the drive pin must not be excessive or we will not have a good contact condition over the full slot width.   In consideration of (S.D.E.) (9.47), this constraint on (T/d) can also be directly imposed on the ratio (T/D) as a regional constraint, i.e., $(T/D) \leq k_3$.   Again we must exercise our practical judgement in deciding on an acceptable upper limit value of k_3, using Table 1 of reference 9-16 as a guide together with perhaps some order-of-magnitude calculations.   Let us now define the *ratio* (T/D) as a limited design parameter x, as follows:

$$x = T/D \qquad \text{(S.D.E.) (9.52)}$$

$$k_7 \leq x \leq k_3 \qquad \text{(L.E.) (9.53)}$$

We are now at a point where the initial formulation equation system can be summarized as follows:

Initial Formulation

$$(\tau_{max}/\omega)^2 \approx k_1{}^2(J_G + J_L)\left(\frac{\partial^2 \beta}{\partial \alpha^2}\right)_{max}/(k\,DTd) \qquad \text{(P.D.E) (9.46)}$$

$$d = k_2 D \qquad \text{(S.D.E.) (9.47)}$$

$$J_G = k_6 T\,D^4 \qquad \text{(S.D.E.) (9.48)}$$

$$x = T/D \qquad \text{(S.D.E.) (9.52)}$$

$$D \leq D_{max}\,;\, T \leq T_{max}\,;\, k_7 \leq x \leq k_3 \qquad \text{(L.E.)s}$$

Specified or Known: J_L, D_{max}, T_{max}, k, k_1 (from equation (9.44); v_1, v_2, E_1, E_2 are also known), k_2, k_3, k_7, $(\partial^2\beta/\partial\alpha^2)_{max}$, k_6 (from equation (9.49); w and k_5 are also known).

Find: Dimensions of Geneva mechanism in Figure 9.29 which minimize $(\tau_{max}/\omega)^2$, for each material combination of interest.

Upon review of our initial formulation equation system, using the notation defined in Chapter 8, we have:

$$N_s = 3$$
$$n_f = 2 \quad \text{(which are } J_G \text{ and } d)$$
$$n_c = 3 \quad \text{(which are } D, T, \text{ and } x)$$

Thus, from equation (8.7) we have $n_v = n_c + n_f = 3 + 2 = 5$. Hence, in reference to Figure 8.2, we have for the initial formulation equation system $n_f = 2 < N_s = 3$, and therefore a case of redundant specifications exists. From equation (8.10) we should have $N_{ff} = N_s - n_f + 1 = 3 - 2 + 1 = 2$ equations in our final formulation system. Also, from equation (8.13) we anticipate $D_{vs} = n_v - N_s + 1 = 5 - 3 + 1 = 3$ as the required number of dimensions for our optimum design variation study.

A simple determinate approach for the optimum design variation study is recognized as follows from the initial formulation summary. Select x as the eliminated parameter, and derive the developed (P.D.E.) of the ideal problem by combining (S.D.E.)s (9.47) and (9.48) with (P.D.E.) (9.46), eliminating free variables d and J_G from the initial formulation equation system. The final formulation equation system so obtained is summarized as follows:

Final Formulation

$$\left(\frac{\tau_{max}}{\omega}\right)^2 \approx \frac{k_1{}^2 \left(\frac{\partial^2 \beta}{\partial \alpha^2}\right)_{max}}{kk_2} \left[k_6 D^2 + \frac{J_L}{TD^2}\right] \tag{I}$$

$$x = T/D \tag{II}$$

where

x = eliminated parameter
D, T = related parameters
(I) = developed (P.D.E.) of ideal problem
(II) = relating equation

From the final formulation equation system we can readily sketch a typical three-dimensional variation diagram as shown in Figure 9.30. For the ideal problem, minimization of $(\tau_{max}/\omega)^2$ would be obtained by placing $T = T_{max}$. From developed (P.D.E.) (I) of the ideal problem, for constant $T = T_{max}$, we have a minimum value of $(\tau_{max}/\omega)^2$, where its partial derivative with respect to D is zero. Thus, point 4 in Figure 9.30 is a possible one for optimum design.

$$\frac{\partial [(\tau_{max}/\omega)^2]}{\partial D} \sim 2k_6 D - 2J_L/(T_{max} D^3) = 0$$

$$D_4 = [J_L/(k_6 T_{max})]^{1/4} \tag{9.54}$$

For a material combination of interest:

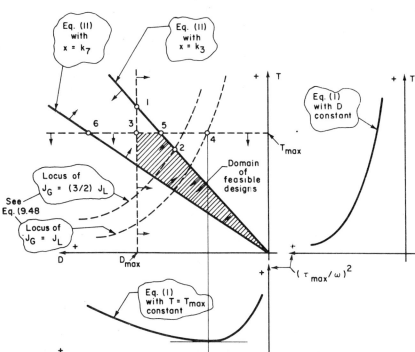

Figure 9.30 Typical three-dimensional variation diagram for design of Geneva mechanism. Possible points for optimum design are 1, 2, 3, 4, 5, or 6. (See final formulation summary for equations I and II.)

For the point 4 optimum design possibility of Figure 9.30 we note the following relationship of interest, referring to equations (9.48) and (9.54):

$$J_L = k_6 T_{max} D_4{}^4 = J_G$$

Hence, for this design at point 4 the Geneva wheel inertia, J_G, would equal the given load inertia, J_L. From the typical variation diagram of Figure 9.30, we readily recognize that it may not be possible to achieve the point 4 design even for the ideal problem. This would be the case if we had $D_4 > D_{max}$, and in that event we would have to settle for point 3 of Figure 9.30 as the ideal problem optimum design.   Next, we find that four other points are possible for optimum design if we impose relating equation (II) of the final formulation, with the regional constraint $k_7 \le x \le k_3$.   The two boundary lines thus determined are typically shown in Figure 9.30, and from the preceding discussion point 6 would be the optimum design if $(D_4$ and $D_{max}) > D_6$.   On the other hand, for the constraints as specifically shown in Figure 9.30, we readily see that it is necessary to investigate the variation of $(\tau_{max}/\omega)^2$ along the boundary line of equation (II) with $x = k_3$.   Hence, along that boundary line we would have $T = k_3 D$, and from equation (I)

$$(\tau_{max}/\omega)^2 \sim k_6 D^2 + J_L/(TD^2)$$

$$= k_6 D^2 + J_L/(k_3 D^3)$$

Thus, along that boundary a minimum exists for

$$\frac{\partial[(\tau_{max}/\omega)^2]}{\partial D} \sim 2k_6 D - 3J_L/(k_3 D^4) = 0$$

This we will designate as the point 2 optimum design possibility in Figure 9.30, and from the preceding we have

$$D_2 = [3J_L/(2k_6 k_3)]^{1/5} \tag{9.55}$$

where

$$T_2 = k_3 D_2$$

For this point 2 optimum design possibility of Figure 9.30 we note the following relationship of interest, referring to equations (9.48) and (9.55):

$$J_L = (2/3)k_6 k_3 D_2^5 = (2/3)k_6 T_2 D_2^4 = (2/3)J_G$$

Hence, for this particular design the Geneva wheel inertia would be 3/2 of the given load inertia, i.e., $J_G = (3/2)J_L$ for the point 2 possibility of Figure 9.30.

From an understanding of the optimum design variation study now made, we conclude that intersection points 1 and 5 of Figure 9.30 are also possibilities for optimum design.   For the specific relationship of constraints as shown in the typical diagram of Figure 9.30, point 2 would be the optimum design.   However, if $T_2 > T_5$ or $T_2 > T_1$ existed, either point 1 or 5 might be the optimum design, depending on their relationship to the other constraints.   Specifically, we will summarize the explicit procedure for determination of the particular point of optimum design in flow-chart form.   This is presented in Figure 9.31, which is derived from an understanding of the typical variation diagram in Figure 9.30.   We will refer to the final formulation for the specific equations unless otherwise stated.   Incidentally, the

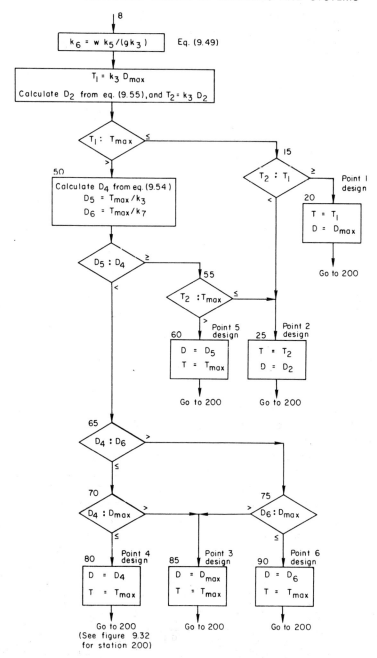

Figure 9.31   Flow chart for determination of optimum design point.   (See Figure 9.30 for variation diagram and Figure 9.32 for total flow chart.)

numbers at the various blocks of Figure 9.31 correspond to the statement numbers in the digital computer program of Figure 9.33. Also, we note in Figure 9.33 that the particular point of optimum design is designated in the print-out, and features like that of course could be readily incorporated if desired in other programs, such as in Figure 9.28. Finally, it should be pointed out that for any material combination of interest there is no possibility for incompatible specifications in this particular problem of optimum design. We readily recognize this from an understanding of Figure 9.30.

Let us now tie together our variation study in the total picture of determining the overall optimum design, where there are M material combinations of interest for drive pin and Geneva wheel. This is summarized in the flow chart of Figure 9.32, where we first read the boundary values of interest. Incidentally, some of the constants are included as material properties, since the value selected might be influenced to some extent by characteristics of the material itself. Next, we execute the steps in Figure 9.31 to determine the optimum values of D and T for the particular point which minimizes $(\tau_{max}/\omega)^2$. Finally, at station 200 of Figure 9.32 we analyze the optimum design to determine what permissible speed of operation, ω_{max}, has been achieved along with the level of τ_{max} which is allowed. These calculations we make using the ideal problem developed (P.D.E.) (I), with allowable τ_{max} determined by the maximum shearing stress theory of fatigue failure, giving for the range of acceptability

$$\tau_{max} \leq S_e/((1 + p) N_e) \qquad \text{(L.E.) (9.56)}$$

Derivation of (L.E.) (9.56) would be similar to derivation of equation (12.34) of reference 9-9, which the interested reader might wish to consult. In the equation, S_e is the fatigue strength of the surface material from a reversed bending test in psi, $p = S_e/S_y$ where S_y is the normal yield strength of the material in psi, and N_e is an appropriate factor of safety to account for the various uncertainties anticipated in design. The properties S_e and S_y can be estimated from readily available sources at the time of design for the materials of interest, and for this reason we have selected the maximum shearing stress theory for application here. Using the right side of (L.E.) (9.56) for τ_{max} in equation (I) of the final formulation, we calculate allowable speed, ω_{max}, as follows.

$$\left[\frac{S_e}{(1 + p)N_e \omega_{max}}\right]^2 = \frac{k_1^2 \left(\frac{\partial^2 \beta}{\partial \alpha^2}\right)_{max}}{kk_2}\left[k_6 D^2 + \frac{J_L}{TD^2}\right]$$

thus

$$\omega_{max} = \frac{S_e}{(1 + p)N_e k_1 \left(\left(\frac{\partial^2 \beta}{\partial \alpha^2}\right)_{max} A_1\right)^{1/2}} \qquad (9.57)$$

Figure 9.32   Flow chart to determine overall optimum design of the Geneva mechanism.

```
1           READ,AJL,DMX,TMX,AK,AK5,D2BA2
2           READ,M
3           WRITE (6,3)M,AJL,DMX,TMX,AK,AK5,D2BA2
4         3 FORMAT('0','SPECS. M,AJL,DMX,TMX,AK,AK5,D2BA2 ='/I4,6E15.6///)
5           J = 1
6        10 READ,ANE,AK2,AK3,AK4,AK7
7           READ,ANU1,ANU2,E1,E2,SE,STY,WD
8         5 WRITE (6,6) J
9         6 FORMAT ('0','MATERIAL COMBINATION',I4)
10          WRITE (6,7) ANE,AK2,AK3,AK4,AK7
11        7 FORMAT ('0',10X,'SPECS. ANE,AK2,AK3,AK4,AK7 =',5E14.5)
12          WRITE (6,8) E1,E2,SE,STY,WD
13        8 FORMAT ('0',10X,'MATL. PROP. E1,E2,SE,STY,WD =', 5E14.5)
14          AK6 = WD*AK5/(386.*AK3)
15          T1 = AK3*DMX
16          D2 = (3.*AJL/(2.*AK3*AK6))**.2
17          T2 = AK3*D2
18          IF (T1 - TMX) 15,15,50
19       15 IF (T2 - T1) 25,20,20
20       20 T = T1
21          D = DMX
22          WRITE (6,21) D,T
23       21 FORMAT ('0',5X,'BEST DESIGN IS PT. 1 WITH D,T =',2E15.6)
24          GO TO 200
25       25 T = T2
26          D = D2
27          WRITE (6,26) D,T
28       26 FORMAT ('0',5X,'BEST DESIGN IS PT. 2 WITH D,T =',2E15.6)
29          GO TO 200
30       50 D4 = (AJL/(AK6*TMX))**.25
31          D5 = TMX/AK3
32          D6 = TMX/AK7
33          IF (D5 - D4) 65,55,55
34       55 IF (T2 - TMX) 25,25,60
35       60 D = D5
36          T = TMX
37          WRITE (6,61) D,T
38       61 FORMAT ('0',5X,'BEST DESIGN IS PT. 5 WITH D,T =',2E14.5)
39          GO TO 200
40       65 IF (D4 - D6) 70,70,75
41       70 IF (D4 - DMX) 80,80,85
42       80 D = D4
43          T = TMX
44          WRITE (6,81) D,T
45       81 FORMAT ('0',5X,'BEST DESIGN IS PT. 4 WITH D,T =',2E14.5)
46          GO TO 200
47       85 D = DMX
48          T = TMX
49          WRITE (6,86) D,T
50       86 FORMAT ('0',5X,'BEST DESIGN IS PT. 3 WITH D,T =',2E14.5)
51          GO TO 200
52       75 IF (D6 - DMX) 90,90,85
53       90 D = D6
54          T = TMX
55          WRITE (6,91) D,T
56       91 FORMAT ('0',5X,'BEST DESIGN IS PT. 6 WITH D,T =',2E14.5)
57      200 A1=(AK6*D**2 +AJL/(T*D**2))/(AK2*AK)
58          AK1 = .2426/((1.-ANU1**2)/E1 + (1.-ANU2**2)/E2)**.5
59          TAUMX = SE/((1. + (SE/STY))*ANE1
60          UMGMX = TAUMX/(AK1*(D2BA2*A1)**.5)
61          RPMMX = UMGMX*9.5493
62          FMX = (AK6*T*D**3 + AJL/D)*UMGMX**2*D2BA2/AK
63          DPN = AK2*D
64          R = AK4*D
65          WRITE (6,201) DPN,R
66      201 FORMAT ('0',10X,'AND DPN,R =',2E15.6)
67          WRITE (6,202) RPMMX
68      202 FORMAT ('0',5X,'RPMMX =',E15.6)
69          WRITE (6,203) TAUMX,FMX
70      203 FORMAT ('0',10X,'WITH TAUMX,FMX =',2E15.6//)
71          IF (J - M) 205,210,210
72      205 J = J + 1
73          GO TO 10
74      210 STOP
75          END
```

Figure 9.33 Digital computer program with application to numerical example 9-3 for optimum design of the Geneva mechanism. (a) Digital computer program written from the flow chart in Figure 9.32.

```
SPECS. M,AJL,DMX,TMX,AK,AK5,D2BA2 =
   4   0.482000E-02   0.500000E 01   0.100000E 01   0.238400E 00   0.506000E-02   0.540900E 01


MATERIAL COMBINATION   1
          SPECS. ANE,AK2,AK3,AK4,AK7 =   0.15000E 01   0.15000E 00   0.15000E 00   0.35000E 00   0.50000E-01
          MATL. PROP. E1,E2,SE,STY,WD =  0.30000E 08   0.30000E 08   0.14100E 06   0.17400E 06   0.28300E 00
     BEST DESIGN IS PT. 2 WITH D,T =   0.454944E 01   0.682415E 00
          AND DPN,R =   0.682415E 00   0.159230E 01
     RPMMX =   0.141851E 04
          WITH TAUMX,FMX =   0.519238E 05   0.132605E 04


MATERIAL COMBINATION   2
          SPECS. ANE,AK2,AK3,AK4,AK7 =   0.15000E 01   0.15000E 00   0.20000E 00   0.35000E 00   0.10000E 00
          MATL. PROP. E1,E2,SE,STY,WD =  0.30000E 08   0.41000E 06   0.40000E 04   0.10000E 05   0.51500E-01
     BEST DESIGN IS PT. 1 WITH D,T =   0.500000E 01   0.100000E 01
          AND DPN,R =   0.750000E 00   0.175000E 01
     RPMMX =   0.539556E 03
          WITH TAUMX,FMX =   0.190476E 04   0.100389E 03


MATERIAL COMBINATION   3
          SPECS. ANE,AK2,AK3,AK4,AK7 =   0.20000E 01   0.15000E 00   0.15000E 00   0.35000E 00   0.10000E 00
          MATL. PROP. E1,E2,SE,STY,WD =  0.41000E 06   0.41000E 06   0.40000E 04   0.10000E 05   0.51500E-01
     BEST DESIGN IS PT. 1 WITH D,T =   0.500000E 01   0.750000E 00
          AND DPN,R =   0.750000E 00   0.175000E 01
     RPMMX =   0.492055E 03
          WITH TAUMX,FMX =   0.142857E 04   0.834908E 02


MATERIAL COMBINATION   4
          SPECS. ANE,AK2,AK3,AK4,AK7 =   0.15000E 01   0.15000E 00   0.15000E 00   0.35000E 00   0.50000E-01
          MATL. PROP. E1,E2,SE,STY,WD =  0.30000E 08   0.30000E 08   0.40000E 05   0.60000E 05   0.28300E 00
     BEST DESIGN IS PT. 2 WITH D,T =   0.454944E 01   0.682415E 00
          AND DPN,R =   0.682415E 00   0.159230E 01
     RPMMX =   0.437104E 03
          WITH TAUMX,FMX =   0.160000E 05   0.125912E 03

COMPILE TIME=    67.94 SEC,EXECUTION TIME=    4.43 SEC,OBJECT CODE=    3992 BYTES
```

Figure 9.33 (continued)   (b) Input and output data for numerical example 9-3.

where

$$A_1 = \frac{k_6 D^2 + J_L/(TD^2)}{k_2 k} \tag{9.58}$$

All terms have been defined in equations (9.57) and (9.58), with long-life speed ω_{max} of the optimum design thus obtained in rad/sec. The corresponding value in revolutions per minute of the drive shaft is designated as RPMMX and calculated after station 200 in the computer program of Figure 9.33. The value of τ_{max} calculated is the allowable value, taken as the right side of equation (9.56). Finally, after statement 200 we calculate peak pin-to-slot contact force, F_{max}, at allowable speed ω_{max} by equations (9.45) and (9.48), giving

$$F_{max} = \frac{(J_G + J_L)\omega_{max}^2}{kD} \left(\frac{\partial^2 \beta}{\partial \alpha^2}\right)_{max}$$

$$= \frac{(k_6 TD^4 + J_L)\omega_{max}^2}{kD} \left(\frac{\partial^2 \beta}{\partial \alpha^2}\right)_{max}$$

thus

$$F_{max} = (k_6 TD^3 + J_L/D)\omega_{max}^2 \left(\frac{\partial^2 \beta}{\partial \alpha^2}\right)_{max} \Big/ k \tag{9.45a}$$

For the optimum design now determined, the value of F_{max} from equation (9.45a) is of interest since we would want to be sure that the mounting design could sustain indefinitely the peak forces encountered, without excessive stress or dynamic deflection. Also, for the optimum design, our digital computer calculates as a last step the pin diameter d by equation (9.47) and cut-out radius $R = k_4 D$, referring to Figure 9.29 for notation.

Let us now illustrate application with a numerical example.

Numerical Example **9-3** A four-station well-lubricated Geneva mechanism is to be designed with the configuration of Figure 9.29 for the objective of maximizing allowable operational speed, as governed by subsurface fatigue failure at the critical region of contact between drive pin and wheel slot. The given inertial load is $J_L = 0.00482$ lb-sec^2-in., which incidentally is equivalent to a solid steel disc 4 in. in diameter and 0.262 in. thick. Space constraints on D and T for the Geneva wheel to be designed are $D \leq 5$ in. and $T \leq 1$ in. Constants k and k_5 for the four-station Geneva mechanism are directly obtained from Table 1 of reference 9-16 as $k = 0.2384$ and $k_5 = 0.00506$. From the same source we obtain $(\partial^2 \beta/\partial \alpha^2)_{max} \approx 5.409$, originally taken from Table 2 in Part 1 of reference 9-15. We wish to consider four material combinations for possible use in the design, so we have $M = 4$.

TABLE 9-3 INPUT DATA FOR VARIOUS MATERIAL COMBINATIONS OF INTEREST[a]

| | Material Combination | | | Proportionality Constants | | | | | | | Material Properties | | | | |
J	Drive Pin	Geneva Wheel	N_e	k_2	k_3	k_4	k_7	v_1	v_2	E_1 (psi $\times 10^6$)	E_2 (psi $\times 10^6$)	S_e (psi $\times 10^3$)[b]	S_y (psi $\times 10^3$)[b]	w (lb/ in.^{3})[c]
1	AISI 2340 steel	AISI 2340 steel	1.5	0.15	0.15	0.35	0.05	0.26	0.26	30	30	141	174	0.283
2	Steel	Delrin[R,d]	1.5	0.15	0.20	0.35	0.10	0.26	0.35	30	0.41	4	10	0.0515
3	Delrin[R,d]	Delrin[R,d]	2.0	0.15	0.15	0.35	0.10	0.35	0.35	0.41	0.41	4	10	0.0515
4	AISI 1020 C.R. Steel	AISI 1020 C.R. Steel	1.5	0.15	0.15	0.35	0.05	0.26	0.26	30	30	40	60	0.283

Geneva mechanism numerical design example 9-3.
For weaker of contacting materials.
For Geneva wheel material.
Delrin® is a trademark of the E. I. du Pont de Nemours & Co., Inc.

The four material combinations of interest with the accompanying input data specifications decided upon are summarized in Table 9-3. For material combinations $J = 1$ and $J = 4$, drive pin and Geneva wheel materials are both of hardened AISI 2340 steel and soft AISI 1020 C.R. steel, respectively, with properties taken from Table 5-1 of reference 9-9. For material combination $J = 2$, we have a steel drive pin with an acetal resin plastic Geneva wheel, with material properties of the latter obtained from references 9-17 and 9-18. For material combination $J = 3$, we have both the drive pin and the Geneva wheel made of the acetal resin plastic, with material properties obtained from the aforementioned references. Appropriate values for factors of safety, N_e, in Table 9-3 were estimated from a consideration of the uncertainties involved in design, including our judgement of confidence in the material properties of significance. Finally, proportionality constants k_2 and k_4 were taken directly from Table 1 of reference 9-16; these values are compatible with the k_5 value previously selected and are in agreement with the plan view proportions of Figure 9.29. Appropriate values for k_3 and k_7 were selected in Table 9-3 based on practical considerations of manufacturing, possibilities of Geneva wheel warpage, and pin-to-slot contact condition anticipated for the four material combinations. Obviously, the values cannot be considered as precise, but they rely heavily on the exercising of our practical engineering judgement, which is the best that we can do with the information available at this stage of design.

Output data from execution of the computer program in Figure 9.33 is listed at the end therein, with a brief summary presented in Table 9-4. The

TABLE 9-4   OUTPUT DATA FROM FIGURE 9.33 FOR THE BEST GENEVA MECHANISM DESIGNS IN EXAMPLE 9-3

	Material Combination		Best Design Dimensions				Long-Life Speed (rpm)
J	drive pin	Geneva wheel	D (in.)	T (in.)	d (in.)	R (in.)	
1	AISI 2340 steel	AISI 2340 steel	4.5494	0.6824	0.6824	1.5923	1418
2	Steel	Delrin[R][a]	5.000	1.000	0.750	1.750	540
3	Delrin[R][a]	Delrin[R][a]	5.000	0.750	0.750	1.750	492
4	AISI 1020 C.R. steel	AISI 1020 C.R. steel	4.5494	0.6824	0.6824	1.5923	437

[a] Delrin[R] is a trademark of the E. I. du Pont de Nemours & Co., Inc.

order of preference happens to be the same as the material combination J numbers, with the hardened steel drive pin and Geneva wheel combination being the first choice.   The maximum long-life operational speed for that overall optimum design is found from Figure 9.33 to be RPMMX = 1418 rev/min for the drive shaft of Figure 9.29, as governed by subsurface fatigue at the contact region.   Of course, the mounting design would have to be adequate to sustain the reactions developed from the accompanying peak dynamic contact force of F_{max} = 1326 lb from Figure 9.33, with adequacy being determined primarily by requirements of adequate bearing life, adequate fatigue strength of the shaft, and acceptable dynamic deflection of the mounting. Satisfaction of these requirements would be an important part in the total optimum design of the system.

Incidentally, from the output data of Figure 9.33 it is interesting to note that the best designs for the material combinations 1 and 4 are both point 2 designs from Figure 9.30.   On the other hand, the best designs for the material combinations 2 and 3 are both point 1 designs from Figure 9.30. Points 3, 4, 5, or 6 of Figure 9.30 might very well be encountered as the best design for another numerical example with different boundary values for the input data.

9-5   REDESIGN OF A DRIVE MECHANISM FOR A HIGH-SPEED MACHINE

In this section we will very briefly present an industrial example of application for the method of optimum design.   The problem was presented to the author in a consulting capacity by the Whitin Machine Works, Inc.   The purpose for inclusion of this example in the book is twofold.   First, it illustrates successful application of the method of optimum design to a complex assemblage of mechanical elements which have interrelated effects on each other. Secondly, it illustrates the very important technique of utilizing the theoretical goals of optimum design in the making of a practical design layout. The work will not be presented in detail because of space limitations in the book as well as the highly specialized and proprietary nature of the example.

The processing of cotton from fibers to fabric consists of many steps, as summarized in Figure 9.34, requiring a system of many machines in a typical mill. One of the important operations in the process is the combing of fibers, which is performed by the Whitin Super J7B Comber shown in Figure 9.35. The combing operation is performed at each station of the machine by a carefully coordinated action between detaching rolls and a nipper frame assembly, shown in extreme positions in Figure 9.36. The detaching rolls extend for the full length of the combing machine, representing a sizeable inertia which must undergo a special type of reversed feeding motion for the proper combing of cotton. Hence, at this stage of design we assume that the peripheral motion of the driving bottom detaching rolls in Figure 9.36 is a specified requirement of known acceleration characteristics.

The problem of redesign was with respect to the detaching roll drive mechanism whose basic configuration is shown in Figure 9.37. This very important mechanism of the machine is located in the head end of the comber, and based on many years of success in the past we would like to retain the same basic configuration for the future. However, the redesign was desired since dynamic forces on the control cam of Figure 9.37 prevented us from operating the comber at the higher speeds which we would like to achieve.

The epicyclic gear train mechanism and detaching rolls inertia of Figure 9.37 is reflected on the cam follower control arm a as an equivalent lumped mass load inertia, $(J_L)_{eq}$, whose value is a function of the mechanism design. The equivalent mechanism is shown in Figure 9.38, and for minimization of critical dynamic cam force F_c a preliminary study revealed that we should in effect minimize load inertia $(J_L)_{eq}$. Hence, the basic objective for optimum design of the detaching roll drive mechanism is selected as minimization of $(J_L)_{eq}$.

The primary design equation expressing $(J_L)_{eq}$ is derived from application of kinematics to the mechanism and basic theory from dynamics to free-body diagrams of the various gears in Figure 9.37. This results in the following relationship, where internal forces other than $F_{a/bc}$ have been eliminated by combination of the equations obtained. The derivation is quite lengthy and will not be presented here.

$$F_{a/bc} = \frac{\ddot{\theta}_a}{R_a} \{[J_y + J_d w^2][1 - uv]^2 + M_{oc} R_a{}^2 + J_{bc}[1 + u]^2\} \quad (9.59)$$

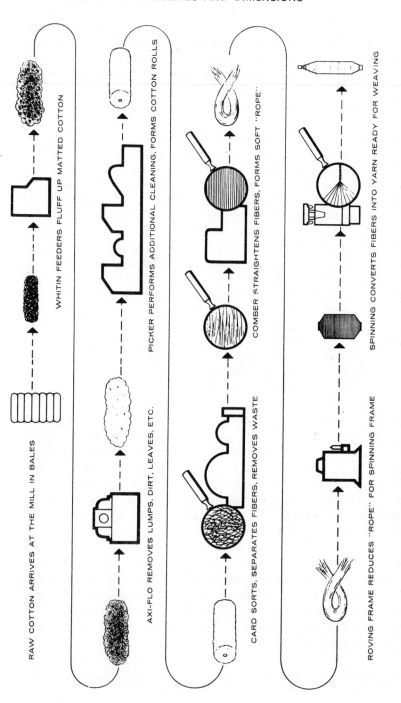

RAW COTTON ARRIVES AT THE MILL IN BALES

WHITIN FEEDERS FLUFF UP MATTED COTTON

AXI-FLO REMOVES LUMPS, DIRT, LEAVES, ETC.

PICKER PERFORMS ADDITIONAL CLEANING, FORMS COTTON ROLLS

CARD SORTS, SEPARATES FIBERS, REMOVES WASTE

COMBER STRAIGHTENS FIBERS, FORMS SOFT "ROPE"

ROVING FRAME REDUCES "ROPE" FOR SPINNING FRAME

SPINNING CONVERTS FIBERS INTO YARN READY FOR WEAVING

Figure 9.34   Process system flow chart for cotton—from fibers to fabric.   *Courtesy Whitin Machine Works, Inc.*

Figure 9.35   The Whitin Super J7B Comber. *Courtesy Whitin Machine Works, Inc.*

In the preceding equation $F_{a/bc}$ is the dynamic force exerted by control arm stud a on compound gear bc, $\ddot{\theta}_a$ is the angular acceleration of control arm a, and R_a is the arm length of the control arm referring to Figures 9.37 and 9.38. The other terms will be defined later.

Assuming excellent lubrication throughout, the only rotational effect which the control arm feels from the epicyclic gear train and from the inertia of the detaching rolls in Figure 9.37 is from force $F_{bc/a}$, whose magnitude and line of action are the same as $F_{a/bc}$.   Also, it can be shown from the derivation that the torque exerted on the control arm from this force, with respect to the rotational axis of the control arm, is $F_{bc/a} R_a$.   Hence, from a knowledge of elementary dynamics of rotating bodies, we could write for a hypothetical equivalent lumped mass on the control arm stud at radius R_a

$$F_{a/bc} R_a = \ddot{\theta}_a (J_L)_{eq} \qquad (9.60)$$

Comparing equation (9.59) with (9.60) we now write the primary design equation expressing $(J_L)_{eq}$ as follows.

$$(J_L)_{eq} = [J_y + J_d w^2][1 - uv]^2 + M_{bc} R_a{}^2 + J_{bc}[1 + u]^2 \quad \text{(P.D.E.)} \,(9.61)$$

In this equation J_y and J_{bc} are mass moments of inertia of gears y and bc respectively about their geometric axes, J_d is the mass moment of inertia sum of the two detaching rolls about their geometric axes, M_{bc} is the mass of

Cushion Plate Position With Knife Open   Cushion Plate Position With Knife Closed

Figure 9.36   Detaching rolls shown with nipper frames in extreme positions for combing operation.   *Courtesy Whitin Machine Works, Inc.*

Figure 9.37   Basic configuration of detaching-roll drive mechanism.

Figure 9.38   Cam mechanism with inertial load $(J_L)_{eq}$ equivalent to that in Figure 9.37.

compound gear bc at radius R_a of Figure 9.37, and u, v, and w are gear ratios of the figure defined as follows:

$$u = N_x/N_b \qquad (9.62)$$

$$v = N_c/N_y \qquad (9.63)$$

$$w = N_y/N_f \qquad (9.64)$$

In the above gear ratio equations, the N's designate numbers of teeth to be selected in design for the gears x, b, c, y, and f as subscripted, referring to Figure 9.37 for notation. Of course, values selected for N_x, N_b, N_c, N_y, and N_f must be positive whole numbers, but this discrete value constraint for each gear covers a very wide range of practical variation which would be acceptable in design.

To be assured that the peripheral motion of the detaching rolls is right for the combing of cotton, two more subsidiary design equations must be included in the initial formulation. The derivations are lengthy and will not be presented here. The equations derived are as follows.

$$D_d(N_d/N_e)uvw = 0.446 \qquad \text{(S.D.E.) (9.65)}$$

$$10.5\,(N_d/N_e)uv = (\Delta\theta_a)[1 - uv] \qquad \text{(S.D.E.) (9.66)}$$

Satisfaction of (S.D.E.) (9.65) will guarantee the proper peripheral advance per cycle of the detaching rolls for feeding the correct amount of cotton.   In that equation D_d is the detaching roll diameter, N_d and N_e are tooth numbers to be selected for gears d and e, respectively, referring to Figure 9.37, and u, v, and w are the gear ratios to be selected in design and defined by equations (9.62), (9.63), and (9.64), respectively.   Next, satisfaction of (S.D.E.) (9.66) will guarantee the proper peripheral reversal motion of the detaching rolls per cycle as required for separating and piecing the fibers in the combing operation.   In that equation, all terms have been defined except $(\Delta\theta_a)$, which is the angle of swing of the cam follower arm as indicated in Figure 9.38.

The need for writing a final subsidiary design equation is recognized from a review of the equations now written, since we realize that detaching rolls inertia J_d and diameter D_d are related mathematically.   Since the desired length of the machine and detaching rolls is a specified constant, and since the desired geometric proportions of the steel detaching rolls are known from past experience, we can express the relationship as follows, based on elementary mechanics:

$$J_d = kD_d{}^4 \qquad \text{(S.D.E.) (9.67)}$$

In this equation k is a readily determined geometric proportionality and material constant from analysis of past designs, and the other two terms have already been defined.

A regional constraint of significance to be adhered to in this problem of design is on cam follower arm swing $(\Delta\theta_a)$, which should not be excessive. Referring to Figure 9.38, based on past cam design experience, we decide that the following range would be acceptable.

$$(\Delta\theta_a) \lesssim 55°$$

Also, we should mention some relatively loose constraints, in that gear ratios u, v, w, and (N_d/N_e) should all be of practical value as should detaching roll diameter D_d.

In brief, our initial formulation consists of (P.D.E.) (9.61), (S.D.E.)s (9.65), (9.66), and (9.67), and the regional constraint on cam follower swing $(\Delta\theta_a)$ assumed at least initially as the only one of significance.   Hence, we have for the number of subsidiary design equations $N_s = 3$, and based on our initial assumptions the number of free variables $n_f \geq 3$.   Thus, referring to Figure 8.2 of Chapter 8 we readily recognize a case of normal specifications.   A relatively easy approach for extraction of the optimum design is to combine the aforementioned three (S.D.E.)s with (P.D.E.) (9.61) by eliminating what we have initially assumed to be free variables, J_d, the ratio (N_d/N_e), and D_d, respectively.   In this way we obtain the developed (P.D.E.) (9.61a) which

d
88T
20P

New departure ball
bearing No. 7209-XR-3-C

H.E. cylinder shaft

y
86T
20P

x
24T
20P

Ⓐ

e
64T
20P

c
18T
20P

b
80T
20P

Inside head
end plate

Tapered spline as on
N9-868 and 853-CII
except I-1/2 diameter
large end

Ⓑ

New departure ball bearing
No. 7211-MR-9-C
(not drawn to scale)

SIDE VIEW SECTION

Figure 9.39 Part of a layout drawing used in optimum design of the detaching-roll drive mechanism.

FRONT VIEW
-GEARING SECTION

Figure 9.39 (continued)

follows, where the terms are assumed to be essentially independent of each other.

$$(J_L)_{eq} = J_y(1 - uv)^2 + \frac{484k}{(\Delta\theta_a)^4 w^2 (1 - uv)^2} + M_{bc} R_a{}^2 + J_{bc}(1 + u)^2$$

$$\text{(P.D.E.)}\ (9.61a)$$

Therefore, from developed (P.D.E.) (9.61a) we conclude the following *goals of optimum design*, for minimization of equivalent load inertia $(J_L)_{eq}$.

1. Design for J_y, M_{bc}, R_a, and J_{bc} all as small as possible (i.e., the epicyclic gear train of Figure 9.37 should be of as small proportions as practically feasible).
2. Cam follower arm swing $(\Delta\theta_a)$ should be as large as practical, which would be $(\Delta\theta_a) \approx 55°$.
3. Choose gear ratio w as large as practical, which from past experience would be approximately 5 in magnitude. Thus, we should strive in design for $w = N_y/N_f \approx 5$.
4. Regarding choices of gear ratios u and v, the decisions explicitly reached were to have both u and v as small as practical, which from past experience would be approximately 0.2 in magnitude. Thus, in design we should strive for $u = (N_x/N_b) \approx 0.2$ and $v = (N_c/N_y) \approx 0.2$. The proof for these optimization conclusions for u and v depends on numerical data; it is quite lengthy, and it will not be presented here.

The preceding "goals of optimum design" were the controlling guide of greatest significance in the making of decisions in the layout drawing for the new design shown in Figure 9.39. Incidentally, the equivalent load inertia $(J_L)_{eq}$ achieved for the new design was only 0.392 times its magnitude in the initial design, representing a 60.8% reduction in inertial load for the cam mechanism. This sizeable reduction as realized by application of the method of optimum design made it possible to operate the comber machine safely at an appreciable increase in speed, far in excess of what was hoped for before starting the work of redesign. Also it should be mentioned that the values determined at the end for gear ratio (N_d/N_e) and diameter D_d were of practical size, thereby validating our initial assumption that these were free variables, resulting in the case of normal specifications.

Derivations of synthesis such as what we have briefly outlined herein, the making of practical design layouts such as presented in Figure 9.39, and in the process the making of theoretical calculations of analysis where appropriate, all go hand-in-hand in the work of mechanical design synthesis. They are all tied together, and for optimum results a careful coordination of effort must be realized. Only a very small part of the total work for the redesign of the detaching roll drive mechanism is presented in this book, having out-lined briefly only the aspects of mechanical design synthesis. Many more drawings were made, several hundred pages of theoretical analysis calculations were carried through, many practical features not to be disclosed were incorporated, and an extensive testing program including many strain gage experiments were undertaken by the Whitin Machine Works on a prototype machine to verify completely the success of design. The design of the detaching roll drive mechanism in the Whitin Super J7B Comber is one which truly exemplifies practical success for the method of optimum design in a complex problem. The author is grateful to the Whitin Machine Works for their foresightedness in allowing him to carry this project through to a successful conclusion! Also, the high quality of workmanship employed by the company for the execution of the latter stages in the Figure 1.3 morphology is greatly appreciated. Without this effort in the total picture the project would not have been a success.

9-6 SYNTHESIS FOR OPTIMUM DESIGN OF A NEW DRIVE MECHANISM FOR A HIGH-SPEED MACHINE

In this section we will very briefly present another industrial example of mechanical design synthesis, where the original problem was presented by the Warner & Swasey Company to the author in a consulting capacity. There are several reasons for including this example in the book. For one, it illustrates the application of some of the techniques of creative synthesis presented in earlier chapters, for satisfaction of the optimization objectives of minimizing shaking force. Secondly, it illustrates another application of the method of optimum design to a complex assemblage of mechanical elements, for the minimization of a critical internal force. Finally, it again illustrates the very important technique of utilizing the theoretical goals of optimum design in the making of a practical design layout. The work of mechanical design synthesis will not be presented in detail because of space limitations in the book as well as the highly specialized and proprietary nature of the example.

In the processing of long staple textile fibers into yarns various drawing or drafting operations are necessary for the eventual obtainment of a high-quality fabric. The Warner & Swasey Company has designed Pin Drafter® machines for performing these operations. A highly efficient machine system flow chart, which incorporates Pin Drafters, is shown in Figure 9.40, known as the "short long-staple system," and taken from reference 9-23. It was first highly successful in the tufted-carpet yarn trade, and later it spread into other areas, such as in the production of medium-weight yarns for upholstery, draperies, etc. It has the characteristics of producing large volumes of stock

SHORT LONG-STAPLE YARN SYSTEM

Figure 9.40 Machine system flow chart for short long-staple system.
Courtesy The Warner & Swasey Company.

with large sales value in a small space of a mill, and thus it appears to be an optimum machine system for the production of coarse-to-medium count yarns. Hence, Figure 9.40 is included at this point as an excellent illustration of a machine system flow chart as generally discussed in section 9-1 of the present Chapter.

In the design of a Pin Drafter machine the stock delivery unit must satisfy the requirements of the particular machine system in which it must function. Hence, single and dual coilers and ballers have been designed, with single or dual automatic or manual doffing features. The Warner & Swasey M-4700

Figure 9.41   The Warner & Swasey M-4700 Pin Drafter® Machine with a dual baller delivery of an earlier design than that in Figure 9.47. *Courtesy The Warner & Swasey Company.*

Pin Drafter machine with dual baller delivery is shown in Figure 9.41.   As the sliver is wound to form a ball, shown at the right of Figure 9.41, it must be twisted and placed by translatory motion at the proper helix angle on the ball to form an acceptable package.   This also requires rapid accelerations of the twister mass during reversal at each end of the traverse stroke.   The problem of mechanical design synthesis which we will consider here is the drive mechanism for reciprocating the twister traverse masses of a dual-baller delivery unit.

Left and right twister masses of a dual-baller delivery unit must be reciprocated through specified stroke S as depicted in Figure 9.42, with a given motion of approximately known acceleration characteristics.   If the given left and right twister masses are moved in opposite directions at any instant, the accelerations of each will be opposite and the accompanying "inertia forces" will cancel.   Hence, the resultant shaking force on the machine frame will be minimized, and this will be a basic objective for optimization in mechanical design synthesis of the drive mechanism.   Also, in our efforts

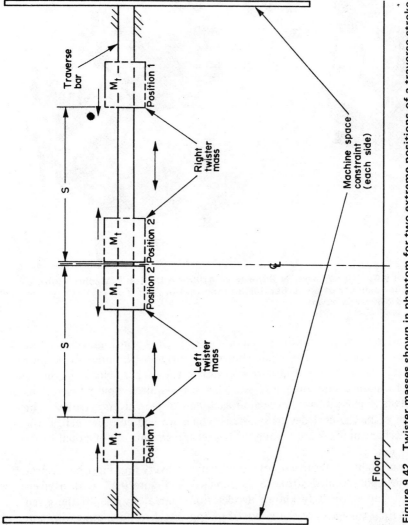

Figure 9.42 Twister masses shown in phantom for two extreme positions of a traverse stroke.

of creative design we should strive for simplicity. The basic technique followed will be a combination of the systematics of linkages with the logical building-block approach from Chapters 3 and 4, respectively. The particular boundary conditions for the mechanism to be designed are indicated in Figure 9.42, and the drive shaft of controlled motion characteristics must be located well, within the area bounded by the machine space constraints shown on each side and the traverse bar and floor in the vertical directions.

The approach for creative design will be first to synthesize a simple mechanism for driving a single twister mass, which will then be used as a building block for the synthesis of the total mechanism configuration in a logical procedure of combining components.   Hence, in reference to the catalog of basic linkages for $F = +1$ in Figure 3.6 of Chapter 3, we start with the simplest possibility and thus attempt to use the four-bar linkage for the synthesis of a suitable basic mechanism for driving one of the twister masses in Figure 9.42.   This we soon find impossible to do for this particular problem, because of the twister mass motion requirements and the space constraints previously described.   Next, from the catalog of basic linkages in Figure 3.6 we attempt to synthesize some suitable mechanisms from Watt's chain, which is in the next category of complexity in the figure, having $L = 6$ links.   In this attempt we are successful, as presented in Figure 9.43 where five basic mechanisms are shown, all compatible with the motion requirements and boundary constraints of Figure 9.42.   Several other possibilities not shown were derived from Stephenson's chain in the same category of complexity of Figure 3.6, but none of these basic mechanism types proved to be of real practical interest for this particular application.

Upon review of the five basic types of mechanisms derived in Figure 9.43, we readily see that two are of primary interest.   They are mechanisms Ⓑ and Ⓓ in that figure, both of which appear to be acceptable for this particular problem.   Mechanism Ⓐ offers no significant features of improvement over mechanism Ⓑ, and in fact is more complicated since it requires an additional link which is the connecting rod 5.   Mechanism Ⓒ of Figure 9.43 was discarded because of the quick-return effect inherent to this type of drive, which it was feared would vary the ball or package helix angle too much.   Mechanism Ⓔ in the figure was discarded because of difficulty in obtaining an adjustable stroke, which was a desired feature because of various size packages to be wound.

In the final review, basic mechanism Ⓑ of Figure 9.43 was chosen as being more desirable than Ⓓ because of the expense involved in the manufacture of a cam for the latter.   For this particular problem it was believed that the motion characteristics of mechanism Ⓑ would be entirely satisfactory for the winding of acceptable packages in the dual baller.   Hence, the next step in the procedure of synthesis was the logical combining of two basic mechanisms

Figure 9.43   Sketches from the synthesis of some basic twister drive
mechanisms derived from Watt's chain.

B, for the driving of right and left twister masses in opposite directions within the boundaries of Figure 9.42. The drive mechanism configuration which naturally followed is shown in Figure 9.44, where gear segments were first employed for connection of the fork arms. The mechanism of Figure 9.44 should have a minimum amount of resultant shaking force for high-speed operation, which was verified in the end by a very smooth operation for the prototype model.

After initial synthesis of the mechanism of Figure 9.44, some feasibility calculations very quickly revealed that dynamic bearing loads at joints A and B of the figure would be critical for the desired speeds of operation. Hence, the method of optimum design from Chapter 8 was applied for the objective of minimizing bearing load F_A for the mechanism in critical position 1 of Figure 9.44. To start, we recognize that specified values can be estimated at this stage of design for twister mass M_t, stroke S with peak twister mass acceleration $\ddot{S}_1$ in critical position 1, fork pivot separation H, gear segment mass moment of inertia J_g, and connecting rod mass M_a, all of which are shown in Figure 9.44. On the other hand, we do have some freedom in design for the selection of dimensions L and R_b, and for location of the drive shaft in Figure 9.44. Let us determine the directions in which to head for these design parameters in order to minimize bearing load F_{A1} at A for critical position 1 of the mechanism.

From our knowledge of basic dynamics, the system of elements in Figure 9.44 can be hypothetically reduced to a single mass rotational system on link b pivoting about O_b, with approximately an equivalent inertia J_{eq} as follows:

$$J_{eq} \approx 2[J_g + M_b L^2/3 + M_t L^2] + M_a R_b{}^2$$

For this equivalent single-mass rotational system, its angular acceleration, $\ddot{\theta}_b$, would be approximately $\ddot{S}/L$, where $\ddot{S}$ is the linear acceleration of twister mass M_t in Figure 9.44. Also, we recognize that connecting rod a is essentially a two-force member with relatively small mass M_a, so the drive torque exerted on link b is approximately $F_{A1}R_b$ for critical position 1 as seen from Figure 9.44. Hence, by the application of elementary dynamics to the equivalent single mass rotational system we obtain the following:

$$F_{A1}R_b \approx J_{eq}(\ddot{\theta}_b)_1$$

$$\approx \{2[J_g + M_b L^2/3 + M_t L^2] + M_a R_b{}^2\}\left(\frac{\ddot{S}_1}{L}\right) \qquad (9.68)$$

Let us now recognize that fork arm mass M_b in Figure 9.44 is not a specified value, but as an approximation varies linearly with L, which is appropriate for this application from feasibility calculations.

$$M_b \approx k_b L \qquad (9.69)$$

Figure 9.44 Sketch of dual baller twister drive mechanism as synthesized from basic mechanism Ⓑ of Figure 9.43.

The constant of proportionality k_b can be estimated at this time, with a sufficient degree of accuracy, from the feasibility calculations assuming the properties of ductile iron desired for the fork arms and avoiding various failure phenomena of significance.   Finally, referring to Figure 9.44, we will define a geometric proportionality variable u as follows:

$$u = R_b/L \tag{9.70}$$

The reason for doing this is not obvious at this stage of explanation, but it will enable us to separate the variables for an easier optimum design variation study to follow.

Combining equations (9.70) and (9.69) with (9.68) by eliminating R_b and M_b we obtain the following developed primary design equation for this case of normal specifications.

$$F_{A1} \approx \left[\frac{2}{u}\left(\frac{J_g}{L^2} + \frac{k_b L}{3} + M_t\right) + M_a u\right](\ddot{S}_1) \quad \text{(P.D.E.)} \tag{9.71}$$

In (P.D.E.) (9.71), J_g, k_b, M_t, M_a, and $\ddot{S}_1$ are constants whose values have been approximately determined at this stage of design as previously discussed. The design variables are u and L, essentially independent of each other as readily seen from Figure 9.44, with $u = R_b/L$ a proportionality ratio. Acceptable values for L and u would be within regional constraints $L_{min} \le L \le L_{max}$ and $u_{min} \le u \le u_{max}$.

From (P.D.E.) (9.71) we readily see that for any specific value of u, F_{A1} has a minimum value for $\partial F_{A1}/\partial L = 0$, as depicted graphically in Figure 9.45(a). Thus, for ideal point L_{opt} we calculate

$$\frac{\partial F_{A1}}{\partial L} \sim -2J_g/L^3 + k_b/3 = 0$$

Therefore

$$L_{opt} = (6J_g/k_b)^{1/3} \tag{9.72}$$

Using the estimated values for J_g and k_b previously discussed, we can calculate ideal value L_{opt} by equation (9.72).   Accuracy of values for J_g and k_b are not too significant in use of that equation, because of the 1/3 exponent and the fact that the curve is of zero slope at ideal point L_{opt} in Figure 9.45(a).   The value of L_{opt} now determined can be compared with estimates of L_{min} and L_{max} in the making of a layout drawing such as in Figure 9.46.   The specific relationship determined between L_{opt}, L_{min}, and L_{max} was as actually shown in the variation diagram of Figure 9.45(a).   Hence, the decision of optimum design reached for minimization of F_{A1} was to place $L = L_{max}$.   Thus, the direction in which to head at this stage in the making of the layout drawing

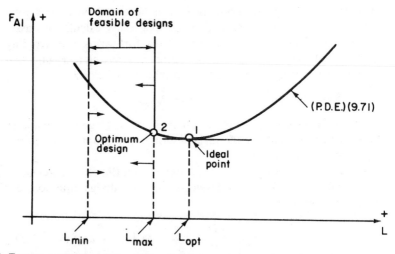

(a) Typical variation of F_{AI} with respect to L for any specific u value

(b) Variation of F_{AI} with respect to u from developed (P.D.E.)(9.71a).

Figure 9.45 Typical variation diagrams used in optimum design of dual baller drive mechanism, for minimization of F_{A1}.

in Figure 9.46 is to make L as large as feasible.   This goal of optimum design was actually incorporated in Figure 9.46 at this time.

Returning now to (P.D.E.) (9.71) with $L = L_{max}$, for the sake of simplicity we define a now determined constant C_1 as follows:

$$C_1 = 2\left(\frac{J_g}{L_{max}^2} + \frac{k_b L_{max}}{3} + M_t\right) \qquad (9.73)$$

Thus, developed (P.D.E.) (9.71) becomes

$$F_{A1} \approx \left[\frac{C_1}{u} + M_a u\right](\ddot{S}_1) \qquad \text{(P.D.E.) (9.71a)}$$

From (P.D.E.) (9.71a), with $(\ddot{S}_1)$ a known constant, we see that F_{A1} has a minimum value for $\partial F_{A1}/\partial u = 0$, as depicted graphically in Figure 9.45(b). Thus, for ideal point u_{opt} we calculate

$$\frac{\partial F_{A1}}{\partial u} \sim -\frac{C_1}{u^2} + M_a = 0$$

Therefore

$$u_{opt} \approx \sqrt{C_1/M_a} \qquad (9.74)$$

Using the estimated values for J_g, k_b, M_t, and L_{max} previously discussed, we could now calculate C_1 from equation (9.73).   Then, with M_a previously estimated we could calculate ideal value u_{opt} from equation (9.74).   However, these specific calculations need not be made since from equations (9.73) and (9.74) we notice that

$$u_{opt} \approx \sqrt{C_1/M_a} > \sqrt{M_t/M_a}$$

and at this stage of design we know that $M_t >> M_a$ exists.   Hence, we conclude that ideal value $u_{opt} > 1$ will be true, and referring to Figure 9.44, for a practical design we require $u = R_b/L < 1$.   Thus, we have without question the relation $u_{max} < u_{opt}$, and the range of acceptability is as actually shown in Figure 9.45(b).   Hence, the decision of optimum design reached for minimization of F_{A1} is to place $u = u_{max}$.   Therefore, the direction in which to head at this time in drawing Figure 9.46 is to have R_b as large as possible, since by equation (9.70) $u = R_b/L$.   This goal of optimum design was actually incorporated in Figure 9.46 at this time, together with the proper location of the drive shaft.

In closing, we again mention the importance of integrating theoretical and practical work in mechanical design synthesis.   The conclusions from theoretical derivations of optimum design as now described were a very important part in the making of practical design decisions in the layout drawing of

Figure 9.46 Layout drawing used in optimum design of the dual baller drive mechanism.

Figure 9.46 (continued)

Figure 9.46. Also, the aspects of mechanical design synthesis as now outlined were only part of the total work for this design project. Many additional calculations of analysis were carried through along the way, many
more drawings were eventually necessary for the presentation of design
specifications, and a number of practical features were incorporated before
reaching the final design. A fairly minor modification not affecting the
basic decisions reached is shown in Figure 9.46. This is the use of the cross-
link *e* instead of the gear segments for connecting the two fork arms, incorporated from a consideration of cost reduction. All the practical and theoretical design decisions made in the complete execution of the Figure 1.3
morphology contributed toward making the design of the dual-baller drive
mechanism a success. Thus, with the incorporation of the decisions of
optimum design as outlined herein, shaking force was insignificant in the
experiments on the prototype and the critical bearing loads were of low
magnitude for long-life operation with high reliability. The completed dual
baller design in a Pin Drafter machine set-up is shown in the photograph of
Figure 9.47, taken at a recent textile machinery show.

**Figure 9.47   The new dual baller design on a Warner & Swasey M-4700 Pin
Drafter** machine set-up at a recent textile machinery show.** *Courtesy The
Warner & Swasey Company.*

9-7 SOME GENERAL GOALS OF OPTIMUM DESIGN FOR HIGH-SPEED MACHINERY

In this chapter we have presented some specific examples of optimum design in high-speed machinery, based on the satisfaction of total system requirements in the best possible way. However, for each of the industrial examples presented, the total design process also incorporated some basic goals of optimum design in the synthesis of construction details. These general goals of optimum design will be briefly explained in what follows. The author has seen many examples in industry where the designer's lack of awareness of these basic principles in the synthesis of construction details resulted in unnecessary failures at the testing stage.

A simplified analytical model of a high-speed mechanism might be approximately depicted by the spring-mass system of Figure 9.7. The given command motion s from a driver causes a driven body to respond with motion y, with mass and elastic properties inherent to the mechanism assemblage. The analytical model analogy would be appropriate as a general approximation for the cam mechanism of Figure 9.13, the Geneva mechanism of Figure 9.29, the detaching roll drive mechanism of Figure 9.37, and the dual-baller twister drive mechanism of Figure 9.44, as specific examples. In each case, in addition to the specific problem previously considered, we would like to have the responding motion of the driven mass as close as possible displacement-wise to the commanding motion of the driver or "rigid-link" ideal motion.

With the preceding in mind, suppose that we look at the analytical model of Figure 9.7 applying Newton's second law to a free-body diagram of mass M without damping. Thus, we obtain $k(s - y) = M\ddot{y}$, which gives

$$\frac{d^2y}{dt^2} + \left(\frac{k}{M}\right)y = \left(\frac{k}{M}\right)s \qquad (9.75)$$

To be general, suppose that we consider the periodic command motion s to consist of N significant and independent Fourier components, expressed as follows:

$$s = \sum_{i=1}^{N} S_i \sin\left(\omega_i t + \phi_i\right) \qquad (9.76)$$

In equation (9.76), S_i, ω_i, and ϕ_i are the amplitude, circular frequency, and phase angle, respectively, of a component of the excitation.

Using this general equation (9.76) for command motion s, the particular integral or steady-state solution of differential equation (9.75) is as follows, for the responding motion y of the driven mass:

$$y = \sum_{i=1}^{N} \frac{S_i \sin (\omega_i t + \phi_i)}{[1 - (\omega_i/\omega_n)^2]} \tag{9.77}$$

where ω_n is the natural circular frequency as follows.

$$\omega_n = \sqrt{k/M} \tag{9.78}$$

From the general analysis now made, we can draw the following "goals of optimum design" with respect to details of construction for two basic types of problems in high-speed machinery design.

1. *For Motion Transmission Applications* (such as the high-speed mechanisms of sections 9.3, 9.4, 9.5, and 9.6). Command motion s is to be transmitted to responding mass with minimum error, so the basic goal is minimization of $|y - s|$. Hence, from equations (9.76) and (9.77) we obtain the primary design equation (9.79).

$$|y - s| = \sum_{i=1}^{N} \left[\frac{1}{1 - (\omega_i/\omega_n)^2} - 1 \right] S_i \sin (\omega_i t + \phi_i) \tag{9.79}$$

Thus, for a given command s with S_i, ω_i, and ϕ_i known, we would minimize discrepancy $|y - s|$ by designing the physical system for ω_n as large as practical, so that we obtain if possible the relation $\omega_n >> \omega_i$ for $i = 1, \ldots, N$. Therefore, from equation (9.78) we see that the general goals of optimum design are:

(a) *Design for high stiffness k between the commanding motion and the responding mass.* This is obtained by incorporating details of rigid mounting construction, such as reasonably large diameter shafts, straddle mountings, bearings well placed and preloaded if necessary, ribbed design for bending members, etc.

(b) *Design for low mass M* of the moving parts, being sure that rigidity of construction is not lost in the process. Hence, mass reduction by decreasing shaft size is often unwise, since total system inertia is generally decreased only a small amount whereas rigidity is generally decreased tremendously by such an approach.

Therefore, in many specific applications the various elements of construction must be viewed in light of their primary contributions to the system (stiffness or mass), with the preceding two goals appropriately applied for minimization of discrepancy $|y - s|$.

2. *For Motion Isolation Applications* (such as the automotive suspension system of section 9.2). Command motion s is *not* to be transmitted to the responding mass in such cases. Hence, the basic goal of optimum design is minimization of response magnitude $|y|$. Thus, from a given excitation or command motion s with S_i, ω_i, and ϕ_i known, we would minimize y by

designing the physical system for ω_n as small as practical, so that we obtain if possible the relation $\omega_n << \omega_i$ for $i = 1, \ldots, N$. This conclusion is drawn from primary design equation (9.77) in this case. This we achieve in the physical system by the following general goals of optimum design, as seen from equation (9.78).

(a) *Design for low stiffness k between the commanding motion and the responding mass.*

(b) *Design for large mass M.*

The preceding goals incorporated in the automotive suspension problem would give the smoothest ride possible, as further discussed in section 9.2. Similarly, these would be the directions in which to head in the design of a suspension system in high-speed machinery where vibration isolation is desired.

Finally, if in a specific application we find that resonance cannot be avoided with respect to a significant component of the excitation because of space limitations or construction requirements, the inclusion of damping or a vibration absorber should be considered for the design. Thus, with some additional expense, it would be possible to mitigate excessive vibrations to a tolerable level in a critical design situation.

BIBLIOGRAPHY

9-1 A. W. Judge, *The Mechanism of the Car*, Robert Bentley, Inc., 993 Massachusetts Avenue, Cambridge, Massachusetts, 1961, 542 pp.

9-2 J. P. Den Hartog, *Mechanical Vibrations*, 4th ed., McGraw-Hill Book Co., Inc., New York, 1956, 435 pp.

9-3 S. Timoshenko, *Strength of Materials*, Part I, 3rd ed., D. Van Nostrand Co., Inc., Princeton, N.J., 1955, 442 pp.

9-4 A. Seireg, *Mechanical Systems Analysis*, International Textbook Company, Scranton, Pennsylvania, 1969, 596 pp.

9-5 L. S. Jacobsen and R. S. Ayre, *Engineering Vibrations*, McGraw-Hill Book Co., Inc., 1958, 564 pp.

9-6 C. Lipson, G. C. Noll, and L. S. Clock, *Stress and Strength of Manufactured Parts*, McGraw-Hill Book Co., Inc., 1950, 259 pp.

9-7 R. E. Peterson, *Stress Concentration Design Factors*, John Wiley & Sons, Inc., New York, 1953, 155 pp.

9-8 A. M. Wahl, *Mechanical Springs*, 2nd ed., McGraw-Hill Book Co., Inc., 1963, 323 pp.

9-9 R. C. Johnson, *Optimum Design of Mechanical Elements*, John Wiley & Sons, Inc., New York, 1961, 535 pp.

9-10 R. C. Johnson, "Predicting Part Failures," Parts 1 and 2, *Machine Design*, Penton Publishing Company, Cleveland, Ohio, Jan. 7, 1965, pp. 137–142 and Jan. 21, 1965, pp. 157–162.

9-11 J. E. Shigley, *Mechanical Engineering Design*, McGraw-Hill Book Co., Inc., New York, 1963, 631 pp.

9-12 G. Sines and J. L. Waisman, *Metal Fatigue*, McGraw-Hill Book Co., Inc., New York, 1959, 415 pp.

9-13 M. F. Spotts, *Mechanical Design Analysis*, Prentice-Hall, Inc., Englewood Cliffs, N.J., 1964, 428 pp.

9-14 F. P. Zimmerli, "Human Failures In Spring Applications," *The Mainspring*, Associated Spring Corporation, Bristol, Connecticut, No. 17, Aug.–Sept. 1957.

9-15 O. Lichtwitz, "Mechanisms for Intermittent Motion," *Machine Design*, Penton Publishing Co., Cleveland, Ohio, Dec. 1951, Jan., Feb., March 1952.

9-16 R. C. Johnson, "How To Design Geneva Mechanisms To Minimize Contact Stresses and Torsional Vibrations," *Machine Design*, March 22, 1956, pp. 107–111.

9-17 "Delrin® Acetal Resins," *Design and Engineering Data*, Plastics Department, E. I. du Pont de Nemours & Co., Inc., Wilmington, Delaware, 1961, 82 pp.

9-18 R. L. Peters, *Materials Data Nomographs*, Reinhold Publishing Corp., New York, 1965, 224 pp.

9-19 C. Lipson and L. V. Colwell, *Handbook of Mechanical Wear*, Chap. 4, "Experimental Load-Stress Factors," and Chap. 5, "Subsurface Fatigue," The University of Michigan Press, Ann Arbor, 1961, pp. 56–107.

9-20 E. Rabinowicz, *Friction and Wear of Materials*, John Wiley & Sons, Inc., New York, 1965, particularly pp. 115–117 and 190–193.

9-21 S. Timoshenko and J. N. Goodier, *Theory of Elasticity*, 2nd ed., McGraw-Hill Book Co., Inc., New York, 1951, 506 pp.

9-22 R. J. Roark, *Formulas for Stress and Strain*, 4th ed., McGraw-Hill Book Co., Inc., New York, 1965, 432.

9-23 R. O. Perrault, "Short Long-staple System," *Textile Industries*, W. R. C. Smith Publishing Co., Atlanta, Georgia, July 1966.

APPENDIX A

EXTENDED CATALOG OF ACCEPTABLE ISOMERS
WITH SINGLE PIN JOINTS

(See Table 3-2 for Link Combinations and Roman Numeral Notation.)

For F = +1.

    I. L = 4: B = 4.

                     See Figure 3.6 for all acceptable isomers.

    II. L = 6: B = 4, T = 2.

    III. L = 6: B = 5, Q = 1.   No acceptable isomers are possible.

    IV. L = 8: B = 4, T = 4.   (See Table in Section 3-3 for derivation and isomer
                                      numbers in parentheses.)

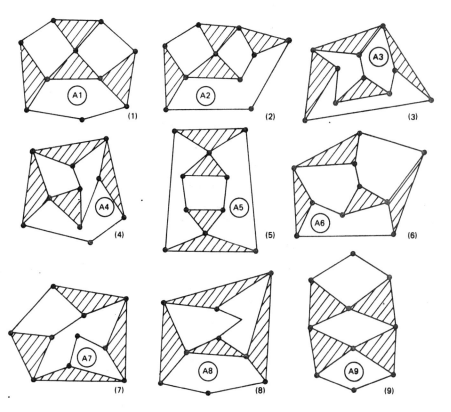

(1)   (2)   (3)

(4)   (5)   (6)

(7)   (8)   (9)

For F = +1 (Cont.)

    V.   L = 8: B = 5, T = 2, Q = 1.

        See Figure 3.9 for the acceptable isomers.

    VI.   L = 8: B = 6, Q = 2.

(1) (2)

    VII.   L = 8: B = 6, T = 1, P = 1.   No acceptable isomers possible.

For F = 0.

    VIII.   L = 3: B = 3.

    IX.   L = 5: B = 3, T = 2.   } See Figure 3.6 for all acceptable isomers.

    X.   L = 5: B = 4, Q = 1.

    XI.   L = 7: B = 5, T = 1, P = 1.

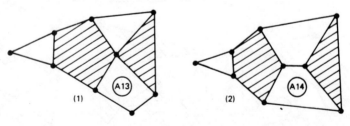

(1) (2)

For F = 0 (Cont.)

XII.   L = 7:   B = 3, T = 4.

(1)

(2)

(3)

(4)

(5)

(6)

(7)

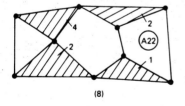

(8)

Number Set Table Used in Derivation of Acceptable Isomers for B = 3, T = 4, for F = 0 Characteristic

T_1	T_2	T_3	T_4	Isomer No.
111	001	001	001	(1)
011	011	011	000	(2)
012	002	001	000	(3)
011	001	011	001	(4), (6), (8)
002	002	001	001	(5), (7)

(To determine the 8 acceptable isomers shown, identical digits in the above table are matched in pairs between the 4 tenary links, avoiding mechanism and preloaded structural components.)

For F = 0 (Cont.)

XIII.   L = 7: B = 4, T = 2, Q = 1.

For F = 0 (Cont.)

XIV.   L = 7:   B = 5, Q = 2.

See Figure 3.10 for the acceptable isomers.

For F = −1.

XV.   L = 2:   B = 2.

XVI.   L = 4:   B = 2, T = 2.

} See Figure 3.6 for all acceptable isomers.

XVII.   L = 4:   B = 3, Q = 1.

XVIII.   L = 6:   B = 4, T = 1, P = 1.

} No acceptable isomers are possible.

XIX.   L = 6:   B = 4, Q = 2.

XX.   L = 6:   B = 3, T = 2, Q = 1.

} See Figure 3.6 for all acceptable isomers.

XXI.   L = 6:   B = 2, T = 4.

See Figure 3.6 for one, and add following two as mentioned therein.

XXII.   L = 8:   B = 6, P = 2.

No acceptable isomers are possible.

XXIII.   L = 8:   B = 5, T = 1, Q = 1, P = 1.

See Figure 3.11 for the acceptable isomers.

For F = −1 (Cont.)

XXIV.   L = 8: B = 4, T = 3, P = 1.

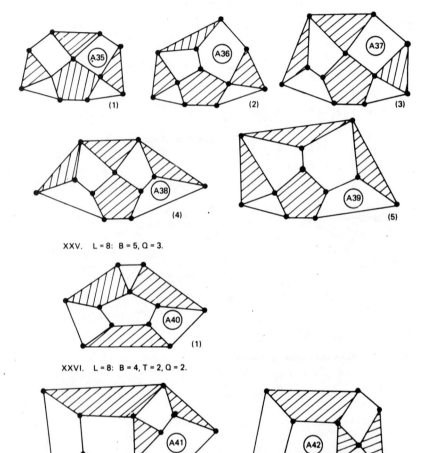

(1)   (2)   (3)

(4)   (5)

XXV.   L = 8: B = 5, Q = 3.

(1)

XXVI.   L = 8: B = 4, T = 2, Q = 2.

(1)   (2)

For F = −1 (Cont.)

(3)

(4)

Plus 11 more not shown.

(5)

XXVII.   L = 8: B = 3, T = 4, Q = 1.

(1)

(2)

(3)

(4)

For F = −1 (Cont.)

(5)

(6)

Plus 16 more not shown.

XXVIII. L = 8: B = 2, T = 6.

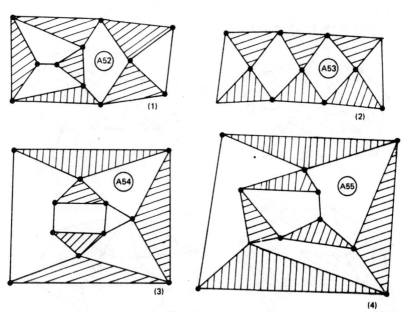

(1)

(2)

(3)

(4)

Plus 11 more not shown.

For F = +2:

XXIX.   For L = 5:   B = 5.

XXX.   For L = 7:   B = 6, Q = 1.   } See Figure 3.6 for all acceptable isomers.

XXXI.   For L = 7:   B = 5, T = 2.

Link Combinations for L = 9, F = +2
(Extension of Table 3-2)

F	L	B	T	Q	P	Designation
+2	9	7	1	0	1	XXXII
	9	7	0	2	0	XXXIII
	9	6	2	1	0	XXXIV
	9	5	4	0	0	XXXV

Derived as
explained in
Section 3-2
using equations
(3.5d) and (3.3)

XXXII.   For L = 9:   B = 7, T = 1, P = 1.

(1)

(2)

(Note: The above acceptable isomers are merely combinations of the
four-bar linkage, Watt's linkage and Stephenson's linkage.)

XXXIII.   For L = 9:   B = 7, Q = 2.

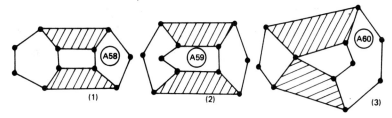

(1)

(2)

(3)

For F = +2 (Cont.)

XXXIV.   L = 9:   B = 6, T = 2, Q = 1.

(1)   (2)   (3)

(4)   (5)   (6)

Plus 10 more not shown.

XXXV.   L = 9:   B = 5, T = 4.

(1)   (2)   (3)

(4)   (5)   (6)

Plus 11 more not shown.

(7)

APPENDIX B

CATALOG OF SOME ACCEPTABLE ISOMERS FOR F = +1 WITH ONE HIGHER JOINT

See Figure 3.12(b) for the simplest possibility with one higher joint.

See Figure 3.13 for four acceptable isomers derived by the partial shrinkage technique, and three of these have one higher joint.

Link Combination B = 6, T = 1 and Q = 1 with One Double Joint Gives F = +1. All possible acceptable isomers of this are:

Figure 3.13(b) (1) plus the following four.

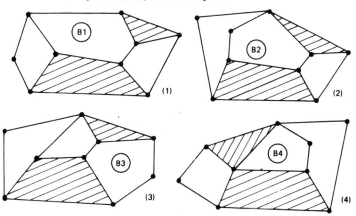

Link Combination B = 6 and T = 2 with One Triple Joint Gives F = +1. Only one acceptable isomer is possible as follows, derived by total shrink-

age of the quaternary link in linkage (A65) of Appendix A.

Link Combination B = 7 and Q = 1 with One Triple Joint Gives F = +1.
Only one acceptable isomer is possible as follows, derived by total shrinkage
of a quaternary link in (A59) of Appendix A.

Link Combination B = 5 and T = 3 with One Double Joint Gives F = +1.
Acceptable isomers are:

Plus 2 shown in Figure 3.13(d).

Plus 10 not shown.

Derivation of Linkages

Shown Here	By Total Shrinkage of Ternary Link in
B7	A69
B8	A70
B9	A70
B10	A70
B11	A71
B12	A73
B13	A73
B14	A73

APPENDIX C

COMPUTERS IN AUTOMATED OPTIMAL DESIGN

C-1 INTRODUCTION

Mathematical programming (MP) optimization methods utilizing digital computers have recently received widespread attention for solving nonlinear problems of the following general type.[1-9]

Find the set $\bar{x}_n$ of variables x_n which minimizes y, where

$$y = f_1(\bar{x}_n), \; n = 1, 2, \ldots, N \tag{C1}$$

while satisfying the inequality constraints

$$g_i(\bar{x}_n) \geq 0, \; i = 1, 2, \ldots, I \tag{C2}$$

and/or the equality constraints

$$h_j(\bar{x}_n) = 0, \; j = 1, 2, \ldots, J \tag{C3}$$

as well as the regional constraints

$$(x_n)_{min} \leq x_n \leq (x_n)_{max} \tag{C4}$$

Exact solution of such an equation system is often possible by the *method of optimum design* presented in Chapter 8, which will guarantee finding of the global optimum. Also, the technique will give a remarkably clear insight into many important aspects of the optimum design solution, unmatched by other methods. Its application to digital computers as a technique for optimal design has been illustrated many times in the examples of Chapters 8 and 9. The computational flow chart so derived is explicit and exact, leading directly to the optimum design with no searching generally required. Hence, once a general program is developed, optimal design is achieved by the computer in very short time. This is true even for fairly large problems, such as in section 9-3 where there are seventeen variables in the initial formulation.

Nonlinear programming techniques are what we will briefly consider in this Appendix. Many algorithms have been developed recently for general application in automated optimal design using the digital computer. However, none of the many algorithmic methods proposed can guarantee finding the solution to specific problems, except in certain relatively restricted problem types.[2]

The algorithmic methods start at an arbitrarily selected point, and based on local properties attempt to move towards the point of optimum design. Eason and Fenton[3] recently evaluated seventeen codes on test problems. Four of these have only regional constraints, with none active at the optimum, whereas the remaining six have inequality constraints on functions as well. No single code was markedly superior to the others on all of the test problems. However, the direct pattern search techniques were in general superior to the gradient-based methods, on these particular problems at least.

Pattern search techniques are particularly good for problems of few variables, but it is found that reliability decreases as the number of variables increases.[4] This difficulty has been alleviated somewhat by combining direct pattern search with special secondary search techniques. This has been done for problems with both equality and inequality constraints.[4,5] Such combined methods, although superior to the basic direct search procedures, are not infallible for general application. For example, the best code in reference 5 solved only eight of the ten test problems of reference 3.

In recognition of this problem, packages of codes have been developed for general use, such as OPTISEP.[5,6] The recommendation is to run several codes in a package to confirm or identify the optimum design solution to a specific problem. Hence, it was found that the OPTISEP package solved all ten of the previously mentioned test problems, which no one code could do by itself. However, the code package approach has the disadvantage of generally being computationally inefficient. Incidentally, as a suggested exercise the interested reader might find it worthwhile to comprise his own code package, such as from the many algorithms of reference 7.

A single, fast and reliable code would be ideal, and efforts are being directed towards achieving this result. As an example of late, a direct search algorithm combined with an improved secondary search strategy is presented in reference 1. This algorithm has been developed in such a way that the weakness of the direct search approach in treating problems of large dimensionality is greatly reduced. It can handle nonlinear problems with or without inequality constraints, and the starting point may be chosen at an infeasible design which violates some constraint. This algorithm has been found to be superior to the other algorithmic methods, which of course is based on the comparison method used and for a relatively small number of test problems.[1]

In the remaining part of this Appendix we will briefly consider a single code example known as PATSH, and a code package example known as OPTISEP.

C-2   PATSH

This is a very successful single code developed by D. E. Whitney at M.I.T.[8], and it is based on the direct pattern search technique developed by R. Hooke and T. A. Jeeves.[9] It is a direct numerical search which makes trial explorations, followed by steps or pattern moves in the direction indicated by the explorations.

For general applications, PATSH has been found to solve optimization problems with a high level of efficiency.[3] It has a better than average solution speed, which is remarkable since its size of only 72 FORTRAN statements is less than half as long as most codes. It has the characteristic of oftentimes converging to a high level of precision. Thus, it is attractive for small computers and repeated use.

The FORTRAN program for PATSH takes the form of a subroutine stored on a disk in a library of a computer. For application a main program must be written to handle input and output data, for specifying parts of the function you want searched if this is necessary and for output of the minumum point solution found by PATSH. Also, it is necessary to write a subroutine for calculating the y value of your function when given the x_n values of the variables. The main program will call PATSH, which in turn will call your function subroutine when necessary. Communication of data between the main program and your function subroutine must be handled by COMMON statements. Let us now consider how to use PATSH.

Suppose we wish to find the values of x_n, (for $n = 1,2, \ldots, N$) which will minimize y in equation (C1). A point in N dimensional space is defined by the values of x_n, and the corresponding value can be calculated for y with your subroutine. Within PATSH the value of y is called OBJ. A size limitation of $N \leq 25$ exists with the present implementation of PATSH given later.

Initially, in your call to PATSH, it is necessary to specify the coordinates x_n of a starting point as well as values for DEL, DLMIN, and ITLIM. Typical starting values are DEL = 1.0, DLMIN = 0.001, and ITLIM = 50. DEL is a step size parameter which is reduced as the search proceeds until it reaches the specified minimum DLMIN. PATSH takes an exploration step size for any x_n of $\Delta x_n = 0.05 * DEL * x_n$, and if $x_n = 0$ it uses $\Delta x_n = 0.05 * DEL$. Upon reaching DLMIN, PATSH returns OBJ (i.e. y) and the corresponding x_n values to the main program. On the other hand, if the number of iterations exceeds ITLIM first, PATSH will at that point return OBJ and the corresponding x_n values to the main program. But, if DLMIN is first reached, PATSH makes one more try for the minimum of y, starting again with the initial DEL and the current best x_n values.

The value of OBJ returned may not be the true minimum for y. If this is suspected, choose a new starting point with perhaps new values for DEL, DLMIN, and ITLIM.

Regarding printed output, two rows of *'s indicate that a pattern move will be attempted through a BASE POINT for improvement in OBJ. To the right of the first row is printed the serial number of the BASE POINT, whereas to the right of the second row is printed the number of pattern moves since the last pattern break. On the other hand, if only a single row of *'s is printed, this indicates that step size DEL has been cut in half or a radical change in direction has occurred for the search.

By permission from Dr. D. E. Whitney of M.I.T., the following summary is given for the Arguments and FORTRAN listing of PATSH, in Figures C1 (a) and (b).

C-3   OPTISEP

This is a very successful package of computer codes developed by J. N. Siddall, as presented in reference 5. It is a group of easily used subroutines requiring very little knowledge of either optimization theory or programming. The designer may rapidly try several optimization techniques of his choice for solving a specific problem. In OPTISEP the subroutines are independent, they have full variable demensioning, and they require relatively small space in the computer memory.

Some of the subroutines from OPTISEP are as follows, taken directly from reference 5.

SIMPLE	Linear programming
SEEK1	Hooke and Jeeves direct search
SEEK3	Hooke and Jeeves direct search
APPROX	Successive linear approximation
RANDOM	Random search with shrinkage
ADRANS	Random search with accelerated pattern moves
SIMPLX	Simplex method of direct search
DAVID	Davidson, Fletcher, Powell gradient method
MEMGRD	Memory gradient method
OPTPLT	Plots contours of objective function
FEASBL	Finds a feasible solution
UREAL	Defines optimization function, user written
EQUAL	Defines equality constraints, user written
CONST	Defines inequality constraints, user written

For specific details on these subroutines together with the FORTRAN program listings, see reference 5.

The codes of OPTISEP can be interchanged easily if one fails or if verification of a solution is desired, and ample documentation is given in the subroutines. Preparation time required for operation is relatively short, and it has better than average speed characteristics.[3]

To use the OPTISEP package, it is necessary to write a main program for dimensioning and a call statement. Also, subroutines must be written for the objective function and constraints. The designer has the freedom to vary the OPTISEP routine from one problem to another, by merely changing the call statements. For example, the strategy might be to try the fastest code first (SEEK1), followed by other more general codes (such as SEEK3, SIMPLX, etc.). In fact, the solution point found by SEEK1 might very well be used as the starting point for the next code. Hence, the package code strategy of OPTISEP gives a degree of reliability and versatility which no single code by itself can match in the field of Computer Automated Optimal Design.

REFERENCES

1. M. Pappas and J. Y. Moradi, "An Improved Direct Search Mathematical Programming Algorithm", TRANSACTIONS OF THE A.S.M.E., *Journal of Engineering for Industry*, Nov. 1975, pp. 1305-1310.
2. W. T. Zangwill, NONLINEAR PROGRAMMING: A UNIFIED APPROACH, Prentice-Hall, Inc., Englewood Cliffs, N. J., 1969.
3. E. D. Eason and R. G. Fenton, "A Comparison of Numerical Optimization Methods for Engineering Design", TRANSACTIONS OF THE A.S.M.E., *Journal of Engineering for Industry*, Series B, Vol. 96, No. 1, Feb. 1974, pp. 196-201.
4. M. Pappas, "Use of Direct Search in Automated Optimal Design", TRANSACTIONS OF THE A.S.M.E., *Journal of Engineering for Industry*, Series B, Vol. 94, No. 2, May 1972, pp. 395-401.
5. J. N. Siddall, "OPTISEP." *Designers' Optimization Subroutines*, ME/71/DSN, REP 1, Faculty of Engineering, McMaster University, Hamilton, Ontario, Canada, 1971.
6. J. N. Sidall, ANALYTICAL DECISION-MAKING IN ENGINEERING DESIGN, Prentice-Hall, Inc., Englewood Cliffs, N. J., 1972, Appendices A-O.
7. J. L. Kuester and J. H. Mize, OPTIMIZATION TECHNIQUES WITH FORTRAN, McGraw-Hill Book Co., New York, 1973, 500 pp.
8. D. E. Whitney, Joint Mechanical and Civil Engineering Computer Facility, M.I.T., 77 Massachusetts Ave., Cambridge, Mass. 02139.
9. R. Hooke and T. A. Jeeves, "Direct Search Solution of Numerical and Statistical Problems", *Journal of the Association for Computing Machinery*, Vol. 8, No. 2, April 1961, pp. 212-229.

ARGUMENTS TO PATSH

X = array of variables, defining OBJ, over which the search will be conducted. In CALL PATSH, X is the starting point of the search. On return from PATSH, X is the point where the minimum was found.

OBJ = minimum returned by PATSH.

N = number of elements of X.

DEL = initial step size.

DLMIN = minimum step size.

ITLIM = maximum number of iterations allowed.

IPT = 1 for detailed printing of explorations and pattern moves; ϕ for printing of base points only; -1 for no printing during execution of PATSH.

NAME = name of your subroutine for calculating OBJ from X.

Your main program must have the following parts:

(control cards)

```
        EXTERNAL NAME
        DIMENSION X(25)
        COMMON A,B, ... }    optional
          .
          . }   Read data such as A, B, initial X, DEL, DLMIN, N
          .
          .

          .
        CALL PATSH(X,OBJ,N,DEL, DLMIN,ITLIM,IPT,NAME)
          .
          . }   Print resulting final X and OBJ
          .
        END
```

Your subroutine must have the following parts:

```
        SUBROUTINE NAME(X,OBJ)
        DIMENSION X(25)
        COMMON A,B, ...  }    optional
          .
          . }   Calculate OBJ from X
          .
        RETURN
```

(a)   Arguments to PATSH.

Figure C1.   Pattern Search Computer Code PATSH.

```
// FOR PATSH - PATTERN SEARCH ROUTINE - D. WHITNEY 2/8/73
*TITLE *** PATSH ***  PATSH *** PATSH ***
      SUBROUTINE PATSH(PSI,SSI,N,DELS,DLMIN,ITLIM,IPT,MRIT4)
      DIMENSION PSI(25),PHI(25),THT(25),XFLG(25)
C.....PRINTER AND HIGH SPEED CONSOLE DEVICE NUMBERS
      DATA IPR,IHZL/5,6/
      DATA ALFA/1.02/
C.....FUNCTION F IS MINIMUM IMPROVEMENT REQUIRED OVER LAST BASEPOINT
      F(SSS)=SSS-ABS(SSS)*.0001*CUT
C
C PSI IS THE CURRENT BASEPT
C THT IS THE PREVIOUS BASEPT
C PHI IS THE TRIAL PT
C S IS THE OBJECTIVE FCT
C IPT= 1  FOR DIAGNOSTIC PRINTOUT
C       0  FOR MINIMAL OUTPUT
C      -1 FOR NO OUTPUT
C
C
C.....INITIALIZATION
      DEL=DELS
      IF (IPT.GE.0) WRITE(IPR,604) DEL,DLMIN,ITLIM,IPT
      DO 705 I=1,N
  705 XFLG(I)=1.
      ITER=0
      CUT=1.
C EVALUATE AT INIT BASEPT
   10 CALL MRIT4(PSI,SSI)
   90 SSITST=F(SSI)
C
C
C.....EXPLORE AROUND CURRENT BASEPOINT
  100 S=SSI
      NPATM=0
      DO 101 I=1,N
  101 PHI(I)=PSI(I)
      ICALL=1
      IF (IPT.LT.0) GO TO 150
      WRITE(IPR,599) ITER
      WRITE(IPR,600) (PSI(J),J=1,N)
      WRITE(IPR,601)  S,DEL
C MAKE EXPLORATORY MOVES
      GO TO 150
C IS PRESENT VALUE < BASEPT VALUE
  160 IF(S.LT.SSITST) GO TO 200
C....CUT STEP SIZE
      DEL=DEL/2.
      IF(DEL.GT.DLMIN) GO TO 100
      IF(IPT.GE.0) WRITE(IPR,704)
      IF(CUT.LT..5) GO TO 702
C  START OVER WITH INITIAL DEL AND CURRENT BASEPOINT
      IF(IPT.GE.0) WRITE(IPR,707)
      DEL=DELS
      CUT=0.
      GOTO 90
C
C
```

Figure C1 (b).   FORTRAN Program Listing of PATSH.

```
C.....SET NEW BASEPOINT.....MAKE PATTERN MOVE.....EXPLORE AROUND PATTERN
  200 SSI=S
      SSITST=F(SSI)
      ITER=ITER+1
      NPATM=NPATM+1
      IF(ITER.GT.ITLIM) GO TO 700
      IF(IPT.LT.0) GO TO 203
      WRITE(IPR,599) ITER
      WRITE(IPR,599) NPATM
      WRITE (IPR,600) (PHI(I),I=1,N)
      WRITE (IPR,601) SSI,DEL
      WRITE(IHZL,601) SSI,DEL
C MAKE PATTERN MOVE
203   DO 201 I=1,N
      THT(I)=PSI(I)
      PSI(I)=PHI(I)
201   PHI(I)=PHI(I)+ALFA*(PHI(I)-THT(I))
      CALL MRIT4(PHI,SPI )
      S=SPI
      IF(IPT.NE.1) GO TO 202
      WRITE(IPR,606) (PHI(I),I=1,N)
       WRITE(IPR,601) SPI,DEL
202   ICALL=2
C MAKE EXPL MOVES
      GO TO 150
C IS PRESENT VALUE < BASEPT VALUE
260   IF(S.LT.SSITST) GO TO 200
      GO TO 100
C
C
C.....INTERNAL SUBROUTINE TO MAKE EXPLORATIONS ABOUT PHI
  150 DO 180 K=1,N
      PHIOLD=PHI(K)
      STEPK=PHIOLD*.05
      IF(STEPK.EQ.0.) STEPK=.05
      STEPK=SIGN(STEPK*DEL,XFLG(K))
      PHI(K)=PHIOLD+STEPK
      CALL MRIT4(PHI,SPI)
      IF(IPT.EQ.1) WRITE(IPR,602) ICALL,K,SPI,(PHI(L),L=1,N)
      IF(SPI.LT.S) GO TO 179
      XFLG(K)=-XFLG(K)
      PHI(K)=PHIOLD-STEPK
      CALL MRIT4(PHI,SPI)
      IF(IPT.EQ.1) WRITE(IPR,602) ICALL,K,SPI,(PHI(L),L=1,N)
      IF (SPI.LT.S) GO TO 179
      PHI(K)=PHIOLD
      GO TO 180
  179 S=SPI
  180 CONTINUE
      GO TO (160,260),ICALL
C
C
700   IF (IPT.GE.0) WRITE(IPR,701)
702   DO 703 I=1,N
  703 PSI(I)=PHI(I)
      IF(IPT.GE.0) WRITE(IPR,607) ITER
      RETURN
```

Figure C1 (b) (continued).

```
599     FORMAT(' **** ',I5)
600     FORMAT('  BASE PT:',1P,4(7E15.6,E15.6,/,T9))
606     FORMAT('  PATTERN:',1P,4(7E15.6,E15.6,/,T9))
601     FORMAT(6X,'OBJ:',1PE15.6,5X,'DEL:',E15.6)
602     FORMAT(1X,2I2,' OBJ: ',1PE14.6,' TRIAL: ',4(6E14.6,E14.6,/,T35))
604     FORMAT('ODEL=',1PE15.6,',      DELMIN=',E15.6,0P,//,
       *' ITLIM=',I6,',    IPT=',I3)
607     FORMAT(' TOTAL NUMBER OF NEW BASEPOINTS (ITERATIONS) :',I5)
701     FORMAT('OSEARCH TERMINATED BECAUSE NUMBER OF ITERATIONS EXCEEDS LI
       *MIT.')
704     FORMAT('OSEARCH TERMINATED BECAUSE STEPSIZE LESS THAN LIMIT.')
707     FORMAT('O STARTING OVER WITH CURRENT BASEPOINT AND INITIAL DEL.')
        END
// DUP
SL PATSH 2
// FOR A TEST OF PATSH
*TITLE OBJECTIVE FUNCTION ***   MERIT   ***
        SUBROUTINE MERIT(X,OBJ)
        DIMENSION X(9)
C ROSENBROCKS GULTCH
        OBJ=100.*(X(2)-X(1)*X(1))**2+(1.-X(1))**2
        RETURN
        END
*TITLE *** MAINLINE FOR TESTING PATSH ***
        EXTERNAL MERIT
        DIMENSION X(2)
        X(1)=1.2
        X(2)=-1.
        N=2
        DEL=1.
        DELMIN=.0001
        ITLIM=200
        IPT=1
        CALL PATSH(X,OBJ,N,DEL,DELMIN,ITLIM,IPT,MERIT)
        WRITE(5,10) X,OBJ
10      FORMAT(' FINAL X:',1P2E15.6,',     OBJ:',1PE15.6)
        END
// XEQ 1
*EL PATSH
```

Figure C1 (b) (continued).

APPENDIX D

SI UNITS IN MECHANICAL DESIGN

D-1 INTRODUCTION

In the United States many practicing engineers are presently most familiar with the English units of measure, because of their extensive background experience with this system. Hence, over the years they have developed a good intuitive understanding for the English units, with respect to the many quantities so essential to the field of engineering design. Therefore, the transition presently taking place to the International System of Units (designated SI) can be a real handicap in the work of design, unless educational steps are taken by the engineer to cope with the problem. Thus, a concentrated effort should be made by design engineers to develop a good intuitive understanding for the SI system of units. This can be done over a period of time through careful study and practice, and the progressive engineer will embark on such a planned program of self-improvement at the earliest possible date.

In what follows we will briefly summarize some of the important items particularly pertinent for a factual understanding of the SI units in the work of mechanical design. To save on space and to avoid excessive duplication of other publications on SI units, we will cover only some of the highlights here. Also, we will introduce some different material showing the relative comparison between magnitudes of various design items in different units. This should be helpful in developing the intuitive understanding so essential to efficient practice in mechanical design synthesis. However, for greater depth in the field of SI units, you should carefully study other publications such as references 1-5.

D-2   SOME SI UNITS AND SYMBOLS

Quantity	Unit	SI Symbol	Formula
1. Base Units			
length	metre	m	—
mass	kilogram	kg	—
time	second	s	—
electric current	ampere	A	—
temperature	kelvin	K	—
amount of matter	mole	mol	—
luminous intensity	candela	cd	—
2. Supplementary Units			
plane angle	radian	rad	—
solid angle	steradian	sr	—
3. Derived Units			
acceleration	metre/second^2	—	m/s^2
angular accel.	radian/second^2	—	rad/s^2
angular velocity	radian/second	—	rad/s
area	square metre	—	m^2
density	kilogram/metre^3	—	kg/m^3
energy	joule	J	$N \cdot m$
entropy	joule per kelvin	—	J/K
force	newton	N	$kg \cdot m/s^2$
frequency	hertz	Hz	cycles/s
power	watt	W	J/s
pressure	pascal	Pa	N/m^2
quantity of heat	joule	J	$N \cdot m$
specific heat	joule per kilogram-kelvin	—	$J/kg \cdot K$
stress	pascal	Pa	N/m^2
thermal conductivity	watt per metre-kelvin	—	$W/m \cdot K$
velocity	metre per second	—	m/s
viscosity, dynamic	pascal-second	—	$Pa \cdot s$
viscosity, kinematic	square metre per second	—	m^2/s
volume	cubic metre	—	m^3
work	joule	J	N

D-3   SI PREFIXES

Multiplication Factor	Prefix	SI Symbol	Multiplication Factor	Prefix	SI Symbol
10^{12}	tera	T	10^{-2}	centi*	c
10^{9}	giga	G	10^{-3}	milli	m
10^{6}	mega	M	10^{-6}	micro	μ
10^{3}	kilo	k	10^{-9}	nano	n
10^{2}	hecto*	h	10^{-12}	pico	p
10^{1}	deka*	da	10^{-15}	femto	f
10^{-1}	deci*	d	10^{-18}	atto	a

*to be avoided where possible

D-4   SOME CONVERSION FACTORS

To convert from	to	Multiply by
Acceleration:		
foot/second^2	metre/second^2 (m/s^2)	3.048 000 E−01
inch/second^2	" " "	2.540 000 E−02
Area:		
foot^2	metre^2 (m^2)	9.290 304 E−02
inch^2	" "	6.451 600 E−04
Bending Moment or Torque:		
pound-force-inch (lbf · in)	newton-metre (N · m)	1.129 848 E−01
pound-force-foot (lbf · ft)	" " "	1.355 818 E+00
Density:		
pound-mass/foot^3 (lbm/ft^3)	kilogram/metre^3 (kg/m^3)	1.601 846 E+01
pound-mass/inch^3 (lbm/in^3)	kilogram/metre^3 (kg/m^3)	2.767 990 E+04
slug/foot^3	kilogram/metre^3 (kg/m^3)	5.153 788 E+02
Energy and Work:		
British thermal unit (Btu) (International Table)	joule (J)	1.055 056 E+03
foot-pound-force (ft-lbf)	" "	1.355 818 E+00
Force:		
pound-force (lbf)	newton (N)	4.448 222 E+00
Force Gradient:		
pound-force/inch (lbf/in)	newton/metre (N/m)	1.751 268 E+02

(D-4 Cont.)

To convert from	to	Multiply by
Length:		
foot	metre (m)	3.048 000 E−01
inch	” ”	2.540 000 E−02
Mass:		
pound-mass (lbm)	kilogram (kg)	4.535 924 E−01
slug	” ”	1.459 390 E+01
Power:		
horsepower (550 ft-lbf/s)	watt (W)	7.456 999 E+02
Pressure or Stress:		
pound-force/foot^2 (psf)	pascal (Pa)	4.788 026 E+01
pound-force/inch^2 (psi)	” ”	6.894 757 E+03
Temperature:		
degrees Fahrenheit ($^\circ$F)	kelvin (K)	$t_K = (t^0{}_F + 459.67)/1.8$
Velocity:		
foot/second (fps)	metre/second (m/s)	3.048 000 E−01
inch/second (ips)	” ” ”	2.540 000 E−02
mile/hour (mph)	” ” ”	4.470 400 E−01
Volume:		
foot^3	metre^3 (m^3)	2.831 685 E−02
inch^3	” ”	1.638 706 E−05

D-5   RELATIVE MAGNITUDE OF SOME ITEMS REFERRED TO SI UNIT

1 joule (J)	= 9.478 × 10^{-4} Btu	= 0.7376 ft-lbf	
1 kilogram (kg)	= 2.205 lbm	= 6.852 × 10^{-2} slug	
1 kg/m^3	= 3.613 × 10^{-5} lbm/in^3	= 6.243 × 10^{-2} lbm/ft^3	
1 metre (m)	= 39.37 inches	= 3.281 feet	
1 metre^2 (m^2)	= 1550 inch^2	= 10.76 ft^2	
1 metre^3 (m^3)	= 6.102 × 10^4 in^3	= 35.31 ft^3	
1 metre/second (m/s)	= 39.37 in/sec	= 2.237 mph	
1 newton (N)	= 0.2248 lbf		
1 newton-metre (N · m)	= 8.851 lbf-in	= 0.7376 lbf-ft	
1 newton/metre (N/m)	= 5.710 × 10^{-3} lbf/in		
1 pascal (Pa)	= 1.450 × 10^{-4} psi	= 2.089 × 10^{-2} psf	
1 Watt (W)	= 1.341 × 10^{-3} hp		

D-6   TYPICAL VALUES OF PROPERTIES FOR SOME MATERIALS IN BOTH UNITS

1. Density, ρ.

Material	(kg/m^3)	(lbm/in^3)
Aluminum	2770	0.100
Cast Iron	7300	0.264
Magnesium	1827	0.066
Steel	7833	0.283
Titanium alloy	4512	0.163
Cast Phenolic	1329	0.048
Nylon	1135	0.041

2. Modulus of Elasticity, E.

Material	(GPa)	(psi)
Aluminum	70	10×10^6
Cast Iron	95	13.8×10^6
Magnesium	45	6.5×10^6
Steel	205	29.7×10^6
Titanium alloy	110	16×10^6
Cast Phenolic	4.90	0.71×10^6
Nylon	4.14	0.60×10^6

3. Tensile Strength, S_t.

Material	(MPa)	(psi)
Aluminum — cast	140	20,000
— plate	483	70,000
Cast Iron — Class 20	138	20,000
— Class 40	276	40,000
Magnesium alloy	276	40,000
Steel — AISI 1020HR	414	60,000
— AISI 1095	745	108,000
— AISI 2340	1944	282,000
Titanium alloy	1034	150,000
Cast Phenolic	68.9	10,000
Nylon	75.8	11,000

D-7 SOME VERY SIMPLE EQUATIONS OF MECHANICS IN SI UNITS

1. Weight, W.

$W = M g$

> where W is in newtons(N)
>
> > M is in kilograms(kg)
>
> and $g = 9.81$ m/s^2

2. Force, F.

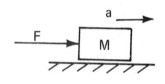

$F = M a$

> where F is in newtons(N)
>
> > M is in kilograms(kg)
>
> > a is in m/s^2

3. Stress,

$\sigma = P/A$

> where σ is in pascals(Pa)
>
> > P is in newtons(N)
>
> > A is in square metres(m^2)

D-8 ILLUSTRATION OF PROBLEM CONVERSION TO SI UNITS

Example 8-6 of Chapter 8 will be converted to SI units, for purposes of illustration. The basic variation study remains the same, starting with equation (8.53) through the final formation of the problem. However, equation (8.55) is modified as follows, where t is now expressed in millimeters (mm) and S_{bu} in pascals (Pa).

$$S_{bu} = [47,000/(t/25.4)^{0.25}] \text{ psi} \times 6.894757 \times 10^3 \text{ Pa/psi}$$

$$\therefore S_{bu} = 7.275 \times 10^8 / t^{0.25}$$

Hence, the initial formulation is the same as before, except change 47,000 to 7.275×10^8 in the limit equation for y. The final formulation remains the same as before, except change the limit value as above in both the variation diagram of Figure 8.23 and the flow chart of Figure 8.24.

For the numerical problem of Example 8-6, input data would be changed as follows:

$P = 15.0 \text{ lbf} \times 4.448222 \text{ N/lbf} = 66.723 \text{ N}$;

$b \leq 1.0 \text{ in} \times 25.4 \text{ mm/in} = 25.4 \text{mm}$;

$25.4 \text{ mm} \leq R \leq 76.2 \text{ mm}$;

Standard stock sizes t_s become, using 1 in = 25.4 mm:

0.010 in = 0.254 mm, 0.014 in = 0.356 mm,

0.018 in = 0.457 mm, 0.022 in = 0.559 mm,

0.025 in = 0.635 mm, 0.032 in = 0.813 mm,

0.042 in = 1.067 mm, 0.058 in = 1.473 mm,

0.072 in = 1.829 mm, 0.095 in = 2.413 mm,

0.109 in = 2.769 mm, 0.120 in = 3.048 mm.

$E = 30 \times 10^6 \text{ psi} \times 6.894757 \times 10^3 \text{ Pa/psi} = 2.068 \times 10^{11} \text{ Pa}$

$= 206.8 \text{ GPa}$ for the modulus of elasticity.

$N_e = 1.5$ factor of safety, and $N = 12$ number of standard stock sizes, both remain the same.

Regarding output data, the overall optimum design would be for:

$\Delta = 0.792$ in $\times$ 25.4 $= 20.117$ mm

$b = 0.861$ in $\times$ 25.4 $= 21.869$ mm

$R = 3.000$ in $\times$ 25.4 $= 76.2$ mm

$t = 0.072$ in $\times$ 25.4 $= 1.829$ mm

As an exercise for becoming familiar with SI units, the interested reader might wish to convert some other numerical problems from Chapters 8 and 9 of the text.

REFERENCES

1. *Metric Practice Guide E380-72E*, American Society for Testing and Materials, 1916 Race St., Philadelphia, Pa., 19103, 1972, 34 pp.
2. *ASME Orientation and Guide for Use of Metric Units*, The American Society of Mechanical Engineers, United Engineering Center, 345 East 47th Street, New York, N.Y., 10017, Third Edition, 1973, 30 pp.
3. E. A. Mechtly, *The International System of Units*, National Aeronautics and Space Administration, Scientific and Technical Information Office, Washington, D.C., 1973, 21 pp.
4. *Engineer's Guide to Modern International Units of Measurement for Rotating Machinery*, Elliott Company, Division of Carrier Corporation, Jeannette, Pa., 15644, 1975, 45 pp.
5. J. L. Meriam, "Using SI Units In Mechanics", *Engineering Education*, April 1977, pp. 682-685.

Index